Chemical and Biochemical Sensing
With Optical Fibers and Waveguides

For a complete listing of the *Artech House Optoelectronics Library*, turn to the back of this book.

Chemical and Biochemical Sensing With Optical Fibers and Waveguides

Gilbert Boisdé and Alan Harmer

Artech House
Boston • London

Library of Congress Cataloging-in-Publication Data
Boisdé, Gilbert.
  Chemical and biochemical sensing with optical fibers and waveguides/Gilbert Boisdé and
Alan Harmer.
        p.   cm.
  Includes bibliographical references and index.
  ISBN 0-89006-737-6 (alk. paper)
  1. Chemical detectors. 2. Biosensors. 3. Optical fiber detectors. 4. Optical waveguides.
I. Harmer, Alan. II. Title.
TP159.C46B65   1996
660'.281–dc20   96-20964
                                                                          CIP

British Library Cataloguing in Publication Data
Boisdé, Gilbert.
  Chemical and biochemical sensing with optical fibers and waveguides
  1. Optical fibers in biochemistry 2. Optical fiber detectors
  3. Waveguides
  I. Title II. Harmer, Alan
      621.3'692

  ISBN 0-89006-737-6

Cover design by Kara Munroe-Brown

International Standard Book Number: 0-89006-737-6
Library of Congress Catalog Card Number: 96-20964

10 9 8 7 6 5 4 3 2 1

To my wife, Mady, for her understanding and patience with this book.
To my children, Isabelle, Marc, and Dominique; to my grandchildren,
Florent, Charlotte, Nicolas, Damian, and Valentin; and to all the others.

G.B.

Especially to all the others.

A.H.

Un livre n'est qu'un pas au fil du temps,
On ne le finit pas, on l'abandonne...

Contents

Foreword   xv

Preface   xvii

Acknowledgments   xix

Chapter 1   Introduction   1
  1.1   Historical Perspective   1
  1.2   Sensors, Transducers, and Optodes   4
  1.3   Relative Merits of Optical Techniques for Chemical Sensing   7
        1.3.1   Advantages of Optical Waveguide Sensors   7
        1.3.2   Limitations of Optochemical Sensing   8
        1.3.3   Further Development   9
  References   9

Part I: Ionic and Molecular Recognition   13

Chapter 2   Chemical Reactions   17
  2.1   Introduction   17
  2.2   Thermodynamic Description of a Reversible Reaction   17
  2.3   Rate Equations of a General Chemical Reaction   19
  2.4   Reversible Sensors   22
        2.4.1   Direct Reactions   22
        2.4.2   Indirect Reactions   28
        2.4.3   Simultaneous Reactions   31
  2.5   Nonreversible Sensors   32
  2.6   Controlled-Reagent Sensors   32
        2.6.1   With Reservoir   32
        2.6.2   With Renewable Reagent   33
        2.6.3   Polymeric Delivery Systems   33

2.7   Exchange Reactions   36
2.8   Intrinsic Properties of the Analyte   39
References   40

Chapter 3   Kinetics and Shape Recognition   41
  3.1   Introduction   41
  3.2   Kinetics   43
         3.2.1   Concepts   43
         3.2.2   Biocatalytic Reactions   45
         3.2.3   Optical Biocatalytic Sensors   49
  3.3   Shape Recognition   50
         3.3.1   Immunological Interactions   50
         3.3.2   Immunoreactions With Enzymes   53
         3.3.3   Immunoassays   54
         3.3.4   Liposome and Enzyme Amplification   56
         3.3.5   Kinetics of Surface Reactions   56
         3.3.6   Molecular Bioreceptors   57
  References   61

Part II: Optical Chemoreception   63

Chapter 4   Indicators, Receptors, and Labels   65
  4.1   Electronic Transitions   65
  4.2   Fundamentals of Indicators   66
  4.3   Common Indicators   69
         4.3.1   Polycyclic Aromatic Hydrocarbons   69
         4.3.2   Carbon-Oxygen Compounds   70
         4.3.3   Carbon-Nitrogen Compounds   76
         4.3.4   Carbon-Nitrogen-Oxygen Compounds   80
         4.3.5   Compounds Incorporating Sulfur   84
         4.3.6   Macrocyclic Compounds—Crown Ethers   86
  4.4   Redox Indicators   90
  4.5   Potential-Sensitive Dyes   91
  4.6   Neutral Ion Carriers and Other Receptor Carriers   94
  4.7   Optical Operating Requirements for Indicators   101
  4.8   Labels and Labeling Protocols   103
         4.8.1   Labels   103
         4.8.2   Protein Labeling Protocols   106
  References   108

Chapter 5   Supports and Immobilization Techniques   113
  5.1   Supports   113

	5.1.1	Materials	113
	5.1.2	Theoretical Aspects of Membranes	117
5.2	Configurations for Sensors		118
5.3	Immobilization Techniques		121
	5.3.1	Physical Methods	122
	5.3.2	Physicochemical Methods	122
	5.3.3	Electrostatic (Ionic) Binding	125
	5.3.4	Covalent Immobilization	125
5.4	Properties of Immobilized Compounds		132
	5.4.1	Characterization	132
	5.4.2	Optical Characteristics in pH Sensors	133
	5.4.3	Coimmobilization of Several Dyes	136
References			138

Part III: Optical Signal Processing and Optical Electrical Transduction   141

Chapter 6   Essential Theory of Optics   143
6.1	Light		143
	6.1.1	Particle Nature	143
	6.1.2	Electromagnetic Waves	144
	6.1.3	Geometrical Optics	149
	6.1.4	Light as Combined Photons and Waves	151
6.2	Interaction of Light With Matter		158
	6.2.1	Scattering	158
	6.2.2	Dispersion	160
	6.2.3	Attenuation and Absorption	161
	6.2.4	Molecular Transitions	162
	6.2.5	Photochemical Effects	163
	6.2.6	Luminescence	164
	6.2.7	Energy Transfer	168
	6.2.8	Nonlinear Effects	170
References			171

Chapter 7   Optical Fibers and Planar Waveguides   173
7.1	Introduction		173
	7.2	Ray Model of Light Propagation	173
	7.2.1	Mode Propagation in Waveguides	174
	7.2.2	Polarization in Optical Fibers	175
7.3	Planar Waveguides		177
	7.3.1	Wave Propagation in Planar Guides	177
	7.3.2	Nonlinearity in Planar Waveguides	180
	7.3.3	Configurations of Planar Waveguides	180

7.4   Types of Optical Fibers   183
7.5   Special Fibers for Sensors   184
        7.5.1   Special Multimode Fibers   184
        7.5.2   Birefringent Fibers   186
        7.5.3   Fibers for Evanescent Wave Spectroscopy   188
        7.5.4   Fibers for Interferometry   189
        7.5.5   Doped Fibers and Lasing Fibers   189
7.6   Fiber Materials   191
        7.6.1 SiO_2-Based Fibers   191
        7.6.2   Infrared Fibers   192
        7.6.3   Plastic Fibers   195
References   195

Chapter 8   Optical Measurement Techniques   199
8.1   Introduction   199
8.2   Optical Modulation Schemes   200
        8.2.1   Intensity (Amplitude) Modulation   200
        8.2.2   Wavelength Modulation   201
        8.2.3   Phase Modulation   202
        8.2.4   Polarization Modulation   203
        8.2.5   Time Modulation   203
8.3   Chemical Measurement Techniques and Signal Characteristics   204
8.4   Absorbance Measurements   208
        8.4.1   Measurements   208
        8.4.2   Fiber-Optic Absorption Spectrometers   208
        8.4.3   Fourier Transform Infrared and IR Spectroscopy   209
8.5   Scattering   211
        8.5.1   Types of Scattering   211
        8.5.2   Reflectance Measurements   211
        8.5.3   Raman Measurements   213
        8.5.4   Surface-Enhanced Raman and IRSpectroscopy   214
8.6   Luminescence   214
        8.6.1   Characteristics of Fluorescence   214
        8.6.2   Practical Measurements   217
        8.6.3   Other Luminescent Sensors   218
8.7   Evanescent Wave Spectroscopy   219
8.8   Surface Plasmon Resonance   222
8.9   Refractometry   225
8.10   Interferometry   226
        8.10.1   Interferometers   226
        8.10.2   Examples of Interferometric Sensors   227
8.11   Photoacoustic Measurements   228

8.12   Other Optochemical Sensing Techniques   229
       8.12.1   Grating Light Reflection Spectroscopy   229
       8.12.2   Ellipsometry   230
       8.12.3   Optical Time Domain Reflectometry   230
References   231

Chapter 9   Light Sources, Photodetectors, and Optical Components   241
9.1   Light Sources   241
       9.1.1   Lamp Sources   241
       9.1.2   Light-Emitting Diodes   243
       9.1.3   Laser Sources   243
       9.1.4   Gas Lasers   243
       9.1.5   Dye Lasers   245
       9.1.6   Solid-State Lasers   245
9.2   Photodetectors   246
       9.2.1   Principles   246
       9.2.2   Photoemissive Detectors   247
       9.2.3   Solid-State Photodetectors   247
9.3   Passive Components   252
       9.3.1   Conventional Components   252
       9.3.2   Fiber-Optic Components   253
9.4   Active Components   255
       9.4.1   Effects   255
       9.4.2   Important Active Fiber Components   256
References   257

Chapter 10   Optodes and Sensing Cells   259
10.1   Classification of Optodes   259
10.2   Extrinsic Optodes   262
       10.2.1   Sensing Cells   262
       10.2.2   Simple Fiber Optodes   263
       10.2.3   Optodes with Immobilized Compounds   265
10.3   Intrinsic Optodes   266
       10.3.1   Refractometric Optodes   266
       10.3.2   Direct Evanescent Wave Spectroscopic Optodes   267
       10.3.3   Intrinsic Coating-Based Optodes   269
       10.3.4   Core-Based Optodes   270
       10.3.5   Hybrid ICB and Core-Based Optodes   272
10.4   Planar Waveguide Devices   272
       10.4.1   Fluorescent Capillary Fill Device   272
       10.4.2   Multianalyte Sensing   274
       10.4.3   Guiding Monolayer Structures   274

10.4.4	Guiding Multilayer Structures	276
10.4.5	Integrated Optical Sensors	277
10.5	Photoelectrode and Optoelectrochemical Sensors	278
References		279

Part IV: Applications 285

Chapter 11 Biochemical Sensors and Biosensors 289
11.1	Introduction	289
11.2	Clinical Diagnostics	289
	11.2.1 Electrolytes	290
	11.2.2 Glucose Biosensor	295
	11.2.3 NH_3/NH_4^+ System	300
	11.2.4 Urea and Creatinine Sensors	301
	11.2.5 Cholesterol and Other Biosensors	302
11.3	In Vivo Monitoring	303
	11.3.1 Oximetry	303
	11.3.2 NADH as Indicator	304
	11.3.3 Principles of Blood Gas and pH Measurements	304
	11.3.4 Extracorporeal Blood Gas Sensors	307
	11.3.5 Invasive Techniques	308
	11.3.6 Transcutaneous Techniques	310
	11.3.7 Noninvasive Techniques	311
11.4	Biotechnology	312
	11.4.1 Biomass, NADH, and ATP	313
	11.4.2 pH, pO_2, and pCO_2 Measurements	315
	11.4.3 Enzyme Activity	316
	11.4.4 Enzyme Optodes With Intermediate Analytes	316
	11.4.5 Other Methods and Techniques	318
11.5	Immunosensors	318
	11.5.1 Immunosensor Characteristics	319
	11.5.2 Reversible and Regenerable Immunosensors	319
	11.5.3 Progress in Immunosensing	322
	11.5.4 Examples of Optical Immunosensing	323
References		325

Chapter 12 Environmental Monitoring, Process Control,
Gas Measurements, and Sensor Networks 339
12.1	Environmental Monitoring	339
	12.1.1 Tracer Dyes in Water	339
	12.1.2 Organic Pollutants	340
	12.1.3 Metal Ions and Anions	343

12.1.4	Biological Oxygen Demand	343
12.1.5	Measurements in Seawater	346
12.1.6	Pesticides and Herbicides	346
12.1.7	Air Pollution	347
12.2	Process Control	347
12.2.1	Nuclear Process Control	347
12.2.2	Other Sensors in Biotechnology	352
12.3	Gas Monitoring	353
12.3.1	Direct Spectrometry	353
12.3.2	Change in Optical Properties of a Sensitive Film	356
12.3.3	Chemically Reactive Film Sensors	356
12.3.4	Other Techniques	358
12.4	Networks	358
12.4.1	General Approach	358
12.4.2	Networks in Chemical Sensing	359
References		361
Glossary		369
About the Authors		375
Index		377

Foreword

The advent of optical light guides has had a tremendous impact on our information society. In the form of optical-fiber transmitters, they have led to a new generation of data communication systems. In addition to this, the introduction of optical-fiber networks was a milestone in that it paved the way for a technology that up to then had been restricted to electrical devices. However, the use of optical fibers is by no means limited to telecommunication applications or to a certain range of wavelengths. In fact, they may be used for numerous other purposes and for transmitting a much greater variety of optical information than is usual in data communications.

Integrated optics may be considered as the younger brother of optical-fiber waveguides, and they have not yet had a commercial success comparable to that of optical-fiber telecommunication systems. However, their potential is as high as that of fiber waveguides, particularly for optical microchips. These are likely to become as powerful as electronic microchips, but with optical rather than electronic circuits and using photons instead of electrons.

One of the application fields of optical waveguide technology is analytical chemistry. Analytical sciences have traditionally relied on numerous kinds of optical information, including absorption and emission spectroscopy, refractometry, and polarimetry. It is surprising, though, how fast optical-fiber technology has been acquired (and adapted to their specific needs) by chemists. Gilbert Boisdé was certainly a pioneer in this area.

The use of fibers for purposes of chemical sensing was initially limited to species possessing an intrinsic color or fluorescence. These included certain colored actinides and transition metals such as the copper ion, but also many gases with a characteristic absorption in the near infrared, such as methane.

An even larger field of application of fiber-optic chemical sensors was provided by combining fiber-optic technology with traditional—sometimes even considered outdated—methods of indicator chemistry, such as the use of dyes that react with a species of interest to undergo a change in their optical properties. This paved the way for sensing (or at least irreversibly "probing") chemical parameters such as pH, oxygen partial pressure, carbon dioxide, heavy metals, and other pollutants, to mention only a few. A further step forward was made by combining biochemical and biological recognition elements with waveguide technology to result in so-called biosensors. The last decade has seen practically all important classes of biomolecules (including enzymes, antibodies, and DNA fragments) and biological components (including tissue and whole cells) immobilized on waveguide structures.

It is obvious that waveguide chemical sensing is an extremely promising and fast growing technology. Not unexpectedly, several sensors have been commercialized in the past few years, a fact that demonstrates the utility of the technology and acts as a thrust for further development in this area. So far, optical-chemical sensors have found application mainly in clinical chemistry and are now routinely used for continuously monitoring blood gases, disposable tests for various clinical parameters using portable or bedside instrumentation, and monitoring environmental parameters such as oxygenation and pollutants such as oil and gasoline. Even more significantly, optical sensors have become an extremely powerful tool in medical research and have provided unique insights into physiological processes.

This book summarizes the state of the art in waveguide-based chemical sensing and biosensing. The index reflects the complexity of the technology, which requires a successful combination of skills not taught in a single study course. Understanding waveguide-based chemical sensing technology requires a substantial background in chemistry, material sciences, immobilization techniques, optics and spectroscopy, optoelectronic components, electronics in general, and data acquisition and handling. In the case of biosensing, numerous aspects of modern biochemistry (including enzymology, immunology, DNA techniques, and cell biology) may be included. Depending on the field of application, a strong interaction will be required with experts in application-oriented areas that include, but are not limited to, clinical chemistry, physiology, surgery, biotechnology, mechanical and chemical engineering, nuclear science, analytical chemistry, geosciences, and space and marine technology.

I really do appreciate the authors' endeavor to write a book on a subject that is growing so quickly that it is difficult to keep pace. In view of its clear structure and the wealth of information it contains, the book represents a most useful source of information for both those who wish to familiarize themselves with the technology and for the specialist. I am very sure that this book will further contribute to the popularity and success of waveguide-based chemical sensing and biosensing.

Otto S. Wolfbeis
Regensburg, July 1996

Preface

We should like to begin with a little history. Professor Jayle, a well-known hormone biochemist at the University of Paris, was blinded by an explosion of ether while analyzing some hormone steroids [1]. This prompted him in 1972 to ask the Instrumentation Group at the Commissariat à l'Energie Atomique (Atomic Energy Commision) in France to construct a remote-reading explosion-proof photometer. A fiber-optic instrument was developed with a bifurcated fiber bundle, specially produced at that time in Switzerland, that enabled measurements to be made at three different wavelengths. The miniaturized probe was a rod of rigid glass fibers immersed in the liquid sample [2]. This was the first of what came to be a series of more than 40 instruments (Telephot), and at the same time, a research group was set up specializing in fiber-optic instrumentation for remote in situ measurements in process control. The group gained momentum from the intense activity in fiber-optic sensors and continued to develop industrial instrumentation over the next 20 years.

Advanced optical techniques are gradually being exploited for industrial applications. The enormous progress in optical-fiber technology, so succesful in long-distance telecommunications, has provided synergy for the new field of optical sensors. Following a suggestion by John Dakin, who coedited two previous volumes in this field [3], the present book was written to complement the work in describing chemical and biochemical optical sensors. This subject, which has already been treated in two previous books (CRC Press and Humana Press, 1991), is evolving rapidly due to its potential in widespread and varied applications.

Information for this book has been drawn from a collection of more than 2,600 publications. Work from the most recent publications (1992 to 1995) is emphasized, perhaps at the risk of a less detailed description of the classic developments in optics and associated instrumentation that are the origins of this discipline. We have endeavored through joint authorship (one of us more of a chemist, the other more a physicist) to cover the gaps in this multidisciplinary field and provide a comprehensive overview for the nonspecialist reader. The book is divided into four parts; the first three describe the concepts of optical sensors. The last part, on applications, was conceived as a separate section so that it can be expanded in the future as the field evolves.

We are extremely grateful to Otto S. Wolfbeis (University of Regensburg, Germany) for his helpful suggestions, his valuable criticism, and his friendly collaboration. Without him this book would not have been written. We offer this book in the hope that it records the debt to all our colleagues worldwide who are patiently developing the present technology and so preparing a promising future.

Gilbert Boisdé
Bures-sur-Yvette, France

Alan Harmer
Geneva, Switzerland

July 1996

References

[1] Jayle, J. M., *Analyse des Stéroïdes Hormonaux*, Paris: Masson, 1962.
[2] Degrelle, H., M. Egloff, J. M. Jayle, G. Boisdé, and J. J. Perez, *IIIème Colloque Biologie Prospective, Pont à Mousson*, Expansion Sci. Fr., Paris, 1975, pp. 463–467.
[3] Dakin, J., and B. Culshaw, *Optical Fiber Sensors*, Norwood, MA: Artech House, 1988 and 1989.

Acknowledgments

In addition to Otto S. Wolfbeis (see preface), we should like to thank all our colleagues and fellow researchers worldwide for their encouragement and for supplying original material. We also gratefully acknowledge the support of the Commissariat Energie Atomique (CEA), Centre d'Etudes de Saclay, and the University of Paris XI, Centre de Bures/Orsay for the use of their excellent libraries. We are especially indebted to Philippe Garderet (Director of Strategy and Evaluation at the CEA), Alain Jorda (Adjoint DEIN, Saclay), Christian Ngo (Chef Service Physique Electronique, Saclay), and to our work colleagues, especially Jean Jacques Perez, Jean Sabatier, Alphonse Denis, Charlotte Bonouvrier, and Jacqueline Piet, who have each contributed in their own way to the realization of this book. Finally, we thank Daniel Trolez and Pierre Lastennet, of STER (Fontenay-aux-Roses), who produced the drawings for this book.

Chapter 1

Introduction

1.1 HISTORICAL PERSPECTIVE

The use of optical fibers in chemical and biochemical analysis began in the 1960s, particularly for oximetry, the measurement of oxygen in the blood [1–3]. Important developments during this period, which inspired new ideas in chemical and biochemical optical sensors, were the characterization of planar guided modes [4] and the principles of internal reflection spectroscopy [5] and surface plasmon resonance [6]. In addition, new concepts were introduced concerning microprobes [7,8], fiber-optic spectroscopic techniques [9,10], and experiments on continuous oxygen sensing [11]. These developments paved the way for major advances that followed in the 1970s:

1. The concept of the *optode* for CO_2 and O_2 measurement [12];
2. Fiber-optic Raman spectrometry [13], remote coherent anti-Stokes Raman spectrometry (CARS) [14], and time-of-flight optical spectrometry [15];
3. The development of fiber-optic refractometers [16,17];
4. The first pH fiber-optic sensor [18,19];
5. Chemiluminescence with an immobilized enzyme [20];
6. Long optical-path fiber-optic cells for the measurement of solutions in process control [21,22] and for a methane gas sensor with a White cell [23];
7. Commercial fiber-optic spectrophotometers connected to large-core (1.0-mm diameter) silica fibers specially developed by the Quartz & Silice Company, France (see Figure 1.1).

Many laboratories and organizations started important research programs on optical-fiber sensors, such as those in the United States at Oak Ridge National Laboratory and Lawrence Livermore National Laboratory under Thomas Hirschfeld and in France at the Commissariat Energie Atomique (Atomic Energy Commission) [24].

The development of low-loss fibers and other components for telecommunications

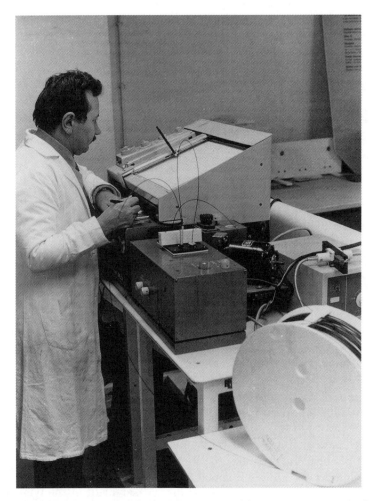

Figure 1.1 One of the authors (G.B.) performing experiments with large-core (1-mm) single silica fibers for spectral analysis measurements at long distances (>50m) for chemical species in solution (1977 to 1978). The fibers are coupled to a commercial double-beam spectrophotometer (DK2 Beckmann). An optical fiber was also used in the reference path to balance the two optical beams. At the same time, cells of various optical path lengths from 1mm to 1m (for uranium solutions) and 30m (for gases) were tried experimentally with this setup. (*Photo*: J. Sabatier, Arch. Commissariat Energie Atomique, Centre de Fontenay-aux-Roses, France.)

spurred research in optical-fiber sensors (OFS) from 1977 onward. The first major impetus for fiber-optic (bio-) chemical sensors occurred at the International Conference on Analytical Chemistry in Gatlinburg, Tennessee in 1980 [25]. An analysis of 17,000 published papers on fiber optics representing about 35% of all work in this field for the period between 1969 and 1994 shows a dramatic growth in activities (Figure 1.2). New areas and ideas developed rapidly, such as the glucose sensor [26], evanescent wave spectroscopy (which included gas detection by surface plasmon resonance [27]), planar SiO_2-TiO_2 waveguides [28], evanescent chemical sensors [29], and immunoassays [30]. The first fiber-optic chemical sensor was commercialized in 1984 by CDI-3M. Since then the device has been used several

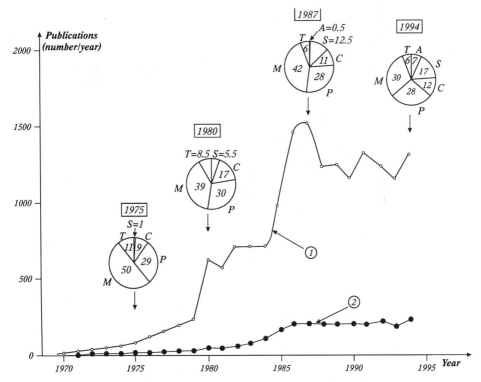

Figure 1.2 Annual publications (papers and patents) between 1969 and 1994 on optical-fiber technology. Curve 1 is for all publications on optical fibers, and curve 2 is for optical-fiber sensors: (A) optical amplification; (S) sensors; (C) components for fiber systems; (P) properties of fibers, fiber measurements, and optical modulation techniques; (M) fiber fabrication, materials, claddings, etc.; (T) testing (environmental, radiation measurements, cable tests, etc.). The distribution for these different classes are shown for 1975, 1980, 1987, and 1994. (Data compiled from an analysis of references from Chemical Abstracts.)

hundred thousand times for monitoring blood gases and pH during cardiopulmonary bypass operations.

The increasing interest and research work is manifest in conference and scientific journal publications. The first Fiber Optic Bio-Chemical Sensor European workshop was organized by Wolfbeis in 1986, and was followed by the biannual Europt(r)ode Congress (see conference proceedings for 1992, 1994, and 1996). Many review articles have been published in journals [31–37], in conference proceedings [38,39], and as contributions to books [40–43]. Two multiauthor books on optical-fiber chemical sensors are valuable sources of reference in this field [44,45].

1.2 SENSORS, TRANSDUCERS, AND OPTODES

Optical (bio-) chemical sensors are devices that convert a (bio-) chemical state, via a transducer, into a signal. The (bio-) chemical state may be in a gaseous, liquid, or solid phase and the output signal is normally electrical, used for measurement or controlling an actuator. This general definition [46], as applied to optical sensors, is represented in Figure 1.3. The device is called an *integrated sensor* if the conditioning

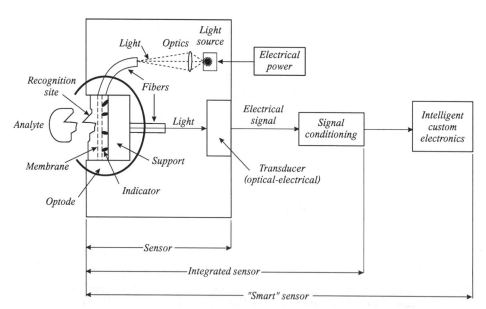

Figure 1.3   Schematic representation of an optical chemical (or biochemical) sensor, an integrated sensor, and a smart sensor (see [46]). Components such as the membrane, light source, and optics can be included in the sensor (shown inside the box). The *optode* is the sensing part, the central element of the *chemical-to-optical* transduction and the site for recognition of the chemical (or biochemical) compounds to be studied.

of the signal is included in the sensing element. Intelligent electronics can also be added to produce a *smart sensor*. A biosensor may be defined as "a device incorporating a biological sensing element either intimately connected to or integrated within a transducer" [47].

In the majority of optical chemical sensors, the transduction steps are:

electrical—optical—chemical—optical—electrical

The first electrical-to-optical step corresponds to the production of light from electrical power by a light source acting as a transducer. In the optical-to-chemical step, the light may influence (e.g., through a photophysical effect such as photobleaching) the chemical compound to be analyzed (the analyte). In most sensors, the light does not effect the chemical reaction. In chemiluminescence and bioluminescence, these two steps are absent because the light itself is generated directly by the chemical or biological reaction. The chemical-to-optical step includes the chemical reactions (Chapter 2), molecular recognition (Chapter 3), and the transfer of the chemical signal into an optical signal, using, for example, indicators (Chapter 4) immobilized on a support (Chapter 5). If no indicator exists for the analyte, another analyte such as oxygen, carbon dioxide, ammonia, or a bioreactive layer, may act as an *intermediate analyte*, which can be used with an indicator or a label. This step can proceed in various ways, as shown in Figure 1.4, in which either the optical waveguiding properties of the light guide form an intrinsic part of the sensing process (for an intrinsic sensor) or, for an intrinsic sensor, the light guide only transmits light to the sensor without using the waveguide characteristics as part of the sensing process (see Chapter 10).

In this book, the sensing element (transducer) including the optical waveguides is called an *optode* when it comprises the two optical-to-chemical-to-optical steps, and an *optrode* when the first electrical-to-optical step is also integrated into the sensing element (see Chapter 10). The last step, the optical-to-electrical transduction, is achieved using the properties of light (Chapter 6), appropriate waveguides and fibers (Chapter 7), optical sensing techniques (Chapter 8), and photodetectors (Chapter 9).

Thus, this book has been divided into four parts: The first part considers ionic and molecular recognition (Chapters 2 and 3); the second part looks at optical chemoreception (Chapters 3 and 4); the third part describes optical signal processing and optical-to-electrical transduction (Chapters 6 to 10); and the last part is on recent applications of optical sensors (Chapters 11 and 12).

In present-day process control systems, the monitoring and control of a chemical process is generally achieved with sensors that measure physical quantities (e.g., temperature, flow, and pressure), not ones that directly sense the chemical species undergoing reactions. Fiber-optic sensors have also been developed for these physical parameters [48], many using principles similar to those of chemical sensors, but they are not described in this book.

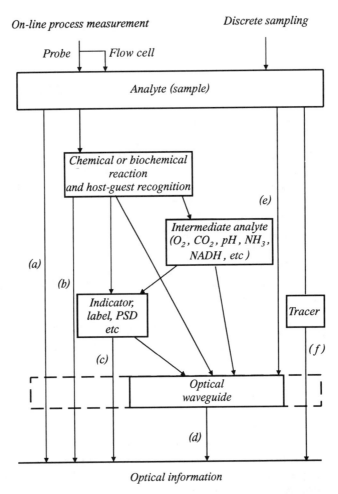

Figure 1.4 The main types of chemical-to-optical transduction schemes showing the transformation of information from the analyte into an optical signal: (a) the spectral properties of the analyte are analyzed directly; (b) chemiluminescence or bioluminescence is used for sensing; (c) typical transduction is performed with an indicator, label, or potential-sensitive dye (PSD), sometimes with an intermediate analyte such as nicotinamide adenine dinucleotide hydrogenated (NADH) (reduced); (d) the intrinsic properties of the optical waveguide are used in the transduction process (intrinsic sensors); (e) the physicochemical properties of the analyte (e.g., refractive index) are measured by an optical waveguide; and (f) the analyte can be a tracer. The dotted lines show extrinsic sensing in which the optical waveguide serves only to transmit the light.

1.3 RELATIVE MERITS OF OPTICAL TECHNIQUES FOR CHEMICAL SENSING

An ideal sensor should measure continuously, reversibly, and in real time (or within a few seconds). It should be able to measure over a large range with high precision and selectivity, and without presenting any problems for the user (e.g., it should have high reliability and be self-calibrating). At present, few sensors possess all these characteristics. Optical techniques for chemical sensing have a number of important advantages over conventional sensing methods, but also include some drawbacks. These relative merits are due, at least in part, to the nature of the optical methods themselves.

1.3.1 Advantages of Optical Waveguide Sensors

Guided-wave optochemical sensors have several notable advantages, which are summarized here.

1. *Optical techniques* are already widely used in analytical chemistry in, for example, absorption and fluorescent spectrophotometry, chromatography, and electrophoresis. The use of fibers and waveguides are an extension of these techniques, with miniaturized sensors for in situ measurement. In addition, the *light produced* as the primary signal in a chemical or biochemical reaction in the presence of a catalyst (chemiluminescence or bioluminescence) leads naturally to the use of optical techniques [49]. The high *sensitivity* (and specificity) of optical spectrometry, particularly fluorescence, is a major advantage.
2. The *low loss transmission* of fibers as light guides and their *adaptability* for optical instrumentation, especially in spectrometry, has encouraged the development of remote and continuous sensing in gases or liquids without sampling. This is becoming very important in real-time techniques (using photodiode arrays, time resolution measurements) for optical Fourier transform infrared (FTIR) spectrometry and chemometrics [50]. Fiber characteristics of high bandwidth, low optical loss, and low temporal dispersion allow remote measurements at long distances (typically 10m to 1,000m).
3. The *inherent properties* of optical fibers offer advantages. In particular, they have excellent electric isolation and immunity to electromagnetic interference, important in many applications, such as medical in vivo monitoring. Their small size and flexibility make miniaturization of optodes possible, and they resist hostile and hazardous environments. Silica is a relatively chemically inert material. Special fibers resist high temperatures (1,000°C), strongly acidic solutions (though not in hydrogen fluoride), and moderate alkaline (pH < 13) environments. Various types of silica fiber have been tested under gamma ray irradiation up to doses of 10^8 Gy (10^{10} rad).
4. Fibers are available in a *variety of materials* (plastic, silica, fluorides,

chalcogenides, tellurides, and halide glasses) with numerical apertures and re-fractive indexes adapted to specific measurements. These materials transmit light over a large spectral range from the ultraviolet (UV) to the near infrared (IR), typically from 0.25 to 12 microns.

5. In *biosensing*, particularly for medical applications, suitable materials can be chosen allowing biocompatibility for plastic optodes, sterilization of silica fibers, and the construction of disposable catheter sensors for in vivo measurements.

6. Light can be *modulated in several ways* (see Chapter 8). Different modulations can be transmitted in a single waveguide without interference between the signals. Optical systems may also have their own *internal reference* as, for example, in multiwavelength measurement methods.

7. *Chemical deposition techniques* for fabricating the chemical part of the sensor have been adapted from work on electrodes and solid-state sensors [51] based on catalytic reactions on surfaces and on the use of membranes. They can be applied to monolayer or multilayer deposition on the surface of the guides by grafting (covalent attachment on the support) or by sol-gel techniques for incorporating specific reagents for ionic or molecular recognition.

1.3.2 Limitations of Optochemical Sensing

1. *Parasitic optical signals* can occur due to light scattering in the medium, unwanted background fluorescence, and stray ambient light. The optical signal may be affected by contamination, fouling of the sensor, and absorption due to suspensions in the analyzed medium, which can lead to a slow reduction in light and long-term instability. Problems are usually avoided by modulating the light, providing a reference signal and signal encoding schemes. Fiber optics is a young technology; components are not yet optimized for sensing applications and high optical losses from multiple links may limit the optical dynamic range.

2. *Interaction of the light and the chemical process* introduces limitations. Photo-degradation leads to a trade-off of signal strength and high light intensity, which may induce photobleaching of the reagent. Self-absorption and inner filter effects (IFEs) of the fluorescence generated inside a solution decreases the measurable signal. Some optical techniques, such as absorption measurements, have a limited dynamic range. Furthermore the relationship between the optical signal and con-centration is generally nonlinear, in contrast to the linear behavior of electrodes. In addition, the type of interaction (adsorption, absorption, ion exchange, chemi-sorption), the morphology of the chemical interface materials, and the encapsu-lation are critical factors for the performance of chemical and biochemical sensors.

3. The *nature of chemical detection* introduces a time delay in the response due to mass transfer between the solid (sensitive film) and the analyte (in gas or solution). The limited lifetime of the immobilized reagent can be a major problem due to the desorption of reagents, the irreversibility of some reactions, and difficulties

encountered in the use of membranes. To overcome this problem requires good control of the immobilization process, particularly of the hydrophobic-hydrophilic characteristics of chemical bindings. New research is needed to improve the sensitivity, reproducibility, and selectivity. In general, these disadvantages are also present in most other types of chemical sensors.

1.3.3 Further Development

As a result of the relative merits and limitations and the difficulty of adopting a new technology like fiber optics, the introduction of optical chemical sensors will be slow and gradual. However, there are some promising niche areas and some successful applications, particularly in biosensing (Chapter 11), environmental monitoring, and remote process control (Chapter 12), that are of growing interest to users.

Planar sensors made by integrated optics offer several advantages [52] in that they can be coupled to fibers, fabricated at low cost, and used with optical amplification. This is an area of growing research. Future developments include the possibility of connecting many sensors into an optical network and the potential of optical computing; these may eventually lead to direct coupling of sensing systems to optical communication systems.

References

[1] Polanyi, M. L., and R. M. Hehir, "In Vivo Oximeter With Fast Dynamic Response," *Rev. Sci. Instrum.*, Vol. 33, 1962, pp. 1050–1054.

[2] Enson, Y., W. A. Briscoe, M. L. Polanyi, and A. Cournand, "In Vivo Studies With an Intravascular and Intracardiac Oximeter," *J. Appl. Physiol.*, Vol. 17, 1962, pp. 552–558.

[3] Kapany, N. S., and N. Silbertrust, "Fibre Optics Spectrophotometer for in Vivo Oximetry," *Nature*, Vol. 204, 1964, pp. 138–142.

[4] Kane, J., and H. Osterberg, "Optical Characteristics of Planar Guided Modes," *J. Opt. Soc. Am.*, Vol. 54, 1964, pp. 347–352.

[5] Harrick, N. J., "Electric Field Strengths at Totally Reflecting Interfaces," *J. Opt. Soc. Am.*, Vol. 55, 1965, pp. 851–857.

[6] Kretschmann, E., and H. Raether, "Radiative Decay of Non Radiative Surface Plasmons Excited by Light," *Z. Naturforsch.*, Vol. A23, 1968, pp. 2135–2136.

[7] Keller, H., "The Photoprobe. An Attempt To Simplify Photometric Measurement by the Use of Fiber Optics," *Z. Klin. Biochem.*, Vol. 7, 1969, pp. 501–504.

[8] Vurek, G. G., and R. L. Bowman, "Fiber Optic Colorimeter for Submicroliter Samples," *Anal. Biochem.*, Vol. 29, 1969, pp. 238–247.

[9] French, C. S., R. W. Hart, N. Murata, and C. Wraight, "Test of Fiber Optics for Fluorescence Spectroscopy," *Canergie Inst. Washington, Yearbook*, 1970, pp. 607–608.

[10] Eysel, H. H., "Raman Scattering of Polycrystalline Substances: Absorption Using a Fiber Optic Cross-Section Converter," *Spectrochim. Acta*, Vol. A27, 1971, pp. 173–177.

[11] Bergman, I., "Rapid Response Atmospheric Oxygen Monitor Based on Fluorescence Quenching," *Nature*, Vol. 218, 1968, p. 396.

[12] Lübbers, D. W., and N. Opitz, "Die pCO_2/pO_2 Optode," Z. Naturforsch., Vol. 30C, 1975, pp. 532–533.

[13] Walrafen, G. E., "New Slitless Optical Fiber Laser Raman Spectrometer," Appl. Spectrosc., Vol. 29, 1975, pp. 179–185.

[14] Eckbreth, A. C., "Remote Detection of CARS (Coherent Anti-Stokes Raman Spectrometry) Employing Fiber Optic Guides," Appl. Opt., Vol. 18, 1979, pp. 3215–3216.

[15] Whitten, W. B., and H. H. Ross, "Fiber Optics Waveguides for Time of Flight Optical Spectrometry," Anal. Chem., Vol. 51, 1979, pp. 417–419.

[16] Kapany, N. S., and J. N. Pike, "Fiber Optics: Part IV, A Photorefractometer," J. Opt. Soc. Am., Vol. 47, 1957, pp. 1109–1117.

[17] David, D. J., D. Shaw, H. Ticker, and F. C. Unterleitner, "Design, Development and Performance of a Fiber Optics Refractometer: Application to HPLC," Rev. Sci. Instrum., Vol. 47, 1976, pp. 989–997.

[18] Peterson, J. I., and S. R. Goldstein, "Fiber Optic pH Probe," U.S. Pat. Appl. 855,384 (1977); U.S. Pat. 4,200,110 (1980).

[19] Peterson, J. I., "Dye Fixation by Copolymerization," U.S. Pat. Appl. 855,397 (1977); U.S. Pat. 4,194,877 (1980).

[20] Freeman, T. M., and W. R. Seitz, "Chemiluminescence Fiber Optics Probe for H_2O_2 Based on the Luminol Reaction," Anal. Chem., Vol. 50, 1978, pp. 1242–1246.

[21] Boisdé, G., and A. Boissier, "Apparatus for Making Photometric Measurement," Fr. Appl. 14753 (1977); Ger. Offen. 2,820,845 (1978); U.S. Pat. 4,188,126 (1980).

[22] Boisdé, G., and A. Boissier, "Photometric Cell With Multiple Reflections," Fr. Appl. 9856 (1978); U.S. Pat. 4,225,232, (1980).

[23] Inaba, H., T. Kobayasi, M. Hirama, and M. Hamza, "Optical Fibre Network System for Air-Pollution Monitoring over a Wide Area by Optical Absorption Method," Electron. Lett., Vol. 15, 1979, pp. 749–751.

[24] Perez, J. J., G. Boisdé, M. Goujon de Beauvivier, G. Chevalier, and M. Isaac, "Automatisation de la Spectrophotometrie du Plutonium," Analusis, Vol. 8, 1980, pp. 344–351; Rep. LA-TR 82–14 (1982).

[25] Borman, S. A., "Optrodes," Anal. Chem., Vol. 53, 1981, pp. 1616A–1618A.

[26] Schultz, J. S., S. Mansouri, and I. J. Goldstein, "Affinity Sensor: A New Technique for Developing Implantable Sensors for Glucose and Other Metabolites," Diabetes Care, Vol. 5, 1982, pp. 245–253.

[27] Nylander, C., B. Liedberg, and T. Lind, "Gas Detection by Means of Surface Plasmon Resonance," Sensors and Actuators, Vol. 3, 1982, pp. 79–88.

[28] Lukosz, W., and K. Tiefenthaler, "Embossing Technique for Fabricating Integrated Optical Components in Hard Inorganic Waveguiding Materials," Opt. Lett., Vol. 8, 1983, pp. 537–539.

[29] Newby, K., M. L. Reichert, J. D. Andrade, and R. E. Benner, "Interfacial Chemical Sensor Using a Single Multimode Optical Fiber," Appl. Opt., Vol. 23, 1984, pp. 1812–1815.

[30] Sutherland, R. M., C. Dähne, J. F. Place, and A. S. Ringrose, "Immunoassays at a Quartz Liquid Interface: Theory, Instrumentation and Preliminary Application to the Fluorescent Immunoassay of Human Immunoglobulin G," J. Immunol. Methods, Vol. 74, 1984, pp. 253–265.

[31] Chabay, I., "Optical Waveguides," Anal. Chem., Vol. 54, 1982, pp. 1071A–1080A.

[32] Stuart, A. D., "Optrodes: Remote Determination of Chemical Concentrations Using Optical Fibers and Optical Sensors," Chem. Aust., Vol. 53, 1986, pp. 350–353.

[33] Wolfbeis, O. S., "Analytical Chemistry With Optical Sensors," Fresenius Z. Anal. Chem., Vol. 325, 1986, pp. 387–392.

[34] Seitz, W. R., "Chemical Sensors Based on Immobilized Indicators and Fiber Optics," Critical Rev. (CRC Press), Vol. 19, 1988, pp. 135–173.

[35] Boisdé, G., and J. J. Perez, "Une Nouvelle Génération de Capteurs: Les Optodes," La Vie des Sciences, Compt. Rend. Acad. Sci. Fr., Vol. 5, 1988, pp. 303–332.

[36] Norris, J. O. W., "Current Status and Prospects for the Use of Optical Fibres in Chemical Analysis: A Review," Analyst, Vol. 114, 1989, pp. 1359–1372.

[37] Wolfbeis, O. S., "Optical Sensing Based on Analyte Recognition by Enzymes, Carriers and Molecular Interactions," *Anal. Chim. Acta*, Vol. 250, 1991, pp. 181–201.

[38] Harmer, A. L., "Guided-Wave Chemical Sensors," *Proc. Electrochem. Soc.*, Vol. 87–9, 1987, pp. 409–427.

[39] Narayanaswamy, R., "Optical Fiber Chemical Sensors," *Proc. Electrochem. Soc.*, Vol. 87–9, 1987, pp. 428–437.

[40] Wolfbeis, O. S., "Fiber Optical Fluorosensors in Analytical and Clinical Chemistry," Chap. 3 in *Molecular Luminescence Spectroscopy: Methods and Applications*, Part II, S. J. Schulman, ed., New York: Wiley, 1988, pp. 129–281.

[41] Krull, U. J., and R. S. Brown, "Fiber-Optic Remote Chemical Sensing," Chap. 7 in *Chemical Analysis: A Series of Monograhs on Analytical Chemistry and its Applications*, R. M. Measures, ed., New York: Wiley, Vol. 94, 1988, pp. 505–532.

[42] Wolfbeis, O. S., G. Boisdé, and G. Gauglitz, "Optochemical Sensors," Chap. 12, Vol. 2, pp. 573–575; O. S. Wolfbeis and G. Boisdé, "Applications of Optochemical Sensors for Measuring Chemical Quantities," Chap. 17, Vol. 3, pp. 867–930; W. Trettnak, M. Hofer, and O. S. Wolfbeis, "Applications of Optochemical Sensors for Measuring Environmental and Biochemical Quantities," Chap. 18, Vol. 3, pp. 931–967, in *Sensors: A Comprehensive Survey*, W. Göpel, J. Hesse, and J. N. Zemel, eds., Weinheim: VCH, 1991.

[43] Lübbers, D. W., "Fluorescence Based Chemical Sensors," in *Advances in Biosensors*, JAI Press, Vol. 2, 1992, pp. 215–260.

[44] Wolfbeis, O. S., ed, *Fiber Optic Sensors and Biosensors*, Vols. I and II, Boca Raton: CRC Press, 1991.

[45] Wise, D. L., and L. B. Wingard, *Biosensors With Fiberoptics*, Clifton, NJ: Humana Press, 1991.

[46] Göpel, W., J. Hesse, and J. N. Zemel, eds., *Sensors: A Comprehensive Survey*, Vols. 1 and 6, Weinheim: VCH, 1991.

[47] Turner, T. E., I. Karube, and G. Wilson, *Biosensors*, London: Oxford University Press, 1988.

[48] Dakin, J., and B. Culshaw, *Optical Fiber Sensors*, Vols. 1 and 2, Norwood, MA: Artech House, 1988 and 1989.

[49] Blum, L. J., and P. R. Coulet, *Biosensor: Principles and Applications*, New York: Marcel Dekker, 1991.

[50] Hirschfeld, T., J. B. Callis, and B. R. Kowalski, "Chemical Sensing in Process Analysis," *Science*, Vol. 226, 1984, pp. 312–318.

[51] Madou, M. J., and S. R. Morrison, *Chemical Sensing With Solid State Devices*, Boston: Harcourt, 1989.

[52] Martelucci, S., A. N. Chester, and M. Bertolucci, *Advances in Integrated Optics*, New York: Plenum Press, 1994.

$P_{art\,I:}$

Ionic and
Molecular Recognition

Chemical and biochemical sensing may be defined as "an information acquisition process in which some insight is obtained about the chemical or biochemical composition of the system in real time" [1]. This information acquisition process requires selectivity of the analyte-receptor interaction. The selectivity is the "ability of a sensor to respond primarily to only one species in the presence of other species." Sensing is achieved by means of a sensor that is sensitive to its chemical surroundings and that can recognize the chemical information and translate it into an optical (or electrical) signal. Thus, the sensing process embodies several steps, as shown schematically in Figure 2.1: the recognition of the specific chemical information, optical chemoreception, optical signal processing, and optical electrical transduction. These different steps may be distinct in a sensor, or several steps may be combined within a single operation. In some sensors the first and/or second steps may be missing; for example, the measurement of refractive index does not require chemical recognition.

Step 1. The ionic or molecular recognition process. It is based on the interaction between the analyte (the chemical species to be measured) and the reagent, or in biosensing between the substrate (a chemical term in enzymology for the chemical analyte to be measured) and the receptor (the sensing element). Neither the analyte nor the reagent need have optical characteristics. This step is governed by either chemical and biochemical reactions, kinetic reactions, or shape recognition. Sometimes these three types of specific interactions coexist and interact within the same sensor. This first step, which is basic to the selectivity of the chemical information, is examined in detail in Part I.

Step 2. The translation of the primary information (the recognition) into an optical

BIO-CHEMICAL SENSING SCHEME

STEP 1	STEP 2	STEP 3	
Ionic / molecular recognition	*Opto-chemoreception* *(Chemical optical transduction)*	*Optical signal processing*	*Optical electrical transduction*
Analyte (A) - receptor (R) interactions	*Optical chemoreceptors*	*Optical techniques* - *Spectrometry*	*Sources* *(electrical → optical)*
Kinetic	- *Indicators* - *Receptors and labels*	- *Refractometry* - *Interferometry* - *Evanescent waves*	
Bio- and chemical reactions	*Supports, membranes*	- *etc ...* *Fibers and waveguides*	*Receivers* *(optical → electrical)*
Shape recognition	*Immobilization techniques*	*Passive components*	*Active components*
Analytical considerations (selectivity, etc...)	*Analytical techniques*		⇓

⇒ *to amplifier*

Figure 2.1   The three steps in biochemical and chemical sensing: (a) *Step 1*: Ionic or molecular recognition; (b) *Step 2*: Optical chemoreception; and (c) *Step 3*: Optical signal processing and optical electrical transduction. Recognition (selectivity of the chemical information) by means of specific receptors is based on chemical reactions, kinetics, and/or shape interaction. Optical chemoreception uses different chemical agents and supports to achieve chemical-optical transduction.

signal by using optical chemoreceptors, a physical support (for immobilization of the chemical receptors), and analytical techniques that control the chemical or biochemical reactions. This step is described as a chemical-optical transduction in Part II.

Step 3. The treatment or processing of the primary optical signal. The third step, in Part III, uses optical hardware in the form of various optical devices and techniques and is based on the properties of interaction of matter with photons. Sometimes a physical process (e.g., the diffusion of a gas in a porous fiber) or a physicochemical process (e.g., the adsorption of the analyte at the surface of the waveguide) can produce directly (without step 1) a change in absorption, a change in the propagation of the evanescent waves, or a change in the refractive index within the sensor. These processes lead to optical signals with nonspecific chemical information. Thus, this third step allows the optical signal to be extracted from the sensor and transformed into an electrical signal by an optoelectronic interface (optical electrical transduction). The electrical signal is subsequently processed by computer and is not described in detail in this book. The optical processing and transduction step is examined in Part III along with optical techniques and associated components.

Part I examines chemical reactions (Chapter 2), kinetic reactions (Chapter 3), and shape recognition (Chapter 3) as the three distinct aspects of the chemical and biochemical information recognition process. Chemical sensing in solutions, solvents, gases, or mixtures relies mainly on chemical reactions and kinetics, whereas biosensing uses kinetics (biocatalysis) and, more importantly, shape recognition, because the molecules are more complex in living organisms than in chemical solutions or gaseous systems.

Chapter 2

Chemical Reactions

2.1 INTRODUCTION

Recognition is the first step in the detection and information acquisition in real time of a chemical composition. This recognition, through a selective chemical reaction, is the basis of sensing for all types of chemical sensors [1,2]. The chemical composition of a chemical species (i) is evaluated in terms of its activity (a) at the concentration C with

$$a_i = f_i \cdot C_i \tag{2.1}$$

where f is the activity coefficient with $\lim f \to 1$ for $C \to 0$. Thus, the activity is the concentration of the chemical species reduced by the factor f, which takes into account the surroundings for a nonideal solution.

Also, in the well-known formal Nernst's equation, the activity is defined from the mobility (μ) of the ion, atom, or molecule subjected to a chemical potential with

$$\mu_i = (\mu_i)^\circ + R \cdot T \ln\{a_i/(a_i)^\circ\} \tag{2.2}$$

where $\mu_i = (\mu_i)^\circ$ for $a_i = (a_i)^\circ$ (standard activity), R is the general gas constant ($8.3144 \ J \cdot T^{-1} \cdot mol^{-1}$), and T is the Kelvin temperature. The concentrations are determined in different units (usually molarity (M), but also normality, molality, and mole fraction, among others) resulting from different standard states.

2.2 THERMODYNAMIC DESCRIPTION OF A REVERSIBLE REACTION

In thermodynamic terms, a chemical reaction occurs when the enthalpy (H) and entropy (S) change. The enthalpy (heat of chemical reaction per mole) corresponds to the internal energy of a system for which heat is absorbed under constant pressure. The entropy is a characteristic of the directionality of processes; $\Delta S > 0$ for an irreversible process. It is a measure in probability terms of the disorder or randomness of a system. A more random system has a higher entropy. Enthalpy and entropy are related through the Gibbs free energy (G) at the absolute temperature (T):

$$G = H - T \cdot S \tag{2.3}$$

When there is no change in the Gibbs free energy $(dG = 0)$ and entropy $(dS = 0)$, the state of the system is an equilibrium. For a simple equilibrium in a reversible chemical reaction

$$A \rightleftharpoons B \tag{2.4}$$

The equilibrium constant K can be expressed in terms of the activities of the two parts of the reaction:

$$K = a_B / a_A \tag{2.5}$$

and the standard free energy change $(\Delta G^\circ$, measured at 1 atm, 298K, and unit concentration of an ideal solution) of the reaction is

$$\Delta G^\circ = -R \cdot T \cdot \ln K \tag{2.6}$$

The equilibrium reaction in (2.4) is composed of two simultaneous processes, forward (subscript f) and reverse (subscript r) reactions, with the rate of the reactions v_f and v_r given in terms of the activities (a):

$$v_f = k_f \cdot a_A = -da_A/dt \tag{2.7}$$

and

$$v_r = k_r \cdot a_B = -da_B/dt \tag{2.8}$$

At equilibrium, the forward and reverse reaction rates are equal and the equilibrium constant is

$$K = k_f/k_r \tag{2.9}$$

The equilibrium condition is reached quickly when the rate constants k_f and k_r are high and slowly for low rate constants. If $k_f \gg k_r$ or $k_f \ll k_r$, the reaction is virtually

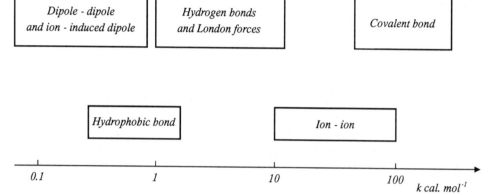

Figure 2.2 Typical ranges of interaction enthalpies in kcal · mol^{-1} for chemical bonds. (See [1].)

irreversible. For instance, with $K > 10^4$ ($\Delta G° < -5.46$ kcal · mol^{-1}), the reaction is irreversible, and the measuring device will generally be considered a dosimeter rather than a sensor [1].

From another standpoint of thermodynamics, the location of binding sites and their associated energy for chemical species is an important characteristic for sensors because of the reaction of molecules through either absorption (bulk interaction) or adsorption (surface interaction). In particular, the type of binding is critical. For example, hydrogen and ionic bonds, hydrophilic and hydrophobic bond interactions, molecular associations, and charge transfer complexes are found inside the binding sites of organic molecules. Figure 2.2 shows typical ranges of interaction energy for common types of chemical bonds.

2.3 RATE EQUATIONS OF A GENERAL CHEMICAL REACTION

The more general form [3] of a chemical reaction is given by the equilibrium equation

$$\alpha \cdot A + \beta \cdot B \rightleftharpoons x \cdot X + y \cdot Y \tag{2.10}$$

The rate (v) is related to the activity (a) of the forward reaction by

$$v_f = k_f \cdot (a_A)^\alpha \cdot (a_B)^\beta \tag{2.11}$$

and for the reverse reaction by

$$v_r = k_r \cdot (a_X)^x \cdot (a_Y)^y \qquad (2.12)$$

where the rate constants k_f and k_r measure the fraction of collisions involved in the chemical reaction. When the reaction is initiated, the rate of the forward reaction is large because A and B are large. As the reaction progresses, this rate diminishes and the rate of the reverse reaction increases. At equilibrium, the forward and reverse reaction rates are equal. Combining (2.11) and (2.12) using the concentration values (in brackets [ ]) expressed as $C_A = [A]$, $C_B = [B]$, $C_X = [X]$, and $C_Y = [Y]$ gives

$$v_f = v_r = k_f \cdot (f_A)^\alpha \cdot (f_B)^\beta \cdot [A]^\alpha \cdot [B]^\beta = k_r \cdot (f_X)^x \cdot (f_Y)^y \cdot [X]^x \cdot [Y]^y \qquad (2.13)$$

The equilibrium constant (K_0) is

$$K_0 = k_f/k_r = (f_X)^x \cdot (f_Y)^y \cdot [X]^x \cdot [Y]^y / (f_A)^\alpha \cdot (f_B)^\beta \cdot [A]^\alpha \cdot [B]^\beta \qquad (2.14)$$

In optical sensors, the chemical potentials cannot be measured, and hence the activities are not determined directly. Thus, all measurements, including the reaction kinetics (reaction rates), are generally evaluated in terms of the concentrations. If K_C is the concentration equilibrium constant (or molar equilibrium constant),

$$K_C = [X]^x \cdot [Y]^y / [A]^\alpha \cdot [B]^\beta \qquad (2.15)$$

with

$$K_0 = K_C \cdot (f_X)^x \cdot (f_Y)^y / (f_A)^\alpha \cdot (f_B)^\beta \qquad (2.16)$$

A large equilibrium constant indicates that the equilibrium will proceed to large concentrations of X and Y, that is, toward the right-hand side of (2.10): $(k_f > k_r)$. Examples are given in Figure 2.3 for small and large K_C with α, β, x, and $y = 1$. It can be seen in Figure 2.3(a,b) that with small K_C, X and Y will never reach concentration A or B, but in the case of Figure 2.3(c,d), concentration A is reached well before equilibrium.

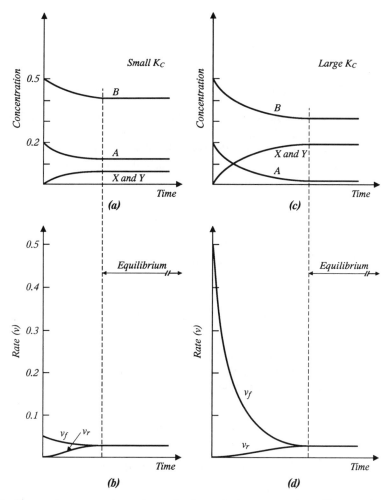

Figure 2.3 Changes in concentration and rate of a chemical reaction to reach equilibrium under idealized conditions. Curves of concentration-time (a,c) and rate-time (b,d) are drawn from (2.15) (concentration) and (2.11) and (2.12) (rate), with α, β, x, and $y = 1$, and $C_A = 0.2$ and $C_B = 0.5$ (in excess) at time $t = 0$: (a) $K_C = 0.1$; (b) $k_f = 0.5$ and $k_r = 5$; (c) $K_C = 10$; (d) $k_f = 5$ and $k_r = 0.5$.

2.4 REVERSIBLE SENSORS

2.4.1 Direct Reactions

2.4.1.1 Single-Step Reaction

In a reversible sensor [4], the analyte (A) and the reagent (R) are in equilibrium, and a direct sensor determines the analyte directly from a measurement of either (A), (R), or (AR). In a stoichiometric reaction, the reaction is complete and depends on a transfer of mass. For a reversible reaction with a molar concentration ratio of 1:1, the chemical equilibrium in solution (subscript s) is

$$A_s + R_s \rightleftharpoons (AR)_s \tag{2.17}$$

where AR is the analyte-reagent association with the molar constant K_s, also called the *binding constant*:

$$K_s = K_{\text{ass}} / K_{\text{diss}} = [(AR)_s] / [A_s] \cdot [R_s] \tag{2.18}$$

K_{ass} and K_{diss} are the association and dissociation constants, respectively. At equilibrium the total concentration of reagent C_R, which may be presumed constant if no photochemical or photophysical effects are present, is

$$C_R = [R_s] + [(AR)_s] \tag{2.19}$$

Thus, the following well-known equations can be obtained:

$$[R_s] = C_R / (1 + K_s \cdot [A_s]) \tag{2.20}$$

and

$$[(AR)_s] = C_R \cdot K_s \cdot [A_s] / (1 + K_s \cdot [A_s]) \tag{2.21}$$

Measuring the ratio $[(AR)_s]/[R_s]$ has the advantage that it is independent of C_R, which is generally an unknown quantity, and it is directly proportional to the analyte concentration (see (2.18)). However, it is important that the optical signals and the concentrations of $(AR)_s$ and R_s be large enough to avoid limiting the dynamic range.

2.4.1.2 Association in Multiple Steps

Many chemical compounds form an association in two steps:

$$A + B \rightleftharpoons AB \tag{2.22}$$

with a binding constant K_1, and

$$A + AB \rightleftharpoons A_2B \tag{2.23}$$

with the binding constant K_2. The total reaction is the sum of the two equilibria:

$$2A + B \rightleftharpoons A_2B \tag{2.24}$$

with the constant K_t

$$K_t = K_1 \cdot K_2 = [A_2B] \,/\, [A]^2 \cdot [B] \tag{2.25}$$

Associations in three or more steps can be handled by the same method of calculation. In addition, depending on the method used for analytic determination, the chemical equilibria can be studied either as an association of molecules (constant K_t) or as a dissociation (inverse constant).

2.4.1.3 Systems With Immobilized Reagent

In practice, in many optical sensors, the reagent in excess is immobilized (phase i) as the solid phase inside or on the surface of a support layer, or directly on the waveguide. As an example, Figure 2.4 shows the different interactions that occur between the analyte in solution and the immobilized reagent [5].

The chemical equilibrium of the analyte between the solution and the solid phase is controlled by the well-known Fick's laws of diffusion and the adsorption conditions on the surface of the waveguide [6]. This is represented by the equilibrium

$$A_s \rightleftharpoons A_i \tag{2.26}$$

where the subscript s refers to the phase in solution and i to the adsorbed or immobilized phase. The constant K_A corresponds to the mass transfer of analyte between the solution and adsorption in the support layer:

$$K_A = [A_i] \,/\, [A_s] \tag{2.27}$$

Also, an interaction is possible between the immobilized reagent and the reagent in

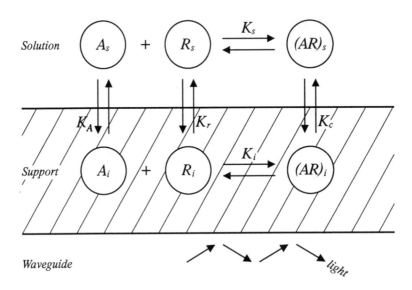

Figure 2.4 Different interactions that occur between the analyte (A) in solution (subscript s) and the reagent (R) immobilized (subscript i) in a support layer on a waveguide.

solution in the presence of the analyte. Taking the state of immobilization as the initial state, the chemical equilibrium is

$$R_i \rightleftharpoons R_s \qquad (2.28)$$

with the constant K_r, which is dependent on the strength of immobilization of the reagent in the surface layer. A smaller constant corresponds to stronger binding, a higher constant to weaker binding and more desorption (or dissolution) of the immobilized reagent.

$$K_r = [R_s] / [R_i] \qquad (2.29)$$

The chemical equilibrium at the surface of and within the support layer is

$$A_i + R_i \rightleftharpoons (AR)_i \qquad (2.30)$$

with the constant K_i

$$K_i = [(AR)_i] / [A_i] \cdot [R_i] \qquad (2.31)$$

The total concentration of reagent C_R is unknown but may be considered constant to a first approximation:

$$C_R = [R_s] + [(AR)_s] + [R_i] + [(AR)_i] \qquad (2.32)$$

In general, the $[A_i]$ value and the K_i and K_A constants are indeterminable. Theoretically, the K_c constant can be used, the initial state being chosen as the immobilization of the $(AR)_i$ species, with the equilibrium

$$(AR)_i \rightleftharpoons (AR)_s \qquad (2.33)$$

and

$$K_c = [(AR)_s] \, / \, [(AR)_i] \qquad (2.34)$$

The relationship between K_c and the other constants can be established from (2.18), (2.27), (2.29), (2.31), and (2.34):

$$K_c = K_s \cdot K_r \, / \, K_i \cdot K_A \qquad (2.35)$$

Hence, depending on the analytical technique, the optical signal is proportional to either $[R_s]$ or $[(AR)_s]$, or alternatively to either $[R_i]$ or $[(AR)_i]$. From (2.18), (2.29), (2.32), and (2.34) the following relations can be established:

$$[R_s] = C_r \, / \, \{1 + 1/K_r + K_s \cdot [A_s](1 + 1/K_c)\} \qquad (2.36)$$

$$[(AR)_s] = C_r \cdot K_s \cdot [A_s] \, / \, \{1 + 1/K_r + K_s \cdot [A_s] \cdot (1 + 1/K_c)\} \qquad (2.37)$$

$$[R_i] = C_r \, / \, \{1 + K_r + K_s \cdot K_r \cdot [A_s] \cdot (1 + 1/K_c)\} \qquad (2.38)$$

$$[(AR)_i] = (C_r \cdot K_s \cdot K_r \cdot [A_s]/K_c) \, / \, \{1 + K_r + K_s \cdot K_r \cdot [A_s] \cdot (1 + 1/K_c)\} \qquad (2.39)$$

Figure 2.5 shows the different variations of $[R_i]$ and $[(AR)_i]$ ((2.38) and (2.39)) as a function of the analyte concentration $[A_s]$.

As for the case of measurement in solution, a large simplification can be obtained in practice if the ratio of the AR and R concentrations is measured. Indeed, it can be shown that this ratio is directly proportional to the analyte concentration. For the analysis in solution, from (2.36) and (2.37), this ratio is independent of C_r and the constants K_r and K_c:

$$[(AR)_s] \, / \, [R_s] = K_s \cdot [A_s] \qquad (2.40)$$

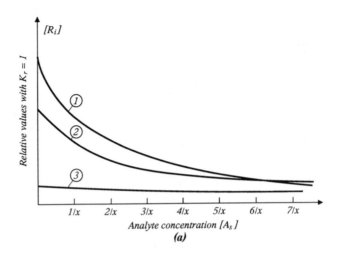

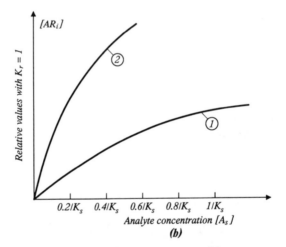

Figure 2.5   Relationship between the analyte concentration and the concentration of compounds measured optically for an immobilized reagent as an indicator (see (2.38) and (2.39)): (a) measurement of the reagent (R_i) for $x = K_r \cdot K_s \cdot (1 + 1/K_c)$ with curve (1) $C_r/K_r = 100$, (2) $C_r/K_r = 10$, (3) $C_r/K_r = 1$; and (b) measurement of the associated form $(AR)_i$ with $C_r/K_r = 10$ and curve (1) $K_r/K_c = 1$, (2) $K_r/K_c = 10$.

This equation, identical to (2.18), indicates that a measurement in solution with an immobilized reagent has some advantages and that the technique using a renewable reagent in a reservoir (or jacket) at the end of a fiber is independent of the values of these constants.

The ratio for immobilized phases is

$$[(AR)_i]/ [R_i] = K_s \cdot K_r \cdot [A_s] / K_c \qquad (2.41)$$

This ratio, proportional to the analyte concentration, is also independent of the C_r concentration, but is dependent on the constants K_s, K_r, and K_c. Consequently, the use of immobilized phases and measurement of $(AR)_i$ and R_i requires rigorous calibration and high stability of the immobilized phase.

It is interesting to note the case for a very large constant K_r. According to the equilibrium (2.28), this represents strong desorption of the immobilized reagent or its absence in the support layer. Thus, the sensor may be regarded as working in solution.

In fact, it is important to consider the ratio K_r/K_c for a reversible sensor. In theory, determination of the ratios $[R_s]/[R_i]$ and $[(AR)_s]/[(AR)_i]$ allows one, by optical methods, to determine the values of these constants.

2.4.1.4 Competitive Association

In some cases, a substance $(S)_s$ interferes by competitive association with the analyte $(A)_s$. In addition to the existing equilibrium of (2.17), this leads to a second equilibrium:

$$(A)_s + (S)_s \rightleftharpoons (AS)_s \qquad (2.42)$$

The molecules $(AR)_s$ and $(AS)_s$ with similar shapes can be adsorbed on active sites and analyzed directly. Indeed, this competitive reaction slows down the rate of the main reaction (with R_s) by *competitive inhibition* and is generally regarded as an interference. This inhibition, frequently observed in catalytic and biochemical reactions, can be reduced by increasing the (R_s) concentration.

2.4.1.5 Homogeneous and Heterogeneous Phases

In summary, reversible and direct chemical reactions between an analyte and a reagent can be studied with fiber optics using two techniques:

1. In the homogeneous phase: The usual optical analytical methods are effective. In this case, (2.20) and (2.21) are valid. Up to now, this has been the main approach by chemists or biochemists using optical fibers for measurement in solution.
2. In the heterogeneous phase (e.g., a solid layer and a liquid phase): One or several reagents normally incorporated in a polymer can be immobilized on the waveguide.

Thus, the fact that a reagent is immobilized means the existence of heterogeneous phases. This is a recent and interesting development of *active optodes* and has a number of advantages [5]. Various reversible sensor types are possible with an immobilized component, generally with the R_i form, and, in particular, examples with the $(AR)_i$ form. In these two cases, the analyte can be measured in solution by the variation of R_s or $(AR)_s$ or their ratio, or alternatively from the decrease of the immobilized species or the increase of the complementary species.

The reversibility of a sensor depends on the values of the K_r and K_c constants, the mass transfer must be as small as possible and in the initial conditions the immobilized form should be in excess.

At present, most chemical sensors measuring ionic species are of the reversible and direct types. In contrast to electrodes, practically all active optodes have nonlinear work functions. Sometimes the logarithmic function of the signal is more appropriate, for example, in the case of the pK value of the binding constant K_s:

$$pK_s = -\log \cdot K_s \tag{2.43}$$

2.4.2 Indirect Reactions

An indirect reaction in a homogeneous phase includes two or more components that interact simultaneously or successively with the analyte. The best-known type of indirect reactions in optical sensors is that of competitive binding.

2.4.2.1 Competitive Binding

In competitive binding, the reagent (R) competes with a ligand (L) for association with the analyte (A). In the homogeneous phase (solution s), the competitive equilibrium is

$$A_s + (LR)_s \rightleftharpoons (AR)_s + L_s \tag{2.44}$$

This equation is the result of two equilibria. One is represented by (2.17) with the constant K_s (2.18), and the other by the equilibrium

$$L_s + R_s \rightleftharpoons (LR)_s \tag{2.45}$$

with the association constant K_b of the L and R species.

$$K_b = [(LR)_s] / [R_s] \cdot [L_s] \tag{2.46}$$

C_{Rs} and C_{Ls} are the total concentrations of reagent and ligand respectively, and the mass balance expressions are

$$C_{Rs} = [R_s] + [(AR)_s] + [(LR)_s] \tag{2.47}$$

and

$$C_{Ls} = [L_s] + [(LR)_s] \tag{2.48}$$

With the elimination of $[R_s]$, the analyte concentration is obtained from (2.18) and (2.46) as

$$[A_s] = (K_b / K_s) \cdot ([(AR)_s] \cdot [L_s] / [(LR)_s]) \tag{2.49}$$

The ratio K_b/K_s represents the relative affinity of L_s and A_s for the reagent R_s.

In this competitive binding reaction, the association constant K_L of A and L species corresponds to the equilibrium

$$A_s + L_s \rightleftharpoons (LA)_s \tag{2.50}$$

where

$$K_L = [(LA)_s] / [A_s] \cdot [L_s] \tag{2.51}$$

A_s can be evaluated from a general expression with one variable, either L_s or $(LR)_s$, by combining the five equations (2.18), (2.46), (2.47), (2.49), and (2.51), including the initial total concentrations C_{Rs} and C_{Ls} and the three constants K_s, K_b, and K_L.

When R_s is very small (with K_s and K_b large), (2.47) may be simplified by eliminating $[R_s]$. In this case, the concentration of A_s relative to L_s is

$$[A_s] = (K_b \cdot [L_s] / K_s) \cdot \{C_{Rs} / (C_{Ls} - [L_s]) - 1\} \tag{2.52}$$

The analyte concentration can also be expressed in a similar form as a function of $(LR)_s$.

A well-known example of a fiber-optic sensor using competitive binding has been investigated by Schultz and coworkers for measurement of glucose [7], shown in Figure 2.6.

With immobilized species, the interactions of ions or molecules are more complicated. The previous equations must be corrected using the terms for the immobilized components R_i, $(AR)_i$, L_i, and $(AL)_i$.

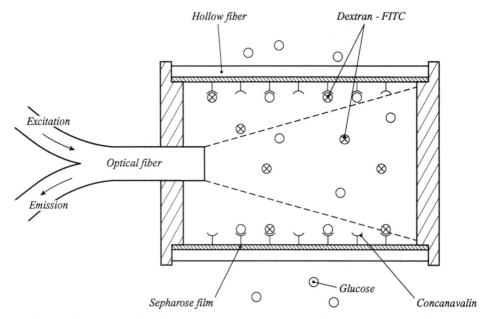

Figure 2.6 Glucose sensor [7]. The reagent is concanavalin A immobilized on a sepharose film, which is coated on the interior walls of a hollow dialysis fiber. The ligand is dextran labeled with fluorescein-isothiocyanate (FITC), a fluorescent compound. The glucose (the analyte) diffuses through the hollow fiber and displaces the dextran from the concanavalin, which migrates into the central area illuminated by light from the input fiber (the dotted lines show the area of the exciting light). The fluorescence (the emitted light) is collected by a second optical fiber. The increase in fluorescence measures the glucose concentration.

2.4.2.2 Successive Reactions

The principle of successive and indirect reactions in two steps is illustrated in solution by the equilibrium

$$A_s + B_s \rightleftharpoons (AB)_s \tag{2.53}$$

associated with a competitive binding reaction:

$$(AB)_s + C_s \rightleftharpoons (AC)_s + (B)_s \tag{2.54}$$

The final reaction can be used to determine A_s from C_s, considered as the reagent. In a generalized way with n successive reactions from reagents R_1 to R_n, the analyte may be determined from the final equilibrium reaction with the reagent R_n. The constant for the total reaction $(K_T = K_1.K_2 \ldots K_n)$ is obtained by multiplying together the individual constants, as many as the number of steps (n). Thus, the reversibility of this type of sensor decreases with the number of steps.

2.4.3 Simultaneous Reactions

Simultaneous reactions can occur between ions in the support (e.g., a membrane), the analyte, and hydrogen ions. Many examples are described later with ion-sensitive optodes (ISO) and coextraction systems (see Chapter 4). A typical example is given in Figure 2.7 for a cation-sensitive optode [8] using a lipophilic ion-selective membrane (see Chapter 4). These reaction equilibria in the aqueous phase, membrane (interface), and organic phase can be analyzed in a similar way, as described in Section 2.4.1.3.

The possibilities of using multireagents immobilized (or coimmobilized) on a support layer are potentially attractive, but have not yet been much exploited. There are some unsolved problems in the immobilization, such as competitive binding (see Chapter 7), and the response time of the sensor is not reproducible at present due to the difficulty of controlling the relative quantities of immobilized reagents. A novel approach has been proposed with three pH-sensitive cones localized by photopolymerization on the face of the same imaging fiber [9].

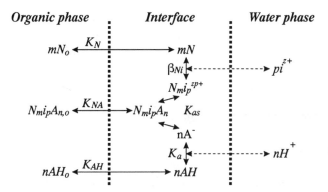

Figure 2.7 Example of successive reactions in a cation-selective optode by ion extraction (see Chapter 4). N is the neutral ionophore, AH a lipophilic anion in the protonated form, and i^{z+} the ion to be measured with a charge z. The subscripts p, m, and n refer to molecules taking part in the chemical reaction, and the subscript o is for organic phases. (*After:* [8] with permission.)

2.5 NONREVERSIBLE SENSORS

If the reagent R_s or R_i is entirely consumed, the reaction stops and the sensor is unusable. However, if the rate of consumption of the reagent is small relative to the total available, the sensor is nonreversible. This occurs when the K_s and/or K_i constants are very large (e.g., in immunosensors) or when the reagent quantity is small compared to that of the analyte. See (2.11), (2.12), and (2.17).

In indirect sensors, the rate of consumption of the second reagent (ligand), such as L_s or L_i and possibly $(LR)_s$ or $(LR)_i$, also results in the irreversibility of the sensor. Thus, in competitive binding reactions the relative quantity of reagent R_s with respect to L_s must also be considered.

Many colorimetric and fluorimetric reactions are irreversible, due to the formation of either a tight binding reagent-analyte complex or a colored additive that undergoes an irreversible reaction. Insoluble compounds, due to a limited solubility, can precipitate on the surface of the waveguide and decrease the reversibility of a sensor. The irreversibility increases when the solubility product (a product of the concentration of the ions of a substance in a saturated solution) decreases. The use of organic precipitating agents, a change of phase (aqueous-organic solvent), or the change of operating conditions (e.g., pH, complexation, ionic strength) can have the same effect.

The support layer (e.g., a membrane) with the immobilized reagent or ligand must be carefully examined to determine whether it remains stable in the operating environment. This is important for most sensors near the limits of their working range (e.g., strong acidity for a pH measurement), in which the membrane can be influenced by the medium and its characteristics can change with time. In addition, the optical window (e.g., a transparent membrane) can produce a loss of selectivity and irreversibility of the sensor.

Conversely, a nonreversible sensor can become reversible if it is used under appropriate conditions during the measurement. This has resulted in the development of fiber-optic sensors with reservoirs, with renewable reagents, or based on reagent delivery systems with controlled-release polymers.

2.6 CONTROLLED-REAGENT SENSORS

2.6.1 With Reservoir

The reservoir is a reagent compartment that is very large compared to the solution volume being examined optically [10]. It has the advantage of maintaining a large quantity of the reagent in the chemical equilibrium zone and avoiding the irreversibility of the sensor through consumption of the reagent. The reservoir can be made with a membrane that adds to the selectivity of the sensor. The reservoir is an interesting alternative to indicator-based sensors (see Chapter 4) with the restriction that the analyte does not exhibit intrinsic luminescence [11]. Indeed, the sensitivity is generally better in solution than in the heterogeneous phase with immobilized reagents. Sometimes the addition of a reagent can make an irreversible reaction between the analyte and the reagent reversible by providing a reservoir with an appropriate medium. The main disadvantages are the limited lifetime when mass transfer occurs with the chemical reaction and the difficulty in manufacturing the sensor. A typical example is given in Chapter 12 for the determination of organochlorides [12].

2.6.2 With Renewable Reagent

The system of a reservoir-based reagent leads to the concept of a fiber-optic sensor with a renewable reagent. The sensor design can be based on irreversible or reversible chemical reactions, since the reaction products are carried away by the reagent flow. Figure 2.8 is an illustration of this concept for the direct control of ammonia with a sensing volume of approximately 400 nl located at the probe tip [13].

The sensor tip consists of a semipermeable microporous hollow fiber, which contains the reagent delivery capillary and three optical fibers: an input fiber, a reference, and a fiber for the detection of the absorbance change of the internal reagent. The reagent is delivered by a precision microinjection pump.

In renewable reagent sensors, the sensitivity and the dynamic range depend strongly on the reagent composition and flow conditions. Schemes have been proposed for multiple reagent delivery capillaries injecting several reagents simultaneously. The major drawbacks of these sensors are the cost and the relative complexity of the sensor.

2.6.3 Polymeric Delivery Systems

Sensors based on controlled reagent release from polymers allow the sustained release of reagents, indicators, or molecules such as enzymes, antigens, antibodies, and hormones. These systems open up new possibilities for chemical or biochemical sensing with analytical reactions that consume reagents. A typical example of a fiber-optic pH sensor illustrating this concept has an ethylene-vinyl acetate copolymer as support for the reagent [14]. The sensing reagent in powder form is added to a polymer solution in an appropriate solvent, mixed, cast in a mold, and dried. Other techniques of immobilization can be used with these types of sensors.

A sensor with a controlled-release polymer, shown in Figure 2.9, consists of three main parts: a reservoir with a polymer matrix, a sensing region with many tiny holes for optimized diffusion of the analyte into the optical measurement zone, and an optical fiber passing through the polymer reservoir and positioned at the top of the sensing region. The design and construction of the sensor is critical in determining its performance. The response time, typically 10 minutes, is generally longer than for

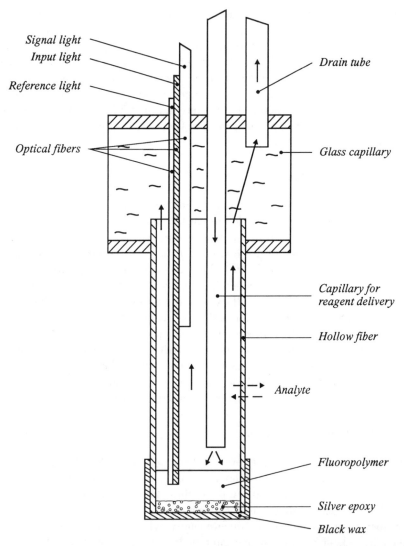

Figure 2.8   An ammonia sensor with a renewable reagent. A syringe pump delivers the reagent via a capillary tube to the bottom of a hollow fiber cavity. The hollow fiber acts as a membrane and allows the analyte to diffuse into the cavity. A drain tube evacuates the waste reactant products. Light from the input fiber is diffused by the fluoropolymer and silver-filled epoxy layers. A second fiber measures the light reflected from the silver particles (reference light), and a third fiber measures the signal from the cavity. (*After:* [13] with permission.)

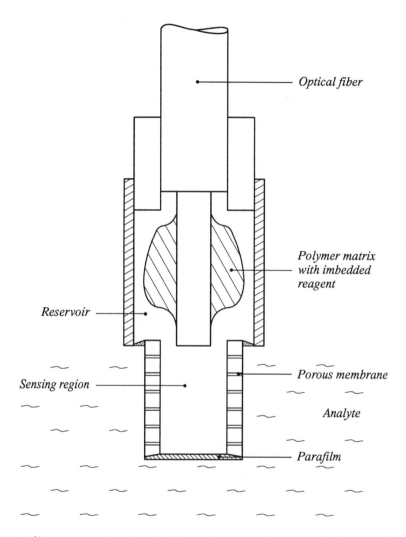

Figure 2.9 A fiber-optic sensor based on the delivery of a reagent from a controlled-release polymer. The reagent is released slowly from the polymer matrix into the reservoir region and reacts with the analyte in the sensing region. (*Redrawn from* [14] with permission.)

other sensors measuring the same analyte, but still acceptable for remote sensing or process control applications. This type of sensor has been used under laboratory working conditions for at least 3 months. However, the kinetics of the rate of release of the reagent from the polymer matrix must be controlled under real working conditions. The interest in this technique is also the possibility of mixing several

Table 2.1
Typical Examples of Direct and Indirect Sensors

Types	Direct	Indirect
Reversible	pH sensor (Section 2.7)	Glucose sensor (Figure 2.7) Biocatalytic sensors (Chapter 3)
Nonreversible	Evanescent wave immunosensor (Chapters 7 and 10)	Competitive binding immunosensor (Chapter 3)
Controlled (reagent release or immobilization control)	UO_2^{2+} sensor [11]	Capillary fill immunosensor (Chapter 10 and [15])

reagents. One reagent could serve as an internal reference to normalize the signal if the variation of its release rate is identical to other reagents.

Table 2.1 summarizes some typical examples of direct and indirect sensors with fiber optics or waveguides. Molecular recognition of large molecules in biosensing generally leads to nonreversible sensors, as opposed to ionic recognition producing reversible sensors.

2.7 EXCHANGE REACTIONS

Various exchange reactions are possible in chemical equilibria, such as the exchange of electrons, protons, ions, and polar molecules, or the exchange between several compounds (ions, protons, and electrons).

1. An exchange of electrons can occur between an oxidizing agent capable of accepting electrons and a reducing agent giving electrons:

$$\text{oxidizing agent} + n \cdot e^- \rightleftharpoons \text{reducing agent} \qquad (2.55)$$

The chemical reaction is generally based on a system with two oxidation-reduction agents. Combining two equations of the form of (2.55) gives

$$Ox_1 + Red_2 \rightleftharpoons Red_1 + Ox_2 \qquad (2.56)$$

In optical sensors the change in the degree of oxidation leads to a change in the absorption or fluorescence spectrum.

2. An exchange of protons can occur between an acid giving a proton and a base accepting the proton:

$$\text{Acid (HA)} \rightleftharpoons \text{Base (A}^-) + H^+ \qquad (2.57)$$

The constant K_a of this equilibrium is

$$K_a = (\gamma_{H^+}) \cdot [H^+] \cdot \gamma_{A^-} \cdot [A^-] / \gamma_{HA} \cdot [HA] \qquad (2.58)$$

where γ_{H^+}, γ_{A^-}, and γ_{HA} are the activities of H^+, A^-, and HA species, respectively, and the brackets [ ] correspond to the concentrations of these species.

The pH is defined from the activity of H^+ ions (a_{H^+}):

$$pH = -\log (a_{H^+}) = -\log (\gamma_{H^+} \cdot [H^+]) \qquad (2.59)$$

Equation (2.58) is normally expressed in logarithmic form as the Henderson-Hasselbach equation:

$$pH = pK_a + \log([A^-] / [HA]) + \log(\gamma_{A^-}/\gamma_{HA}) \qquad (2.60)$$

In optical pH sensors, an indicator that is immobilized on the fiber or into an inert support at its end is used as the reagent. Sörensen's definition (with $\gamma_{H^+} = 1$) is applied for the pH. Also, a constant K_a' is defined as

$$pK_a' = pK_a + \log(\gamma_{A^-} / \gamma_{HA}) \qquad (2.61)$$

Hence,

$$pH = pK_a' + \log([A^-]/[HA]) \qquad (2.62)$$

The general form of the signal S, measured in terms of the fluorescence or absorbance, is a sigmoid as shown in Figure 2.10, with

$$pH = pK_a' - \log\{(S_{max} - S)/(S - S_{min})\} \qquad (2.63)$$

where S_{max} is the maximum signal of the completely dissociated form, S the signal at any given pH, and S_{min} the signal of the totally undissociated (conjugated) form (HA).

Thus, the pK_a' is generally affected by the ionic strength (I_s) of the medium and especially the microenvironment of the sensing part of the optode:

$$pK_a' = pK_a - M \cdot (Z_{HA}^2 - Z_{A^-}^2) \cdot I_s^{1/2} / (1 - B \cdot d \cdot I_s^{1/2}) \qquad (2.64)$$

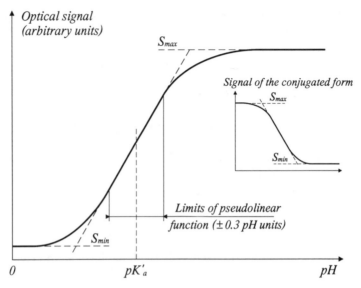

Figure 2.10 Response curve for pH sensing measurements. The optical signal (in arbitrary units) is the absorbance measurement of an indicator and has a general sigmoid form for acid-base reactions. A pseudolinear function is obtained within a narrow zone ($\pm$0.3 units) around pK'_a. S_{max} and S_{min} are the signals for the completely dissociated and totally undissociated forms, respectively. The general form of the signal is reversed for the conjugated form, shown in insert.

where Z_{A^-} and Z_{HA} are the charges of the fluorophore components A^- and HA, respectively, B is a dielectric constant of the solvent, d is an ion-size parameter, and M a constant that varies with the temperature.

3. The exchange of a charged particle can occur between ions or polar molecules in which a complex ion, the donor, gives a charge to an acceptor according to the following general reaction:

$$\text{acceptor}_1 + \text{donor}_2 \rightleftharpoons \text{donor}_1 + \text{acceptor}_2 \qquad (2.65)$$

This can result in the formation or disappearance of a colored complex, allowing the detection of specific ions or molecules. Many metal-ligand complexes are exploited in complexometry, for example, in titration systems to indicate the equilibrium end point.

4. Molecular recognition in optical sensors can be based on the equilibrium conditions between several different components, such as ions and protons (complexation or

changes in acidity), electrons and ions (oxidation-reduction reactions and complexation), and electrons and protons (oxidation-reduction and acidity changes).

2.8 INTRINSIC PROPERTIES OF THE ANALYTE

In some optical sensors, steps 1 and/or 2 of Figure 2.1 may not be needed and the intrinsic properties of the analyte (e.g., optical spectrum, color) are sufficient for a quantitative and direct determination after calibration. For instance, many compounds can be detected directly in transmission, reflectance, or fluorescence with a bifurcated fiber probe (Figure 2.11). Also, fibers can serve as simple light transmission guides to optical measurement cells (extrinsic optodes) such as probes of flow-through cells (see Chapter 10). A large number of examples of these simple sensors can be found in various applications, such as the determination of copper in electroplating baths [16], the analysis of nuclear compounds in process control [17], oximetry [18] (see Chapter 11), measurement of gases (see Chapter 12), or the determination of bilirubin in blood based on its strong fluorescence (excitation at 430 nm; emission at 520 nm) [19].

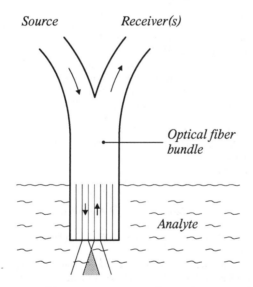

Figure 2.11 Direct measurement of the intrinsic properties of an analyte with a bifurcated fiber bundle. For fluorescent measurements (shown here), fluorescence is collected from the overlap area (shown shaded for two fibers) between the cone of exciting light and the acceptance cone for the collected fluorescence. A bundle with many fibers improves the collection efficiency and randomization of the light. For absorption, a reflection element (e.g., a mirror) is added. For multiwavelength analysis, several sources and detectors can be connected to the input and output fibers.

References

[1] Janata, J., *Principles of Chemical Sensors*, New York: Plenum Press, 1989.

[2] Edmonds, T. E., ed., *Chemical Sensors*, Glascow, U.K.: Blackie and Sons, 1988.

[3] Christian, G. D., *Analytical Chemistry*, 4th ed., New York: Wiley, 1986.

[4] Seitz, W. R., "Chemical Sensors Based on Fiber Optics," *Anal.Chem.*, Vol. 56, 1984, pp. 16A, 18A, 20A, 22A, 24A, 33A, 34A.

[5] Wolfbeis, O. S., and G. Boisdé, "Optochemical Sensors," Chap. 17 in *Sensors: A Comprehensive Survey, Vol. 3, Chemical and Biochemical Sensors*, W. Göpel, T. A. Jones, M. Kleitz, J. Lundström, and T. Seiyama, eds., Weinheim: VCH, 1991, pp. 887–930.

[6] Narayanaswamy, R., and F. Sevilla III, "Reflectometric Study of the Acid-Base Equilibria of Indicators Immobilized on a Styrene/Divinylbenzene Copolymer," *Anal. Chim. Acta*, Vol. 189, 1986, pp. 365–369.

[7] Schultz, J. S., S. Mansouri, and I. J. Goldstein, "Affinity Sensor: A New Technique for Developing Implantable Sensors for Glucose and Other Metabolites," *Diabetes Care*, Vol. 5, 1982, pp. 245–253.

[8] Hisamoto, H., K. Watanabe, E. Nakagawa, D. Siswanta, Y. Shichi, and K. Suzuki, "Flow Through Type Calcium Ion Selective Optodes Based on Novel Neutral Ionophores and a Lipophilic Anionic Dye," *Anal. Chim. Acta*, Vol. 299, 1994, pp. 179–187.

[9] Barnard, S. M., and W. R. Walt, "A Fibre Optic Chemical Sensor With Discrete Sensing Sites," *Nature*, Vol. 353, 1991, pp. 338–340.

[10] Wolfbeis, O. S., "Fiber Optical Fluorosensors in Analytical and Clinical Chemistry," Chap. 3 in *Molecular Luminescence Spectroscopy: Methods and Applications*, Part II, S. J. Schulman, ed., New York: Wiley, 1988, pp. 129–281.

[11] Milanovich, F. P., and T. Hirschfeld, "Process, Product and Waste Stream Monitoring With Fiber Optics," *Adv. Instrum.*, Vol. 38, 1983, pp. 407–418.

[12] Milanovich, F. P., D. G. Garvis, S. M. Angel, S. M. Klainer, and L. Eccles, "Remote Detection of Organochlorides With a Fiber Optic Based Sensor," *Anal. Instrum.*, Vol. 15, 1986, pp. 137–147.

[13] Berman, R. J., and L. W. Burgess, "Renewable Reagent Fiber Optic Based Ammonia Sensor," *Proc. SPIE: Int. Soc. Opt. Eng.*, Vol. 1172, 1989, pp. 206–214.

[14] Luo, S., and D. R. Walt, "Fiber Optic Sensors Based on Reagent Delivery With Controlled-Release Polymers," *Anal. Chem.*, Vol. 61, 1989, pp. 174–177.

[15] Badley, R. A., R. A. L. Drake, I. A. Shanks, A. M. Smith, and P. R. Stephenson, "Optical Biosensors for Immunoassays: The Fluorescent Capillary Fill Device," *Phil. Trans. R. Soc. London, Ser. B*, Vol. 316, 1987, pp. 143–160.

[16] Freeman, J. E., A. G. Childers, A. W. Steele, and G. M. Hieftje, "A Fiber Optic Absorption Cell for Remote Determination of Copper in Industrial Electroplating Baths," *Anal. Chim. Acta*, Vol. 177, 1985, pp. 121–128.

[17] Boisdé, G., F. Blanc, and J. J. Perez, "Chemical Measurements With Optical Fibers for Process Control," *Talanta*, Vol. 35, 1988, pp. 75–82.

[18] Polanyi, M. L., and R. M. Hehir, "In Vivo Oximeter With Fast Dynamic Response," *Rev. Sci. Instrum.*, Vol. 33, 1962, pp. 1050–1054.

[19] Coleman, J. T., J. F. Eastham, and M. J. Sepaniak, "Fiber Optic Based Sensor for Bioanalytical Absorbance Measurements," *Anal. Chem.*, Vol. 56, 1984, 2249–2251.

Chapter 3

Kinetics and
Shape Recognition

3.1 INTRODUCTION

In a sensor, ionic and molecular recognition is based on kinetics and shape recognition of molecules. This forms part of the first step of the sensing process and complements chemical reactions as discussed in Chapter 2. Kinetic measurement is well-adapted for enzymatic reactions and improves the selectivity of analysis. However, this selectivity varies from enzyme to enzyme.

Shape recognition is another method of increasing selectivity for determination of the substrate. The substrate is defined as the analyte in a bioprocess. Figure 3.1 shows the formation of an enzyme-substrate complex by shape recognition, represented schematically as a key (substrate) and lock (enzyme). Shape recognition has advantages in immunological determinations, for example, in the binding of an antibody (Ab) to an antigen (Ag). Antibodies normally occur with two common forms, Y and T, as shown in Figure 3.2.

Stereospecificity in relation to shape recognition is demonstrated in Figure 3.3 for an enzyme (lock), which can distinguish the key (substrate) from a foreign body, in a system where the water of hydration is eliminated from the binding site under the action of hydrophobic forces [1].

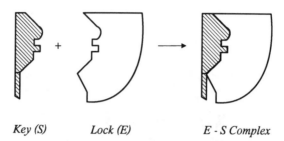

Key (S)   *Lock (E)*   *E - S Complex*

Figure 3.1   Shape recognition of molecules, such as an enzyme. The substrate S, in the form of a key, reacts with the enzyme E, in the form of a lock, to create an enzyme-substrate complex.

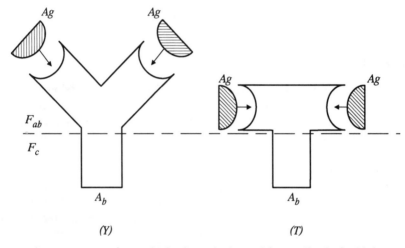

Figure 3.2   Shape recognition of an antibody. The antibody can deform itself at the flexible hinge region to occur in two shapes, the Y and T forms. Antigens bind to the antibody in the F_{ab} upper region.

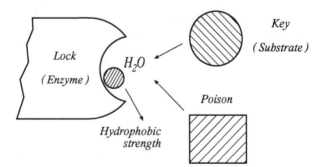

Figure 3.3   Stereospecific binding of an enzyme (lock) with a substrate (key) in the presence of a poison and water molecules. The poison, as a foreign body, cannot become locked to the enzyme due to its different three-dimensional shape.

3.2 KINETICS

3.2.1 Concepts

Kinetics deals with reaction rates, which are commonly determined by optical techniques in chemical measurement. The method is automatically self-referencing because it takes into account the initial state of the analyte. It also yields information about the dynamic state of a system.

3.2.1.1 First-Order Reactions

A reaction is classed as first-order when the reaction rate is directly proportional to the concentration of a single substance. For an analyte (A), which undergoes a chemical reaction to produce a product (P), the reaction rate, equal to the rate of disappearance of A $(-dA)$ during a time (dt), is proportional to the concentration of A:

$$-dA / dt = k \cdot [A] \tag{3.1}$$

where k is the specific rate constant. Integrating this equation yields

$$\log([A] / [A_0]) = -k \cdot t / 2.303 \tag{3.2}$$

where A_0 is the initial concentration of the analyte. The rate of the reaction decreases exponentially with time. The half-life of the reaction, $t_{1/2}$, is the time for one-half of the quantity of the substance to react:

$$t_{1/2} = 0.693 / k \tag{3.3}$$

3.2.1.2 Higher Order Reactions

When the rate of reaction is equal to the disappearance of either A or B components, the reaction is first-order with respect to A or B, but second-order overall, since the exponential term is the sum of two concentration terms:

$$-dA / dt = -dB / dt = k' \cdot [A] \cdot [B] \tag{3.4}$$

If $B >> A$, the concentration of B remains virtually constant and the equation is reduced to a first-order rate.

3.2.1.3 Catalysis

A catalyst is a substance that changes the rate of a reaction without altering the chemical equilibrium and without being modified itself. When two substances A and

B react slowly, the presence of a catalyst can accelerate (or decelerate) the reaction rate. In certain reactions, the catalyst can become attached to the reaction products or the substrate, and its concentration can change and be used as a measure of analyte concentration. For example, in enzyme-controlled reactions, the enzyme activity, defined as transforming 1 µmol of substrate per minute [2], can change during the process and be determined by optical means [3]. The variation of catalyst concentration is measured in two basic ways, as shown by the optical absorption curves in Figure 3.4. The first method measures the concentration of one species at a given constant time (*t*), and the calibration curve is obtained from the absorbance at this time against the catalyst concentration (vertical dotted line in Figure 3.4). The second method takes the time required to reach a constant absorbance. This time increases proportionally to the reciprocal of the catalyst concentration (horizontal dotted line in Figure 3.4).

There is a large interest in catalytic reactions. Biocatalysis is frequently found in living processes, such as enzymatic reactions. However catalytic reactions are not

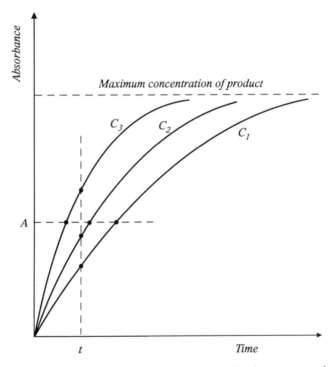

Figure 3.4 Method of measurement of catalyst activity (*C*) by the absorbance (*A*) as a function of time (*t*). The calibration curve for different concentrations C_1, C_2, and C_3 can be determined either at a fixed time or at fixed absorbance. The rate depends on the catalyst (see Figure 3.6).

specific in nature, because several catalysts can have the same effect on a substance and other foreign substances can inhibit or accelerate a reaction by changing the reaction rate.

3.2.1.4 Photokinetics

In chemical reactions caused by photons leading to photophysical or photochemical effects, the rate of reaction can be correlated with the concentration of interfering substances produced by the action of the light. The photochemical effect takes place in a fluorophore, which is influenced by the surrounding medium and hence can be changed by the analyte or a substance reacting with this analyte. The photochemical process changes the nature of the fluorophore and may compete with the deactivation of molecules by the emission of fluorescence. In this case, the absorption or fluorescence spectrum of the fluorophore can be repeatedly recorded during the change of the photoreaction to determine the analyte [4]. Recording with photodiode arrays is preferable because the signal can be scanned rapidly in a few milliseconds.

Practically all organic and metal organic chromophores are sensitive to photons, and light induces chemical transformations by photodecomposition. This decomposition, particularly under the effect of UV light, is a disadvantage for online sensors, since it influences their stability and reproducibility. Thus, at present, no optode or waveguide sensor uses the principle of photodecomposition of chromophores.

3.2.2 Biocatalytic Reactions

3.2.2.1 Enzymes

Enzymes are proteins of high molecular weight (10^4 to 2.10^6) that catalyze reactions of chemical or biochemical substances called *substrates*. As catalysts, enzymes are fully regenerated after a reaction and have a reversible response. An enzyme is characterized by the turnover number (TN). This is defined as the ratio of the converted substrate (S_c) to the enzyme concentration $[E]$, to the number of active sites per molecule (N), and to unit conversion time (t) [2]:

$$TN = S_c \ / \ ([E] \cdot t \cdot N) \tag{3.5}$$

The selectivity of enzymatic reactions varies from enzyme to enzyme in the same way as the binding constant (10^2 to 10^8) and the turnover number (TN = 10^{-1} to $10^5 \sec^{-1}$). In general, the action of enzymes is very fast with a high turnover number, up to $6.10^5 \sec^{-1}$ for a typical sensor. Enzymes are involved in body biochemistry and in most living processes (e.g., plant growth and biotechnology). Making use of enzymatic reactions, optical biocatalytic sensors are kinetic devices based on the establishment of a steady-state concentration of an optically detectable species at the surface of the transducer. An optical enzymatic sensing scheme is shown in Figure 3.5.

Typically, the reaction consuming a substrate (S) in the presence of an enzyme (E) gives a product (P) as described by the Michaelis-Menten expression:

$$S + E \underset{k_{-1}}{\overset{k_{+1}}{\rightleftharpoons}} ES \overset{k_2}{\rightleftharpoons} P + E \tag{3.6}$$

where ES is the activated enzymatic complex, k_{+1} is the rate constant for the forward reaction for step 1, k_{-1} is for the reverse step 1, and k_2 is the reaction rate constant for step 2. The reaction rate (v) of the (ES) compound, evaluated in terms of its concentration, is

$$v = k_2 \cdot [ES] \tag{3.7}$$

When the substrate concentration is high, all the available enzyme (E_T) is complexed with the substrate, and the reaction rate is a maximum (v_{max}) and proportional to the total enzyme concentration:

$$v_{max} = k_2 \cdot [E_T] \tag{3.8}$$

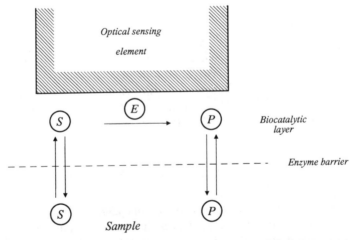

Figure 3.5   A biocatalytic optical sensing scheme. The substrate S reacts with the enzyme E in the biocatalytic layer to form the product P. The enzyme barrier prevents the enzyme from escaping into the sample. The optical sensor measures the reaction by absorbance or fluorescence from the variation of the enzyme, the substrate, or the product.

with

$$[E_T] = [E] + [ES] \tag{3.9}$$

At the steady-state condition, the concentration of the enzymatic complex is constant $(d([ES])/dt = 0)$, and

$$k_1 \cdot [S] \cdot [E] = (k_{-1} + k_2) \cdot [ES] \tag{3.10}$$

The Michaelis-Menten constant k_M is defined by

$$k_M = (k_{-1} + k_2) / k_1 = [S] \cdot [E] / [ES] \tag{3.11}$$

Using (3.7) through (3.9), this becomes

$$k_M = (v_{\max} / v - 1) \cdot [S] \tag{3.12}$$

or the well-known rate function

$$v = v_{\max} \cdot [S] / ([S] + k_M) \tag{3.13}$$

If $v = v_{\max}/2$, $k_M = [S]$ and the Michaelis-Menten constant gives the measurement of the specific enzyme activity. This specific activity is defined as the number of units of an enzyme per milligram of protein. Figure 3.6 is a graphical representation of the rate function for various values of k_M.

When k_M is small, the enzyme becomes saturated at small concentrations of the substrate. For large k_M, high concentrations of substrates are necessary to achieve a maximum reaction rate. When $[S] >> k_M$, the response of the sensor is independent of the substrate concentration. When $[S] << k_M$, the reaction rate is first-order with respect to substrate concentration. Hence, the generation of the product and the response of the sensor are also first-order. For low enzyme loading, with $[E]$ small, the k_M value gives the upper limit of detection (see (3.11) and Figure 3.6). Thus, the dynamic range is dependent on the enzyme loading.

Physically, an enzyme has a stereospecific binding site. This binding site contains amino acid residues that catalyze the formation or the dissociation of covalent bonds. The exact orientation of the protein surface influences the enzymatic reaction with the substrate. From this point of view, an enzyme, in contrast to most catalysts, is selective for a given substrate. However, other chemical species can affect the enzymatic activity, and the information acquisition process scheme is not specific in nature [1].

Enzymes normally refer to the reaction type or compound and their names end in *ase*. For instance, luciferase is an enzyme from luciferin and penicillinase from

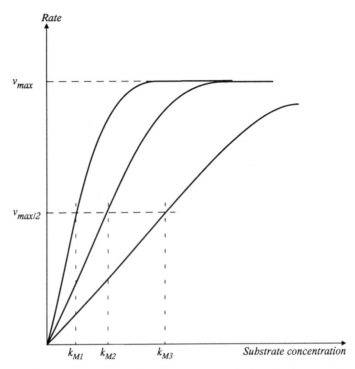

Figure 3.6   The Michaelis-Menten constant, illustrated on a graph of reaction rate against substrate concentration, corresponds to the concentration at $v = v_{max}/2$. The curves with constants k_{M1}, k_{M2}, and k_{M3} are for different reactions.

penicillin. They are divided into six main groups [5,6] corresponding to their catalytic action: removal of water (hydrases and hydrolases), oxidation reduction (oxidases and reductases), transfer of functional radicals (transferases), addition onto double bindings (lyases and synthases), isomerization (isomerases), and combination of two coupled molecules (ligases and synthetases). More detailed information about enzyme functions can be found in specialized texts [5–10].

3.2.2.2 Inhibitors and Coenzymes

Sometimes other substances called *inhibitors*, which have a size and shape similar to enzymes, can be adsorbed on the active sites at the same time as the enzymes. They can slow down the rate of a catalyzed reaction by competitive inhibition. If the inhibition depends only on the concentration of the inhibitor, the inhibition is noncompetitive. This is the case if the inhibitor is adsorbed at a site other than the active site, but which is important for the activation process. Also, substrate inhibition can occur when excessive amounts of substrate are present and block the active sites.

The formation of an organic stereochemical complex with a metal (e.g., a magnesium ion) may be required for the activation of an enzyme. These nonprotein activators are normally called *cofactors*. They are organic substances of relatively low molecular mass (usually less than 1 kilodalton (kDa)) that link two reactions by transferring a group from one substance to another. Conversely, a substance complexed with the same metal can become an inhibitor decreasing the activity of a coenzyme (such as flavine adenine dinucleotide (FAD) covalently bound to an enzyme).

3.2.2.3 Measurement Methods

There are two methods for measuring enzymatic substrates. The first technique measures the complete conversion of the substrate. The product or the depletion of a reactant originally in excess is analyzed before and after completion of the reaction. Calibration is done relative to a known concentration or quantity of the substrate.

The second technique, the most common method, measures the rate of the catalyzed reaction either by determining a single point under well-defined experimental conditions (end point measurement) or by continuous measurement. In the former case, the analysis of the production of a preset amount of the product or the consumption of a preset amount of substrate is determined after a given time. In the latter case, the continuous determination of the variation of P or S concentration with time can be made when the rate is a pseudo-first-order reaction.

Enzymes can be analyzed by measuring the amount of transformed substrate or product produced in a given time. If the substrate is in excess, the rate reaction is dependent only on the enzyme concentration. Inhibitors and coenzymes may be analyzed by the same techniques, particularly by studying the decrease or increase of the reaction rate.

3.2.3 Optical Biocatalytic Sensors

Optical biocatalytic sensors can be divided into two classes [11]. One class includes sensors in which the intrinsic optical properties of the enzyme, substrate, cosubstrate (a substrate that can activate several enzymes), and product are directly measured in the biosensing process. These optically detectable compounds produce a direct change in the absorbance or the luminescence of the optical spectrum. These compounds can be cofactors, such as the pair NADH-NAD$^+$. Furthermore, photons produced by the enzymatic reaction can be measured directly by chemiluminescence or bioluminescence techniques [12].

Sensors in the second class require a secondary reaction, an optochemical transduction to convert the substrate or the product into an optically detectable compound. These can be described as transducer-based optical biocatalytic sensors. Biosensing, using this secondary chemical transduction, is normally based on the measurement of an intermediate analyte such as oxygen, ammonia or ammonium ion, H_2O_2, carbon dioxide, NADH, or pH.

The immobilization of the enzyme is an important aspect of the fabrication of a sensor [13]. Normally it is done on a membrane serving as an enzymatic barrier and placed on the sensitive part of the waveguide, such as the distal tip of an optical fiber or on the surface of a waveguide for evanescent wave spectroscopy. Sometimes a second membrane is used for selective diffusion of the reaction gases. The most common steps in the immobilization procedure are physical trapping or attachment, protein cross-linking, and covalent attachment (see Chapter 5). Enzyme loading is important for the performance of the sensor. Immobilization must maintain a large enzymatic activity, but the amount of enzyme is contained on a small available surface area. This favors a planar geometry.

3.3 SHAPE RECOGNITION

3.3.1 Immunological Interactions

In living organisms, a highly selective chemical sensing mechanism is based on the stereospecific binding of an antigenic determinant (antigen, hapten, epitope) with an antibody. An antigen is a foreign substance that invades the host and stimulates the synthesis and the production of antibodies for the self-defense of the organism [14,15]. Haptens are small chemical molecules unable to induce antibody formation, but can react with its specific antibody once formed. An epitope is a simple single antigenic determinant. Antibodies, protein molecules (about 70Å in diameter) having a high molecular weight, typically 100 to 1,000 kDa, are formed from the B cells by the immune system to detect foreign matter (e.g., nucleic acids, hormones, vitamins, and viruses) by formation of a complex. Figure 3.7 shows the principle of an immunological reaction with the formation of an Ag-Ab complex, the basic step in immunosensing [16].

3.3.1.1 Structure of Antibodies

Antibodies are composed of serum proteins, the most common being immunoglobulin G (IgG), which has a molecular weight of 146 kDa. The structure of IgG is shown in Figure 3.8. It is made up of hundreds of individual amino acids arranged in a highly ordered sequence. The area of binding sites is estimated to be 1 to 4.10^{-6} μm^2. A single binding site corresponds to about 5,000 Da, with 25 to 50 amino acids.

The IgG molecule consists of two heavy and two light chains composed of segments of amino acids and linked together by disulfide (-S-S-) bridges [17]. These bridges contribute to the stability of the molecule and form loops of folded peptides interacting mainly through hydrophobic interactions. Disulfide linkages with

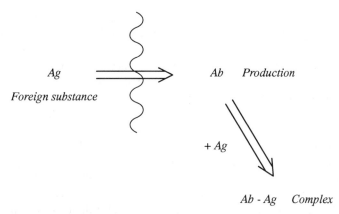

Figure 3.7 Principle of immunological reactions. A foreign antigen Ag penetrates across the boundary B of a living organism or cell and stimulates an immune reaction with the creation of an antibody Ab. The antibody then eliminates the antigen by reacting to form the complex Ab-Ag.

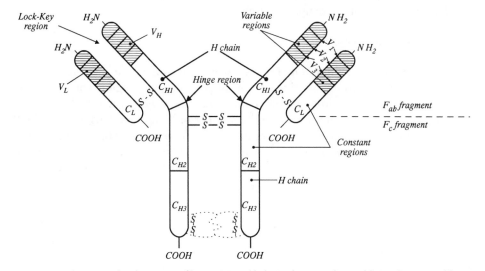

Figure 3.8 The IgG molecule consists of heavy (H) and light (L) chains, with variable (V), hypervariable (V_1, V_2, V_3) and constant (C) domains. The H chains are held together by double sulfide bonds. The upper fragment F_{ab} (antigen binding fragment) is flexible and can bend around the hinge region. The molecule can be broken into fragments F_c (crystallizable fragment), for example by enzymatic action. The tips of the L and H chains are linked by NH_2 groups and the lower ends with COOH functions.

hydrophobic interactions also hold the two H chains together. The hinge region is particularly susceptible to protein splitting. This zone also gives the molecule a certain flexibility so that it can change from the Y form to the T form (or intermediate forms) (see Figure 3.2). The hinge region of the molecule is the dividing line between the F_{ab} and the F_c fragments. At the end of the H and L chains, the variable regions have about a hundred amino acids placed in a variable-sequence characteristic of the antibody. The variable domains are divided up into three regions (V_1, V_2, V_3) of the H and L chains (e.g., HV_1, HV_2, HV_3) separated by a framework region that maintains the three-dimensional structure of the molecule. The variable domains are called the *complementary determining residues*. In the constant domains (C_L, C_{H1-3}), the amino acid sequences are practically invariant for different antibodies.

There are five distinct types of immunoglobulins, called G, A, M, D, and E. The molecules IgG, IgA, IgD, and IgE, but not IgM, are also divided into subclasses such as 1 and 2 (e.g., IgG_1). The IgG molecule has a double symmetry and four double-symmetry regions. Two sites on this molecule can fix an antigen by a combination of subunits in the V_H and V_L regions, each site being located at the tip of the F_{ab} fragment (on the tip of the variable regions in Figure 3.8).

3.3.1.2 Antigen-Antibody Interaction

An antigen and a specific antibody interact together through their three-dimensional structures like a key fitting a specific lock. Antibodies produced against antigens react with the specific areas of proteins called *antigenic determinants*. Except for IgM, the valence of the antibody determined by the number of bound antigens is generally 2, one for each tip of the arms on the immunoglobulin molecule. The approach of an antigen to an antibody produces a fast association with high interaction strength, quickly establishing a thermodynamic equilibrium. In terms of rate constants, the equilibrium equation is

$$Ag + Ab \underset{k_{diss}}{\overset{k_{ass}}{\rightleftharpoons}} Ag \cdot Ab \qquad (3.14)$$

with

$$K = k_{ass} / k_{diss} = [Ag \cdot Ab] / [Ag] \cdot [Ab] \qquad (3.15)$$

The Ag-Ab complex is bound by noncovalent binding forces, allowing a certain reversibility of the Ab-Ag interaction. Nevertheless, the values of the K constant are high (10^5 to 10^{11} mol^{-1}) resulting in partial nonreversibility. This explains the interest in regenerable sensors [18].

Affinity is a thermodynamic parameter that measures the interaction strength between one Ab site and one Ag site. In contrast, *avidity* expresses the strength with which a collection of antibodies (an antiserum with different affinities) is bound to an antigen. The equilibrium constant can be considered as the affinity constant. If it is large, the antibody shows a high affinity for the antigen. Frequently, affinity is used to refer to a monoclonal antibody (an antibody produced by daughter clones of a single B cell), which recognizes a single epitope, and avidity refers to polyclonal antibodies.

Polyclonal antibodies, coming from clones of a number of separate B cells, are heterogeneous, react with different epitopes, and can have a high specificity and a high avidity [2]. However, their preparation (e.g., from the blood of an appropriately immunized animal) must be carefully controlled to maintain a cross-reactivity (binding of different antigens to the same antibody) at an acceptable level. Polyclonal antisera may have a higher avidity and specificity for a given epitope than a monoclonal antibody. This is due to a bonus effect [19]. When two antibodies are bound to two epitopes of the same antigen, the complex is dissociated only when both antibodies detach themselves simultaneously. Thus, the overall Ag-Ab complex appears intact even when only a few antibodies are attached to the antigen, explaining an apparent specificity bonus.

The chemical interaction of an antigen-antibody involves four factors: electrostatic or coulombic forces at ionized sites ($COOH$ and NH_2 groups) and less strongly attached dipoles, hydrogen bonds (e.g., with amino acid function groups OH and NH_2 interacting with the water molecule), hydrophobic attractions resulting from apolar atomic groups, and van der Waals interactions coming from the dipole attractions.

3.3.1.3 Hapten-Antibody Interaction

To produce antibodies to haptens, an immunogenic carrier molecule, called a *partial antigen* (haptin) or *incomplete antigen* (haptene) can also be coupled to them and induce a helper T-cell response. For small antigens, such as haptens, the k_{diss} dissociation constant exhibits much greater variations (10^{-5} to 10^3 sec^{-1}) than the k_{ass} constant (10^7 to 10^8 mol^{-1}). Thus, the K constant is largely influenced by k_{diss} (see (3.15)). When two antigens have similar epitopes, the half life of interaction ($t_{1/2} = 0.693/k_{diss}$) can be long (>100 sec) or very short (10^{-2} sec) for high- or low-affinity antibodies, respectively.

3.3.2 Immunoreactions With Enzymes

Enzyme immunoassay (EIA) is based on the discriminatory power of antibodies to exhibit an affinity for a specific antigen or hapten, and on the high catalytic power and specificity of an easily detectable enzyme [2]. The use of enzymes covalently coupled to an antibody or an antigen have many advantages for high-sensitivity investigation of immunoassays. The enzyme is used as a tag that can be linked to either an antibody or an antigen. One enzyme molecule can react with thousands of substrate molecules, giving an *enzyme amplification* with a higher sensitivity. In these sensors, detection of the enzyme incorporated in the substrate is usually carried out by analyzing the change in color, luminescence, or refractive index.

Another advantage is that many reactions used to immobilize enzymes on substrates can be adapted for the immobilization of antibodies. However, the F_{ab} antibody portion that binds to the antigen must not be blocked by the immobilization.

3.3.3 Immunoassays

Immunoassays can be divided into three classes: direct, competitive, and noncompetitive methods [16]. In direct assays, a simple incubation of a naturally fluorescent antigen can be measured in the presence of an excess of antibody.

In competitive immunoassay (see Figure 3.9), a labeled antigen competes with an unlabeled antigen for reaction on a limited number of antibody sites, the antibody being immobilized on the support. After the substrate has been washed to eliminate the excess of the unbound labeled antigen, it is measured to determine the amount of labeled antigen that has been captured by the antibody. If the original concentration

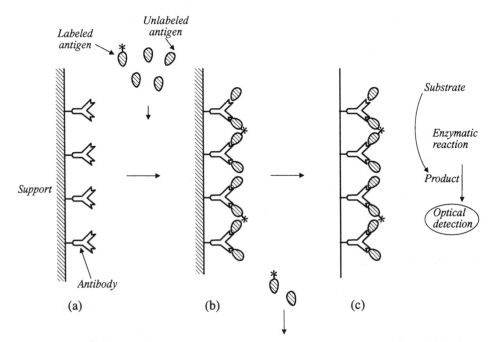

Figure 3.9 Competitive immunoassay: (a) *Step 1*: Labeled and unlabeled antigens compete for antibody binding sites on the support surface. (b) *Step 2*: After reaction, excess antigens are removed by washing. (c) *Step 3*: The concentration of unknown antigen is deduced from the measurement of the labeled antigen (e.g., by monitoring the rate of an enzymatic reaction).

of labeled antigen is known, the concentration of unlabeled antigen can be determined from the relative proportions of captured-to-uncaptured labeled antigen.

Noncompetitive binding assay, using the sandwich technique, is illustrated in Figure 3.10. A first antibody is immobilized, and an antigen is added that becomes immobilized on the antibody. There are a greater number of antibody sites than antigen to ensure complete adsorption of the antigen. After this, another antibody with an enzyme (or fluorescent) label is added. After eliminating the excess of the labeled antibody, the reaction between the enzyme-labeled antibody and the substrate takes place. Optical detection is made by measuring either a change in the absorption of the substrate (by a wavelength scan) or the fluorescence from the enzyme marker. The concentration of labeled antibody is a direct indication of the antigen concentration. This technique is known as the enzyme-linked immunosorbent assays (ELISA) method.

A cascade amplification may be obtained when two enzymes are used as markers, one attached to the labeled antibody and the second used as a catalyst in the reaction with the substrate to produce an optical signal proportional to the antigen concentration and the time squared [20].

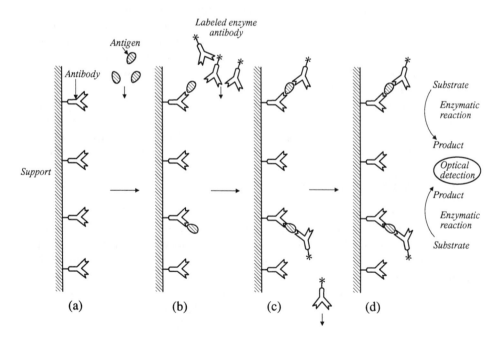

Figure 3.10   Noncompetitive immunoassay illustrated by the sandwich technique (Ab-Ag-Ab): (a) *Step 1*: The antigen to be measured reacts with the antibody on the support surface. (b) *Step 2*: A second enzyme-labeled antibody reacts with the trapped antigens. (c) *Step 3*: Excess antibodies are removed by washing. (d) *Step 4*: The antigen concentration is determined by measurement of the labeled antibody by an enzymatic reaction or by measuring a fluorescent marker.

In these techniques the signal is directly proportional to the antigen concentration and to the time from the start of the immunoassay test, subject to the available number of binding sites on the support. The disadvantages of the enzyme immunoassay techniques are related to problems of labeling antibodies with enzymes, the inhibition of the enzyme that may occur by the immunochemical reaction, and the increase in noise with time from the nonspecific reaction of the substrate.

3.3.4 Liposome and Enzyme Amplification

Liposomes are artificial vesicles of concentric shells of lipid bilayers surrounding an interior aqueous compartment [21]. They can be loaded in aqueous interspaces with an easily detectable optical marker such as a chromophore or enzyme. They possess natural specific antigenic determinants on their surface.

In the presence of a protein complement (a system of nine serum proteins reacting in a specific order), the lysis (dissolution) of an enzyme-sensitized liposome induces the release of a marker in high concentrations by a double amplification scheme. This scheme uses the liposome effect, which is the direct chemical release by leakage from the liposome, plus enzyme amplification. This amplification effect (factor 2 to 5) has been demonstrated with a Clark electrode for the determination of pO_2 using NADH [22], which can also be measured by the usual optical methods. However, this double amplification technique requires the stabilization of both the protein complement and the liposomes in order to avoid an inadequate leakage of the marker through the lipid bilayer.

An advantage of the liposome approach is the possibility of manipulating the interior composition of the liposome, such as making probes with various fluorescent liposomes [23]. Optical detection can be accomplished with the usual enzyme-immunoassay technique, for example, by liposome injection flow immunoassay [24].

3.3.5 Kinetics of Surface Reactions

Heterogeneous immunoassays require a separation step that is slow and time-consuming and introduces errors. Hence, efforts have been made to develop immunoassays with surface reactions in real time [25]. The antibody (or antigen) is immobilized on the surface of the optical substrate and the Ab-Ag interaction is monitored as the binding occurs. In such a system, the kinetics of the reaction are controlled either by the addition of the reagent itself or as a function of the surface properties of the substrate (e.g., surface charge, hydrophobicity). The boundary layer of liquid at the surface acts as an important barrier, affecting the diffusion of the antigen, and can change the reaction kinetics. These diffusion effects are concentrated in a boundary layer of thickness δ (a value relative to the cell dimension, the flow velocity, and the viscosity of the analyte):

$$\delta = (D \cdot t_d)^{1/2} \qquad (3.16)$$

where D is the diffusion coefficient and t_d the diffusion time. For reactions in which diffusion is the rate-limiting step (and not the Ag-Ab interaction), the value of δ can have an important influence on the equilibrium time of the reaction. Usually, in immunoassays at continuous surfaces, the equilibrium times are determined by the diffusion rates.

To understand the kinetics [25] based on rate measurements requires a knowledge of:

- The rate-limiting step (binding reaction or diffusion of the analyte);
- In kinetic assays, the diffusion step when the reaction rate factor dominates;
- The relative and absolute rates for k_{ass} and k_{diss} (see (3.14));
- The quantity of excess antigen/antibody (normally Ag, although sometimes Ab).

Figure 3.11 shows an example of dynamic measurements of human IgG on continuous surfaces using total internal reflection fluorescence (TIRF). A fluorescent marker, FITC, is attached to the antibody, and the detected fluorescence increases as a function of time as the binding takes place.

Kinetic measurements have the advantage of fast measurement times and are less affected by the reaction cell volume, which is very small. Many optical techniques (e.g., TIRF and surface plasmon resonance (SPR)) are possible with thin films on a continuous surface (see Chapter 8). However, problems exist for real-time biosensors, particularly if the sampling volume is very large and the flow of the analyte is not well defined.

3.3.6 Molecular Bioreceptors

Shape recognition is the ability of a molecular bioreceptor to identify and interact with a substrate by means of its form and size. Many different types of bioreceptors can be exploited in sensors and probes:

a. *Antibodies* (and antigens), especially immunoglobulins, represent a large number of different compounds. In addition, there are derived forms such as haptens, monoclonal antibodies, and hybrid antibodies that are a combination of two F_{ab} sites of antibody molecules. These sites can be identical with the same specificity, or different with a dual specificity.

b. *Enzymes* combine the transformation and recognition of molecules and allow amplification in biosensors. They are normally used as catalysts to accelerate the kinetic reaction and/or as marker tags.

c. *Lectins* are proteins and bind to oligosaccharides or single-sugar residues. The binding constants are typically 10^2 to 10^8 M for both reversible and irreversible sensors. The addition of saccharides competitively bound with a protein increases

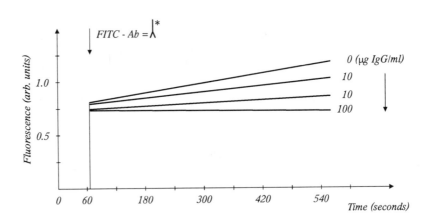

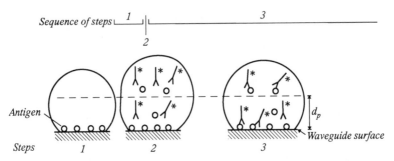

Figure 3.11 Immunoassay at a continuous surface showing binding curves of a fluorescent-labeled antibody (FITC-Ab) to immobilized antigen in the presence of a standard antigen solution. *Step 1*: No fluorescence is obtained with immobilized antigen on the waveguide surface. *Step 2*: There is an immediate jump in the fluorescent signal after introduction of fluorescent-labeled antibodies, due to their presence within the penetration depth d_p of the evanescent wave at the waveguide surface. *Step 3*: The fluorescent signal increases with time as the labeled antibody binds to the immobilized antigen. The fluorescent signal depends on the concentration of free antigens in solution, which reduces the quantity of labeled antibodies available to bind to the waveguide surface. (*After:* [26] with permission.)

the dissociation. Among the best known forms are jacalin, agglutinins, ricin, and concanavalin. The latter molecule (55 kDa for the A form) has been extensively studied because it is specific to both α-D-mannose and α-D-glucose residues (e.g., glycogen, dextrans, and mannans). A typical example of a fiber-optic sensor using concanavalin was given in Figure 2.6. Lectins can also bind to glycoproteins as immunoglobulins in a way similar to Ag-Ab interactions. This property is used in affinity purification and as a model for immunological interactions.

d. *Neuroreceptors* (neurological active compounds) indicate to an organism the presence of a compound that they recognize in the environment. These receptors, which include hormones and neurotransmitters, act as messengers. The membrane-embedded proteins, or protein aggregates, transduce the chemical signal from the extracellular to the intracellular environment via an interaction with a ligand. Their hydrophobic region interacts with the lipids in the membrane. The most studied are insulin (460 kDa), which consists of a glycoprotein with two α (135 kDa) and two β (95 kDa) units, and acetylcholine, the best neurotransmitter receptor in terms of both structure and function [27,28]. The receptor must be labeled with a fluorescent component in order to give an optical signal.

e. *Other molecular receptors* can be used in optical sensors. Avidin is a glycoprotein that selectively binds to four equivalents of the β-group vitamin biotin. The transport proteins form another class of molecular receptors, which are subdivided into three according to their transport function [29]: extracellular transport proteins found in the bloodstream (e.g., lipoproteins, albumin, and transferin), membrane-transport proteins found in the cell membrane (e.g., ion-channel proteins, which open and close to allow the flow of specific ions into or out of the cell), and intracellular transport proteins (e.g., the protein A, which binds to immunoglobulins without interacting with the antigen binding F_{ab} part of these molecules).

f. *DNA* and *DNA-binding proteins*: Deoxyribonucleic acid (DNA) is a long-chain polymer built from nucleotide units. The DNA molecule is unique in that it is present in the nucleus of the biological cell to control the production of proteins, and it carries essential genetic information. The fine structural detail of the DNA monomer is a phosphate group ($-O-PO_3^{2-}$) covalently bonded to a carbon atom attached to a pentose ring (see Figure 3.12) [30]. One of four bases, which can be adenine (A), thymine (T), guanine (G), or cytosine (C), is attached to a carbon atom of the deoxyribose ring, covalently bonded through the nitrogen of the (NH-) group forming a nucleoside. The polymers are formed in a structure known as a nucleic acid. The DNA molecule binds to another molecule in a double helix (duplex form) in which two long nucleic acids strands are wound around each other. In this helix, the bases fit together via hydrogen bonds formed by base pairs (AT or GC pairs). The specificity and the genetic code of DNA are determined by the sequence of these base pairs, which allows recognition of the complementary DNA. Thus, the sequence ATGC only binds to TAGC, the complementary sequence of another DNA molecule. In ribonucleic acid (RNA), the uracil base is equivalent to thymine.

DNA can also bind to complementary polycyclic molecules of RNA by means of hydrogen bonds between bases. Thus, polycyclic aromatic compounds metabolically activated to electrophilic intermediates can bind covalently to DNA and be detected in trace quantities by optical methods. A recent example of fluoroimmunoassay was developed for the detection of DNA-adduct products of carcinogens down to 10^{-15} concentration levels [31].

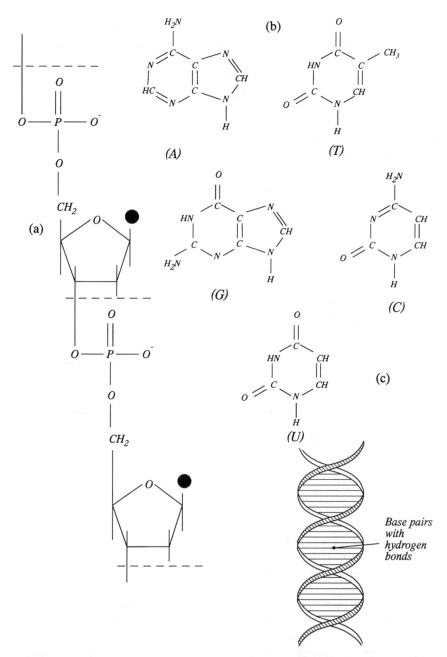

Figure 3.12 Deoxyribonucleic acid (DNA): (a) backbone of the molecule showing the place (black point) where the base binds onto it through the nitrogen of the NH groups; (b) base molecules adenine (A), thymine (T), guanine (G), cytosine (C), and for ribonucleic acid (RNA), uracil (U) replaces thymine; and (c) DNA double helix with hydrogen bonds between base pairs (AT or GC).

References

[1] Janata, J., *Principles of Chemical Sensors*, New York: Plenum Press, 1989.

[2] Tijssen, P., "Practice and Theory of Enzyme Immunoassays," in *Laboratory Techniques in Biochemistry and Molecular Biology, Vol. 15*, R. H. Burdon and P. H. van Knippenberg, eds., Amsterdam: Elsevier, 1985, p. 39.

[3] Wolfbeis, O. S., "Fiberoptic Probe for Kinetic Determination of Enzyme Activities," *Anal. Chem.*, Vol. 58, 1986, pp. 2874–2878.

[4] Gauglitz, G., and H. Mauser, *Principles and Applications of Photokinetics*, Amsterdam: Elsevier, 1991.

[5] Barman, T. E., *Enzyme Handbook, Vols. 1 and 2*, Berlin: Springer Verlag, 1969.

[6] Mouranche, A., and C. Costes, *Hydrolases et Depolymerases*, Paris: Gauthier-Villars, Bordas, 1985.

[7] Mosbach, K., *Methods in Enzymology, Vol. 44*, New York: Academic Press, 1976.

[8] Carr, P. W., and L. D. Bowers, *Immobilized Enzymes in Analytical and Clinical Chemistry*, New York: Wiley, 1980.

[9] Guilbault, G. G., *Analytical Uses of Immobilized Enzymes*, New York: Marcel Dekker, 1984.

[10] Mosbach, K., "Immobilized Enzymes and Cells," Part D in *Methods in Enzymology, Vol. 137*, New York: Academic Press, 1988.

[11] Arnold, M. A., and J. Wangsa, "Transduced-Based and Intrinsic Biosensors," Chap. 16 in *Fiber Optic Chemical Sensors and Biosensors, Vol. 2*, O. S. Wolfbeis, ed., Boca Raton: CRC Press, 1991, pp. 193–216.

[12] Blum, L. J., and S. M. Gautier, "Bioluminescence and Chemiluminescence Based Fiberoptic Sensors," Chap. 10 in *Biosensors Principles and Applications*, L. J. Blum and P. R. Coulet, eds., New York: Marcel Dekker, 1991.

[13] Madou, M. J., and S. R. Morrison, *Chemical Sensing With Solid State Devices*, Boston: Academic Press and Harcourt, 1989.

[14] Paul, W. E., *Fundamental Immunology*, New York: Raven Press, 1989.

[15] Freifelder, D., *Biologie Moleculaire*, Paris: Masson, 1990.

[16] Vo-Dinh, T., M. J. Sepaniak, G. D. Griffin, and J. P. Alarie, "Immunosensors: Principles and Applications," *Immunomethods*, Vol. 3, 1993, pp. 85–92.

[17] Vo-Dinh, T., G. D. Griffin, and M. J. Sepaniak, "Fiber Optics Immunosensors," Chap. 17 in *Fiber Optic Chemical Sensors and Biosensors, Vol. 2*, O. S. Wolfbeis, ed., Boca Raton: CRC Press, 1991, pp. 217–257.

[18] Bowyer, J. R., J. P. Alarie, M. J. Sepaniak, T. Vo-Dinh, and R. Q. Thompson, "Construction and Evaluation of a Regenerable Fluoroimmunochemical Based Fibre Optic Biosensor," *Analyst*, Vol. 116, 1991, pp. 117–122.

[19] Roitt, I., *Essential Immunology*, 4th ed., Oxford, U.K.: Blackwell Scientific, 1980.

[20] Thomson, R. B., and F. S. Ligler, "Chemistry and Technology of Evanescent Wave Biosensors," in *Biosensors With Fiber Optics*, D. L. Wise, and L. B. Wingard, eds., Clifton: Humana Press, 1991, pp. 111–137.

[21] Seitz, W. R., "Optical Ion Sensing," Chap. 9 in *Fiber Optic Chemical Sensors and Biosensors, Vol. 2*, O. S. Wolfbeis, ed., Boca Raton: CRC Press, 1991, pp. 1–17.

[22] Haga, M., S. Sugawara, and H. Itagaki, "Drug Sensor: Liposome Immunosensor for Theophilline," *Anal. Biochem.*, Vol. 118, 1981, pp. 286–293.

[23] Morgan, J., "Application of Fluorescent Liposomes," *Appl. Fluoresc. Technol.*, Vol. 2, 1990, pp. 14–16.

[24] Locascio-Brown, L., A. L. Plant, V. Horwath, and R. A. Durst, "Liposome Flow Injection Immunoassay: Implications for Sensitivity, Dynamic Range and Antibody Regeneration," *Anal. Chem.*, Vol. 62, 1990, pp. 2587–2593.

[25] Place, J. F., R. M. Sutherland, A. Riley, and C. Mangan, "Immunoassay Kinetics at Continuous

Surfaces," in *Biosensors With Fiber Optics*, D. L. Wise and L. B. Wingard, eds., Clifton: Humana Press, 1991, pp. 253–291.

[26] Place, J. F., R. M. Sutherland, and C. Dähne, "Optoelectronic Immunosensors: A Review of Optical Immunoassay at Continuous Surface," *Biosensors*, Vol. 1, 1985, pp. 321–353.

[27] Rogers, K. R., J. J. Valdes, and M. E. Eldefrawi, "Acetylcholine Receptor Fiber Optic Evanescent Fluorosensor," *Anal. Biochem.*, Vol. 182, 1989, pp. 353–359.

[28] Krull, U. J., R. S. Brown, K. Dyne, B. D. Hougham, and E. T. Vandenberg, "Acetylcholine Receptor Mediated Optical Transduction From Lipid Membranes," Chap. 22 in *Chemical Sensors and Microinstrumentation*, R. W. Murray, R. E. Dessy, W. R. Heinemann, J. Janata, and W. R. Seitz, eds., Washington, D.C., ACS Series, Vol. 403, 1989, pp. 332–344.

[29] Wingard, L. B., and J. P. Ferrance, "Concepts, Biological Components and Scope of Biosensors," in *Biosensors With Fiber Optics*, D. L. Wise and L. B. Wingard, eds., Clifton, Humana Press, 1991, pp. 1–27.

[30] Atkins, P. W., and J. A. Beran, *General Chemistry*, 2nd ed., New York: Scientific American Books, 1992.

[31] Vo-Dinh, T., T. Nolan, Y. F. Cheng, M. J. Sepaniak, and J. P. Alarie, "Phase Resolved Fiber Optics Fluoroimmunosensor," *Appl. Spectr.*, Vol. 44, 1990, pp. 128–132.

Part II:

Optical Chemoreception

Part I of this book has clearly demonstrated the importance of chemical reactions, kinetics, and shape recognition as the first step in the sensing process for a chemical sensor. The interaction of light with matter (see Chapter 6) is the main basis of optical sensors. In particular, light can be used to measure organic molecules having characteristic functional groups (e.g., chromophores) and can produce a temporary change in their optical properties, such as a change of their absorption or fluorescence spectrum.

Molecules that are optical chemoreceptors, such as indicators, dyes, receptors, and labels, can themselves be used to recognize particular molecules and ions or merely serve as markers. These compounds for optical sensing are examined in Chapter 4. Furthermore, the ways in which these indicators and labels are employed in optical sensors are extremely diversified. In particular, their arrangement (immobilization) on a support is also an essential part of the chemical-optical transduction mechanism as described in Chapter 5.

An excellent approach that provides a global overview of the subject [1] has been incorporated into this Part II.

Chapter 4

Indicators, Receptors, and Labels

4.1 ELECTRONIC TRANSITIONS

A molecule absorbs light at particular wavelengths or energies ($h\nu$) corresponding to transitions between well-defined internal energy levels. These energy levels correspond to quantized states for three different types of transitions: rotational, vibrational, and electronic. Many energy levels occur for each type. The result is a complex spectrum of absorption peaks corresponding to sharp lines from discrete transitions, and broad bands from individual transitions that are broadened by interactions within the molecule.

In the optical spectrum, electronic transitions are caused by the absorption by specific bonds and functional groups within the molecule. The absorption wavelength corresponds to the energy of transition between two states (electronic configurations); shorter wavelengths are equivalent to higher energies. The intensity of the absorption band, represented by the molar absorptivity (ε), depends on the probability of the transition and the polarity of the excited state.

In the UV and visible regions, the functional groups of a molecule that absorb light are called *chromophores* [2]. In general, they are unsaturated atomic groups, such as $\rangle C=C\langle$, $\rangle C=O$, $\rangle C=N-$, $-N=N-$, $O=N-$, and $\rangle C=S$, and include aromatic ring and quinoid structures. Other groups called *auxochromes*, such as hydroxy, amino groups, and halogens, that do not absorb radiation are attached to the chromophore and can enhance the absorption (hyperchromism) or shift the wavelength (bathochromic effect toward longer wavelengths, hypsochromic effect toward shorter wavelengths). These possess unshared (n) electrons interacting with the π electrons in the chromophore ($n - \pi$ conjugation).

In principle, the spectral effects of two isolated chromophores are independent and are simply added together if they are separated by at least two single bonds.

However, the spectrum is altered when two chromophores interact. When multiple bonds (e.g., double or triple bonds) are separated by one single bond, they are conjugated. The overlap of π orbitals of conjugated chromophores induces a bathochromic shift and generally increases the absorption intensity. When the double (or triple) and single bonds occur alternatingly along a carbon chain, the degree of conjugation increases. This is the case with aromatic compounds, containing phenyl or benzene groups. Also, the addition to the benzene ring of substituted groups such as amino ($-NH_2$), nitro ($-NO_2$), and aldehydic ($-CHO$) groups has a bathochromic effect and increases the intensity. Polynuclear aromatic compounds, having several benzene rings, have greater conjugation and absorb at longer wavelengths. Thus, naphthacene (with four rings) and pentacene (with five rings) have absorption maxima at 470 and 575 nm, respectively. By comparison, benzene, with one ring, has absorptions at 184 ($\varepsilon = 46,700$), 202, and 250 nm; and naphthalene, with two rings, has absorptions at 220 ($\varepsilon = 112,000$), 275, and 312 nm.

Indicator dyes, used widely in acid-base and redox titrations, and more recently in indicator-based sensors, are extensively conjugated systems that absorb in the visible region. The loss or addition of an electron or proton changes the electronic configuration and hence the color. Many molecules do not absorb directly in the visible region, but derivatives can often be prepared (e.g., complexes with metals) that have strongly colored absorptions.

4.2 FUNDAMENTALS OF INDICATORS

A large number of ions and molecules in chemical reactions (analytes or substrates) do not have suitable optical characteristics that are easily measurable. Indicators are chemical systems whose electronic spectrum changes under the effect of certain ion concentrations and thus can serve to *indicate* a distinct step in a chemical reaction and to determine these ion concentrations. In a general way, an indicator consists of an acceptor-donor system (see (2.65)) in which the acceptor and the donor have different colors or optical spectra under the influence of a proton, electron, or metal ion.

Hence, the indicator acts as a chemical transducer for species that are not directly measurable by optical techniques. Certain conditions for the use of indicators in sensors must be met: The indicator should be highly sensitive; a large change in the spectrum should occur for small concentration changes and should be easily detected; and the indicator concentration must be very small to avoid changing the concentrations of acceptor and donor. This implies that a high molar extinction coefficient and fluorescent quantum yield are desirable for an indicator. Also, the indicator should be specific for a given determination.

In acid-base reactions, the indicators exchange protons. According to the Brönsted and Ostwald concepts, indicators are weak acids that have a given color in the undissociated form in acidic solutions (HI, where I is the indicator and H the proton), and another color in the dissociated form (I^-). They are commonly used to provide a visible indication of the end point (at the completion of a chemical reaction) or of the equivalent point (the point at which a stoichiometric amount of reagent has been added).

In aqueous solutions, an indicator is in equilibrium with the acid-base system H_3O^+/H_2O:

$$HI + H_2O \rightleftharpoons H_3O^+ + I^- \qquad (4.1)$$

The concentration of the ion product of water is 10^{-14}, and for an acid indicator, the K'_{al} constant can be written in a simplified form:

$$pK'_{al} = pH - \log([I^-]/[HI]) = 14 + \log([I^-]/[HI]) - \log[OH^-] \qquad (4.2)$$

As an example of a typical indicator, the absorption spectrum of a dye in solution at various pH values is shown in Figure 4.1. The acid and basic forms of the 3,4,5,6-tetrabromophenolsulfonephthalein (TBPSP) indicator (molecular weight 670) and the isobestic points occur in the UV and visible regions. The isobestic point is the point on the spectrum (the wavelength) at which there is no change in absorption with changes in pH. It should be noted that the usable range of pH values with one indicator is relatively narrow (around 3 pH units) because of the sigmoid form of the optical signal curve with pH (see Figure 2.10).

The most common reaction of indicators with metals is chelation, in which an

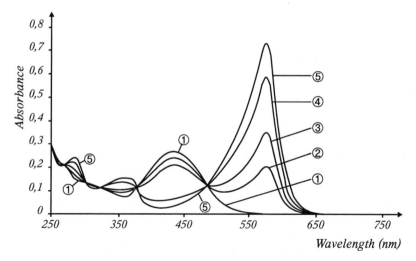

Figure 4.1 The absorption spectra of TBPSP (1.65 μM) in a trisodium phosphate buffer (50 mM) for different pH levels: (1) pH of 5.0; (2) 7.32; (3) 7.72; (4) 8.41; (5) 10.52.

excess of ligand, as L^-, reacts with M^{n+} metals, forming a series of products such as $M^{(n-1)+}L$, $M^{(n-2)+}L_2$, including the neutral chelate ML_n. Photometric measurement is possible when the chelating reagent, such as HL, contains chromogenic or fluorogenic ionophores, which are ligands that selectively bind ions, as well as chelating groups [3]. The majority of ligands used as indicators (I) for metal ions combine with hydrogen (H), with the displacement of a proton:

$$M^{n+} + HI \rightleftharpoons M^{(n-1)+} I + H^+ \qquad (4.3)$$

With an $H_n I$ indicator, the successive equilibria are

$$H_n I \rightleftharpoons H_{(n-1)} I^- \rightleftharpoons H_{(n-2)} I^{2-} \ldots \rightleftharpoons I^{n-} \qquad (4.4)$$

Each equilibrium has its specific pK'_{ai} value (i is from 1 to n), and each species corresponds to a specific wavelength $(\lambda_i)_{max}$. The metal atom is coordinately bonded to a functional group, which can be weakly or strongly basic, to form a five- or six-membered ring structure, that of the metal chelate. Typical coordinating groups are =O, -OH, -NH$_2$, -NH-, =N-, -SH, and =S. The ring formation is often found in crown ethers, so called because of their ability to *crown* cations by complexation. The cation selectivity is related to the molecular structure, and hence to the variation of ether oxygen atoms, of aliphatic and aromatic groups, and of the length of $(CH_2)_n$ bridges and the ring size.

Furthermore, the equilibrium of an indicator system based on ion pair formation and measured in an organic medium [4] is

$$M^+ + I + F^- \text{ (aq)} \rightleftharpoons MI^+F^- \text{ (org)} \qquad (4.5)$$

A large number of indicators are known [5], but only a few are specifically useful in sensing. The majority are ionizable or neutral dyes, and can be categorized by their characteristics:

- Optical properties (absorbance mediators, fluorophores, or quenchable fluorophores);
- Chemical composition (e.g., azo, phthalein compounds, and crown ethers);
- Electronic exchange mechanism (acid base, redox indicators, PSDs, ion-association complexes);
- Application to optical sensors (pH, metals, anions, and organic molecular recognition).

Further information about the optical properties of indicators and their use in sensors can be found in the literature [1,6,7]. The description of indicators in this chapter concentrates on their chemical composition and reactions [3].

4.3 COMMON INDICATORS

4.3.1 Polycyclic Aromatic Hydrocarbons

The polycyclic aromatic hydrocarbons (PAH), such as pyrenes and their derivatives, fluoranthene, diphenylanthracene, benzoperylene, and decacylene, are fluorescent indicators that can serve as oxygen sensors. However, their excitation spectrum occurs only in the UV region and their emission is generally below 450 nm, apart from fluoranthene (480 nm) and rubrene (552 nm), in a region where interference is present from background fluorescence of biological materials. In PAH-based oxygen optodes, CO, CO_2, nitrogen, and noble gases do not interfere. Major interference is produced by SO_2, chloride ions, halothane, and some heavy metals, which quench the fluorescence of PAHs. Naphthalenes and pyrenes have two distinct absorption maxima and are excitable at two wavelengths with only one emission band. Benzo[a]pyrene (BP) is one of the most-used indicators as a test reagent in immunology. In immunosensors, it is linked to a large protein structure and employed as a standard label (see Section 4.6) in the form of a chemically reactive BP derivative [8].

4.3.2 Carbon-Oxygen Compounds

4.3.2.1 Phthaleins

Phthaleins are part of the family of triphenylmethane dyes, resulting from condensation reactions between phenols and phthalic anhydride, and are used principally for titrations. Some familiar dyes are given in Table 4.1, and the chemical structural formula of phenolphthalein is shown in Figure 4.2(a). At low pH, the phthaleins exist as their colorless lactones. Successive removal of protons yields a tautomer with a molecular rearrangement—a quinoid structure with an intensive red color [9]. The maximum absorption wavelengths (around 552 nm for phenolphthalein, 660 nm for α-naphtholphthalein) are convenient for sensors, but their pKs are generally high (around 8 to 10) and these dyes are only suitable in a few special cases.

Table 4.1
Chemical Composition and pK_a'
of Some Phthaleins

Name	Formula	pK_a'
α-Naphtholphthalein	$C_{28}H_{18}O_4$	8.4
o-Cresolphthalein	$C_{22}H_{18}O_4$	9.4
Phenolphthalein	$C_{20}H_{14}O_4$	9.5
Thymolphthalein	$C_{28}H_{30}O_4$	9.7
Fluorescein	$C_{20}H_{12}O_5$	2.2/4.4/6.7

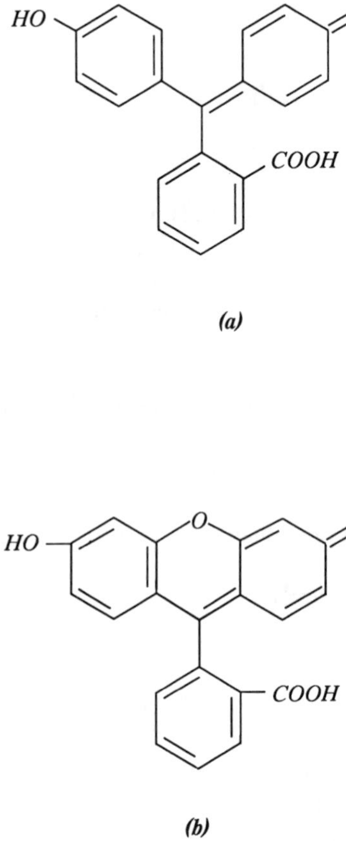

(a)

(b)

Figure 4.2 The structural formula of phthaleins: (a) phenolphthalein in quinone form and (b) fluorescein.

Fluorescein (also called resorcinol phthalein) is a fluorescent phthalein (Figure 4.2(b)) in which the single-band excitation wavelength (492 nm) coincides with the 488-nm line of the argon laser. It is used as a pH probe, but has a limited photostability, and its three pK'_as overlap each other. In geology, fluorescein and its derivatives are well known as tracers for water flow and can be detected in streams with a simple extrinsic fiber probe. The commonest derivative is FITC, which is employed in labeling (see Section 4.8).

Other derived compounds, such as fluorescein partly converted to eosin (2,4,5,7-tetrabromofluorescein), fluoresceinamine and 2,7'-bis-carboxy-ethyl-5,6-carboxy-fluorescein have been covalently immobilized on solid polymeric supports by different workers (see Chapter 5). Carboxynaphtofluorescein has been used in cellulose and sol-gel-based pH sensors [10]. Useful derivatives such as seminaphthofluoresceins

(SNAFL) or seminaphthorhodafluors (SNARF) have been developed (by Molecular Probes, Inc., of Portland, Oregon) as dual fluorescence indicators and used for pH and carbon dioxide analysis in physiological applications [11]. These fluorophores have two distinct emission maxima, one in acidic solutions, the other in alkaline solutions. At the isobestic point, the emission intensity is independent of pH and serves as the reference value, allowing a normalized signal measurement from the ratio of the intensity at two wavelengths. The vita blue dye, with $\lambda_{ex} = 524$ and 610 nm, and $\lambda_{em} = 570$ and 665 nm at pH 5.2 and 10, respectively, has similar properties [12].

4.3.2.2 Sulfophthaleins

Sulfophthaleins, which are also part of the triphenylmethane dyes, are obtained from a condensation reaction between phenols and o-sulfobenzoic anhydride; they are absorbance (or reflectance) indicators used for acid-base reactions in fiber-optic sensors. They can have two pK_a's (e.g., thymol blue has a pK_a' of 1.65 in acid form and 9.2 in basic form) and o-cresol purple 1.51 and 8.32, respectively. The different dyes can be categorized, as in Figure 4.3(a), by four radicals, R_1, R_2, R_3, and R_4. The

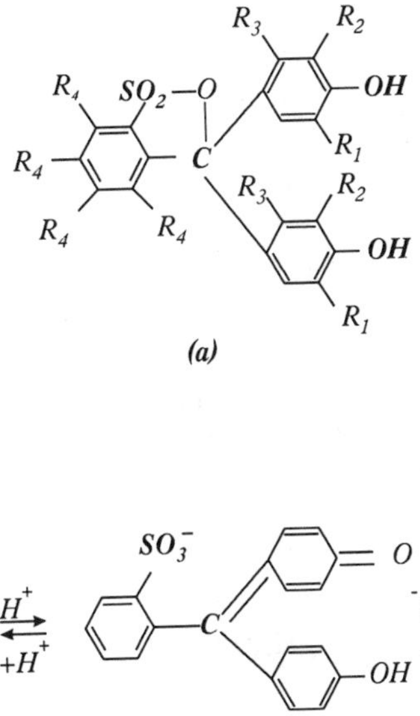

(a)

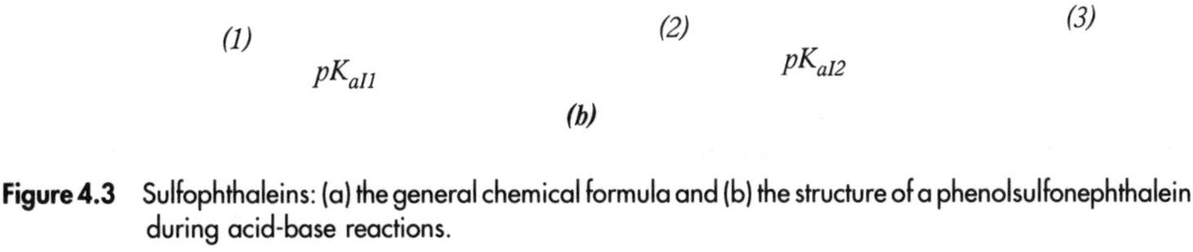

(b)

Figure 4.3 Sulfophthaleins: (a) the general chemical formula and (b) the structure of a phenolsulfonephthalein during acid-base reactions.

simplest form (where R_1 to R_4 = H) is phenol red (phenolsulfonephthalein), whose acid-base functions are shown in Figure 4.3(b).

The different forms that have been tried in optodes by various authors are summarized in Table 4.2, giving the pK'_{a1} and the absorption wavelength λ_{max} (taken from [5,9,13] and commercial literature). The molecular extinction coefficients (ε) have been measured with a monobeam diode-array spectrophotometer equipped with optical fibers and a 4-cm-path-length probe [14]. The whole range of pH can be covered by the sulfophthaleins, which are also easy to immobilize on various supports.

Other sulfophthaleins, such as cathechol violet and eriochrome cyanine R, possess OH groups in R_1 and R_2 adjacent to the -C=O group and can form highly colored chelates ($\varepsilon = 3.10^4$ to 7.10^4) with a large number of metals in weakly acidic or basic solution. However, their nonselective nature in forming chelates and the small color changes with pH are disadvantages for their use in sensors.

Table 4.2
Selected Sulfophthaleins for pH Measurement With Optodes

Indicator	Abbreviation	Formula	R_1	R_2	R_3	R_4	pK'_{a1}	λ_{max} (nm)	ε (at λ_{max})
Thymol blue (acid form)	TB	$C_{27}H_{30}O_5S$	iPr	H	Me	H	1.65	546	31,500
Tetrabromophenol blue	TBPB	$C_{19}H_6O_5Br_8S$	Br	Br	H	Br	3.56	605	64,000
Tetrabromothymol blue	TBTB	$C_{27}H_{26}O_5Br_4S$	iPr	Br	Me	Br	3.8	612	—
Bromophenol blue	BPB	$C_{19}H_{10}O_5Br_4S$	Br	Br	H	H	4.1	592	79,000
Bromocresol green	BCG	$C_{21}H_{14}O_5Br_4S$	Br	Br	Me	H	4.9	618	—
Bromophenol red	BPR	$C_{19}H_{12}O_5Br_2S$	Br	H	H	H	6.16	574	35,000
Chlorophenol red	CPR	$C_{19}H_{12}O_5Cl_2S$	Cl	H	H	H	6.25	572	35,000
Bromocresol purple	BCP	$C_{21}H_{16}O_5Br_2S$	Br	Me	H	H	6.3	588	—
Tetrabromophenol-sulfonephthalein	TBPSP	$C_{19}H_{10}O_5Br_4S$	H	H	H	Br	7.03	578	42,000
Tetrachlorophenol-sulfonephthalein	TCPSP	$C_{19}H_{10}O_5Cl_4S$	H	H	H	Cl	7.04	575	—
Bromothymol blue	BTB	$C_{27}H_{28}O_5Br_2S$	iPr	Br	Me	H	7.3	617	41,000
o-Cresol tetrachloro-sulfonephthalein	CTSP	$C_{21}H_{14}O_5Cl_4S$	Me	H	H	Cl	7.51	590	—
Phenol red	PR	$C_{19}H_{14}O_5S$	H	H	H	H	8.0	559	—
Cresol red	CR	$C_{21}H_{15}O_5S$	Me	H	H	H	8.2	572	—
Thymol blue (basic form)	TB	$C_{27}H_{30}O_5S$	iPr	H	Me	H	9.2	598	22,000

Note: pK'_{a1}, λ, ε values are given in solution for a temperature of 15° to 30°C and zero ionic strength. For R_1 to R_4 groups: H = hydrogen, Me = methyl group, iPr = isopropyl group, Br = bromide, Cl = chloride.

4.3.2.3 Hydroxy Aromatic Compounds

In hydroxy aromatic compounds, one or several OH groups are present. In chelation, the hydrogen can be replaced by a metal ion or complex, which binds to the =O, -OH, and -COOH.

The simplest classes are:

- Hydroxybenzenes (Figure 4.4 (a)) such as pyrocathechol with OH groups for R_1 and R_2 and pyrogallol with three OH groups for R_1, R_2, and R_3;
- Hydroxynapthalenes (Figure 4.4(b)) with OH groups for R_1 and R_2, and SO_3H groups for R_4 and R_7;
- Pyrene derivatives (Figure 4.4(c)).

8-Hydroxypyrene-1,3,6-trisulfonic trisodium salt (HPTS), with one OH group in R_1 and three SO_3H groups in R_2, R_3, and R_4, although sensitive to ionic strength, has been found to be one of the best fluorescence indicators for pH and CO_2 sensors because of its high quantum yield. In addition, it has a pK_a' (7.3) within the normal range for medical applications and is easily immobilized. During immobilization, it is important to ensure that the chromophore groups maintain their active function as an indicator (see Chapter 5). The fluorescence emission ($\lambda_{em} = 520$ nm), shown in Figure 4.5 as a function of the pH, may be excited at two wavelengths, 403 and 470 nm [15]. A similar derivative, 1,3-dihydroxypyrene-6,8-disulfonic acid (DHPDS), with $pK_a' = 7.41$, is also used as an indicator for pH measurements.

Among the hydroxyanthraquinones, alizarin red (Figure 4.6(a)) with three pK_a's

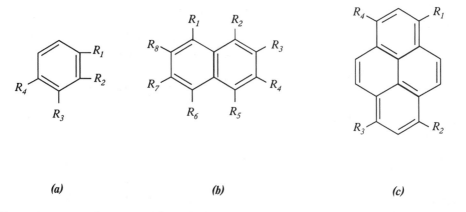

(a)   *(b)*   *(c)*

Figure 4.4 PAHs. The structural formulas for (a) hydroxybenzenes, (b) hydroxynaphthalenes, and (c) hydroxypyrenes. Some of the substitutions for R are OH groups. When R is not indicated, then hydrogen is present.

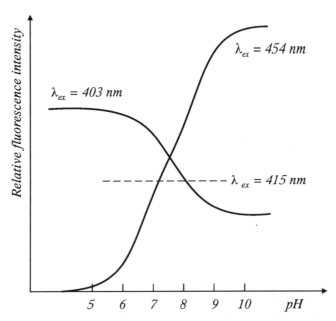

Figure 4.5 The fluorescent emission intensity at 520 nm of HPTS as a function of pH for excitation wavelengths (λ_{ex}) at 403, 415, and 454 nm (*After*: [15]).

is valuable for pH determination over a wide range by absorbance measurement. In chelate reactions, alizarin is nonselective.

Some derivatives of the furanes, such as Fura 2, and of the fluorones, such as phenylfluorone (Figure 4.6(b)), and hydroxyflavones, whose parent is flavone (Figure 4.6(c)) with a bridging oxygen, have several OH groups and act as fluorophores. During chelation, the 5-hydroxyflavone forms six-membered rings and 3-hydroxyflavone five-membered rings. These have been tried as sensors; for example, morin (3,5,7,2',4'-pentahydroxyflavone-sulfonaphthalein), which has five hydroxy groups, has been covalently immobilized on cellulose for Al^{3+} detection. However, the pK_a' varies with pH. Hence, the selectivity depends on the pH; in the 4- to 5-pH range the equilibrium favors reaction with Al^{3+}, and at higher pH favors Be^{2+} [16].

Among compounds having a bridging oxygen, the 7-hydroxycoumarins (HC), which have an electron-donating group in position 7, form an important class of indicators (Figure 4.6(d)) known for their strong fluorescence. Thus, 4-methyl-umbelliferone (4-MU) was employed in the first optodes for CO_2 and pH sensors [17]. Its pK_a' (7.8) is well adapted for pH measurement in physiological applications. Coumarin 7-hydroxycoumarin-3-carboxylic acid (HCC), with an electron acceptor in position 3 (R_1), has also been used as a pH sensor [18].

Synthesis of a new class of fluorescent coumarins having (*N*-arylsulfonyl)-amino groups in R_3 (Ph-SO_2-NH-) have been studied for an extended pH range, typically 2 to 9, for compounds having several pK'_as [19,20]. For example, the 4-pyridyl amino coumarin has two pK'_as (4.74 and 7.64). The presence of a cyano (-CN) group in R_2 shifts the wavelength by about 50 nm in absorption, by the bathochromic effect, and 100 nm in fluorescence, but decreases the pK'_a (0.4 to 0.7 units). Also, some 3-aryl-coumarins have been investigated for acid-base titrations [21]. Table 4.3 gives some

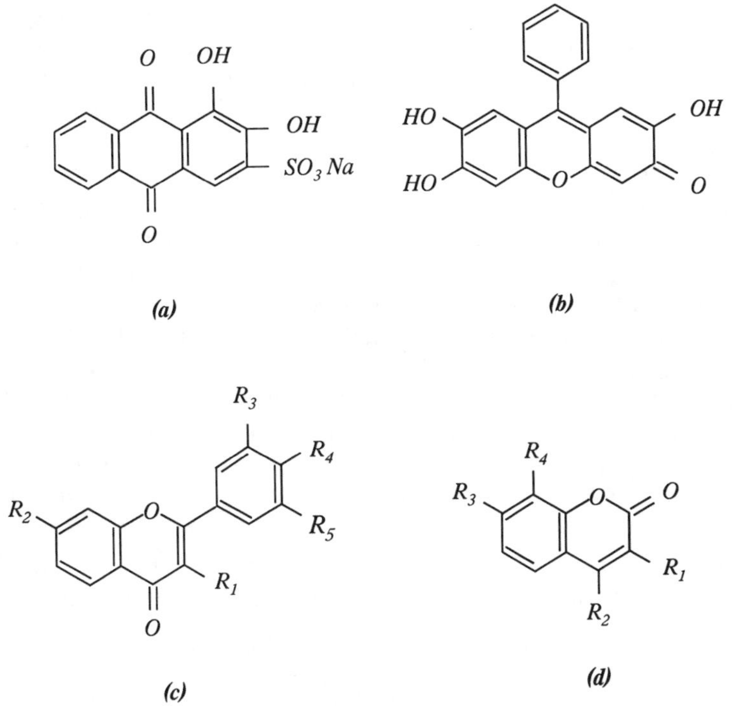

Figure 4.6 Hydroxyaromatic compounds: (a) hydroxyanthroquinone (alizarin red), (b) trihydroxyfluorone, (c) hydroxyflavones, and (d) hydroxycoumarins.

Table 4.3
Typical 7-Hydroxycoumarin Indicators

Name	Abbreviation	R_1	pK'_a (at 22°C)	λ_{ex} (nm)	λ_{em} (nm)
4-Methyl-umbelliferone	4MU	CH_3	7.78	355 – 410	445
4-Trifluoromethyl-umbelliferone	4TMU	CF_3	7.26	355 – 410	503
7-HC-3-carboxylique	HCC	Carboxy	7.04	360 – 400	448
4-Pyridyl-amino-coumarin	4PAC	4-Pyridyl	7.64 and 4.74	377 – 400	470

typical coumarin indicators with their $pK_a's$, fluorescent excitation (λ_{ex}) and peak emission (λ_{em}) wavelengths, as well as their usual abbreviation.

4.3.3 Carbon-Nitrogen Compounds

4.3.3.1 Azo Compounds

The azo group (-N=N-) is a strong chromophore and can be linked either to two aryl (monoazo and biazo compounds) or three aryl groups, or to a heterocyclic compound as represented in Figure 4.7(a). The R_1 and R_2 radicals can be aryl, benzyl, or naphthyl groups with -H, -OH, -NO$_2$, -SO$_3$H, or -AsO$_3$H$_2$ functions. In the group of heterocyclic compounds, 1(-2-pyridylazo)-2-naphthol, called PAN, has one -OH group as the R_1 radical, and 4(-2-pyridylazo)-resorcinol (PAR) has -OH groups as R_1 and R_2 radicals.

Some azo compounds can be used for acid-base reactions at appropriate wavelengths as seen in Table 4.4 (see also [22]). The existence of several $pK_a's$ (e.g., with the solochrome violet RS) allows a large pH range to be covered with the same indicator. All azo indicators are nonfluorescent.

Of considerable interest is the possibility of azo compounds forming rings with metals, making various ligand/metal or ligand/metal-hydroxy compounds, as shown in Figure 4.7(b) for different valence states. Typical examples are thorin with thorium, eriochrome black T with magnesium, and arsenazo I with uranium.

For bis-azo compounds, the resulting chelate structures can be complex with a three-dimensional form surrounding the metal ion, as shown in Figure 4.8(a) for an

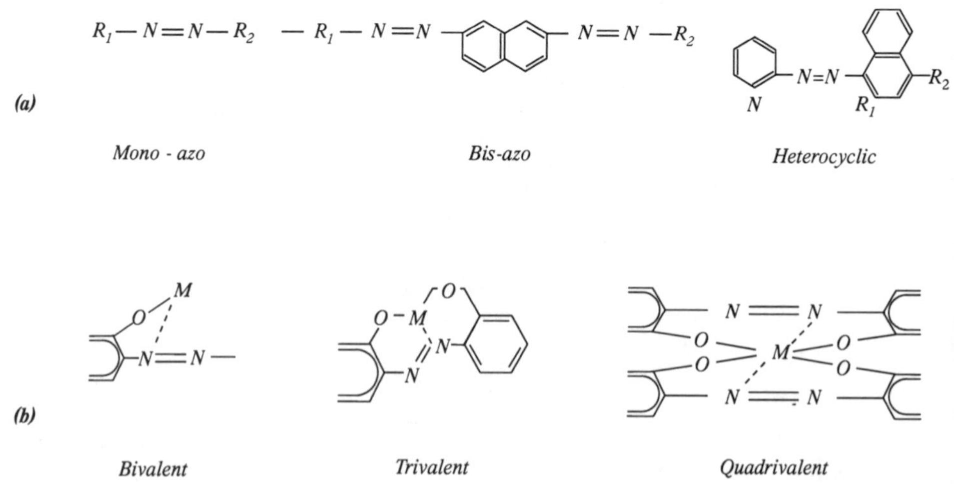

Figure 4.7   Azo compounds: (a) general formula for the mono-, bis-, and heterocyclic azo compounds and (b) chelate compounds with bivalent, trivalent, or quadrivalent metals (M).

Table 4.4
Selected Azo Compounds for pH Measurement With Optical Fibers

Name	Abbreviation	Formula	pK'_a	λ_{max} (nm)
Methyl red	MR	$C_{15}H_{15}N_3O_2$	5.0	526
Nitrazine yellow	NY	$C_{16}H_8N_4O_{11}S_2Na_2$	6.5	590
Palatine chrome black	PCB	$C_{20}H_{13}N_2O_5SNa$	7.4	643
Solochrome violet RS	SVRS	$C_{16}H_{11}N_2O_5SNa$	4.35/7.4/9.35	562

Note: pK'_a and λ_{max} values in solution.

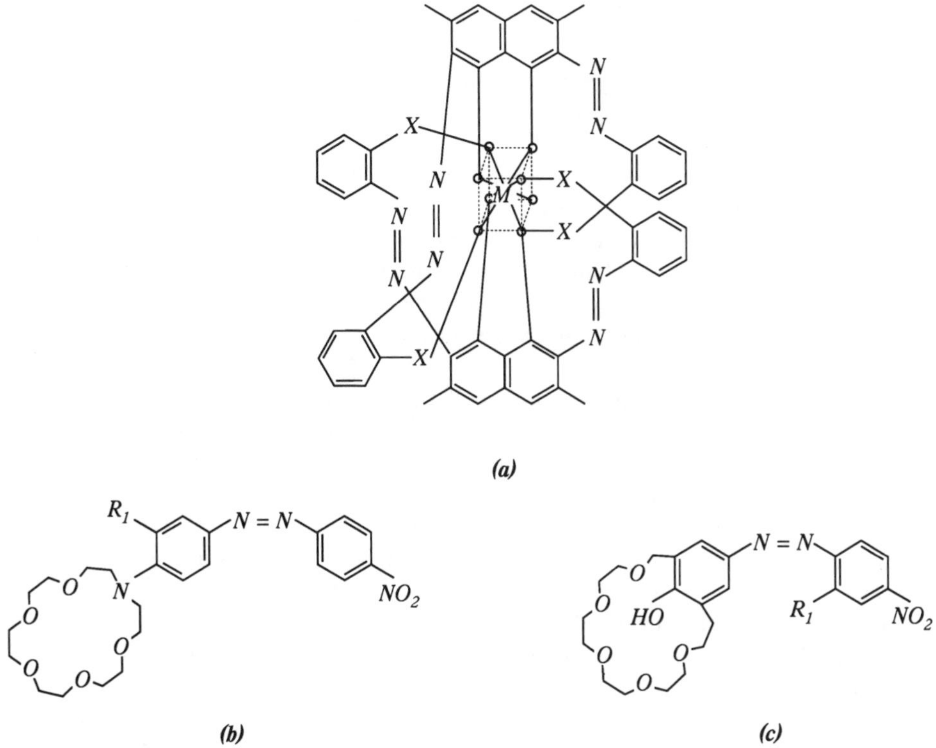

Figure 4.8 Azo compounds: (a) the cubic structure with a 1:2 molar ratio and a coordination number of 8, in which X = SO_2 or AsO(OH) or PO_2(OH) and (b,c) two examples of azo compounds bound to a macrocyclic polyether.

eightfold coordination. New chromogenic crown ethers including the azo form and a macrocyclic polyether, shown in Figure 4.8(b), have been synthesized for K^+ determination [23]. A second form is given in Figure 4.8(c). Variations in the structure induce a change in the pK'_a; for example, the pK'_a is 9.48 when R_1 is H and 7.42 when R_1 is an additional NO_2 group.

4.3.3.2 Other Compounds

Some trimethylmethane dyes forming R^+ or RH^+ ions and derived from aromatic amines can form ion associations with halides or thiocyanates of anionic metal complexes. These are chelating agents. The R^+ form, in Figure 4.9(a), generally predominates in neutral solutions, the RH^{2+} form in acid solutions, and the ROH form in alkaline mediums. Table 4.5 lists some typical compounds with radicals in the R_1 to R_4 positions.

Various other compounds are photometric reagents forming cationic complexes with many metals. For example:

- Hydrazones (Figure 4.9(b)), which can lose the H of the imino group (=N-N<) to give uncharged complexes;
- Ferroins, which in an aromatic system form five-membered chelate rings with metals, such as 1,10-phenanthroline (Figure 4.9(c)), a reagent for Fe^{2+};
- Triazines, such as tripyridyltriazine (TPTZ) (see Figure 4.9(d)), a highly colored reagent specific for Fe^{2+} at low concentrations in water.

Some derivatives of oxines, such as 8-hydroxyquinoline (Figure 4.10(a)), also have chelating properties. Hence, quinolines are indicators for heavy metals [24]. Also, the fluorescence of some quinolinum compounds such as 6-methoxyquinoline can be dynamically quenched by thiocyanates and halogenides, as in a halide sensor [25].

The indophenols are mainly used as redox indicators. They also have acid-base characteristics. Indophenol (Figure 4.10(b)) with a $pK'_a = 8.1$ is the best known. For instance, 2,6-dichlorophenolindophenol ($pK'_a = 5.7$) has been immobilized on XAD-type amberlite™ resins for sulfide measurement [26]. However, these compounds have poor stability, as in the case of indolines, aminoindophenols, and indamines. In contrast, the hydrophobic indicator 7-decyl-2-methyl-4-(3′,5′-dichlorophen-4′-one) indonaphth-1-ol (MEDPIN) has an anionic form that gives a stable ternary complex with valinomycin and potassium [27]. The change of absorbance of MEDPIN induced by potassium ions was detected from the quenching of a fluorescent dye (1,1′-dioctadecyl-3,3,3′,3′-tetramethyl-indo-dicarbocyanine perchlorate) by energy transfer.

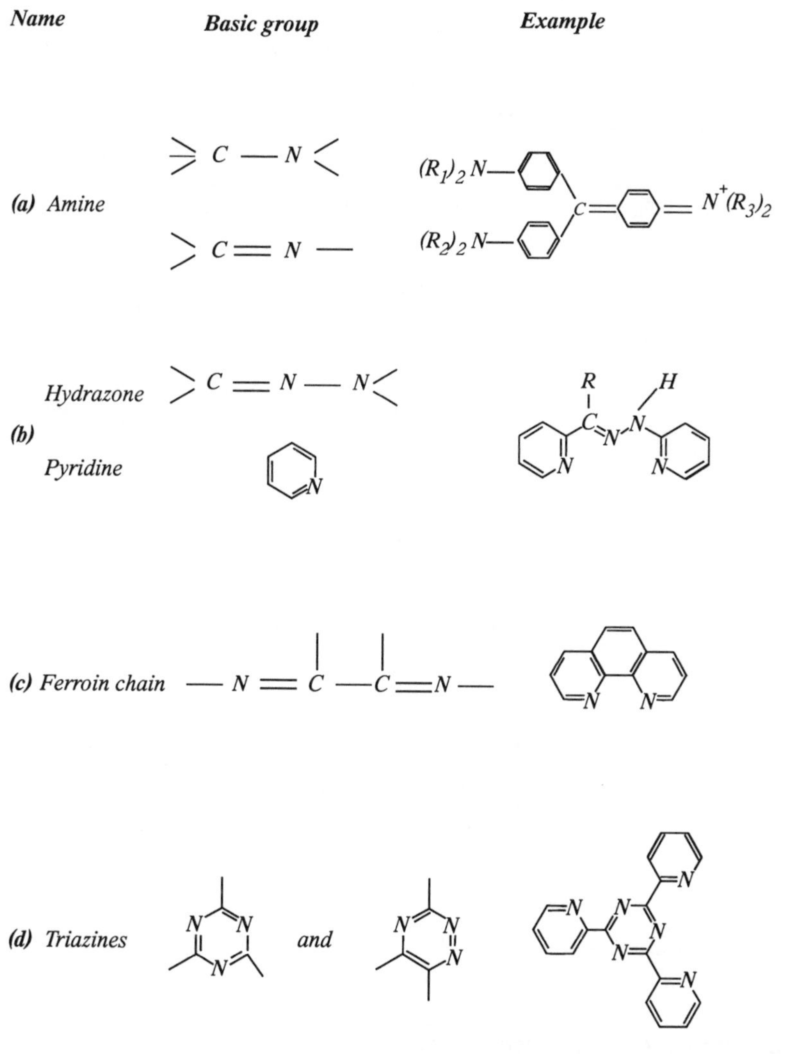

Name	Basic group	Example

(a) Amine

(b) Hydrazone

Pyridine

(c) Ferroin chain

(d) Triazines

Figure 4.9 Examples of indicators with =N-C≡ or -N=C⟨ bonds for different basic groups: (a) aromatic amines with the R⁺ form with $R_3 = R_4$, (b) pyridylhydrazone, (c) 4,10-phenanthroline, and (d) TPTZ.

Table 4.5
Trimethylmethane Dyes Forming Halogenide Ion
Association Complexes

Name	R_1	R_2	R_3	R_4	λmax (nm)
Rosaniline	H	H	H	CH_3	547
Crystal violet	CH_3	CH_3	CH_3	H	591
Brilliant green	C_2H_5	H	C_2H_5	H	615
Victoria blue	CH_3	C_6H_5	CH_3	C_6H_5	635

Note: R_1 to R_4 are radicals, and λ_{max} is for the R^+ form.

Name	*Basic group*	*Example*
Oxine		

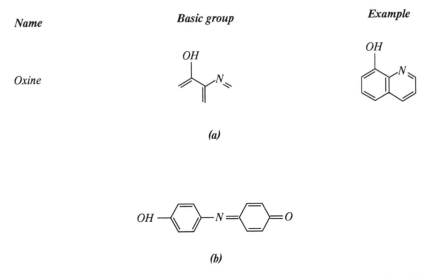

(a)

(b)

Figure 4.10   Further examples of indicators with a -N=C⟨ bond: (a) 8-hydroxyquinoline and (b) indophenol.

4.3.4 Carbon-Nitrogen-Oxygen Compounds

4.3.4.1 Rhodamines

Rhodamines are well-known compounds with very strong fluorescence and are used in dye-lasers and as tracers in hydrology. The signal intensities for 1 ppm of rhodamine 6G used as a tracer with a remote fiber fluorimeter is 2.5 times greater than for fluorescein [28]. The absorption maximum of rhodamine B, occurring at $\lambda_{max} = 556$ nm for the violet form (RH^+), is intense ($\varepsilon = 1.1 \times 10^5$) and varies with pH. Rhodamine 6G (Figure 4.11(a)) has the same characteristics but is more

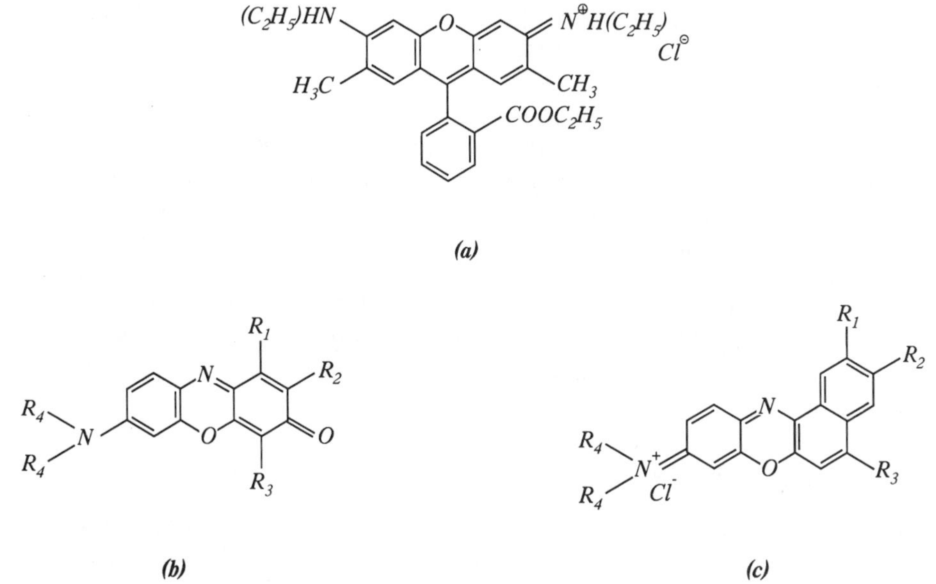

(a)

(b)

(c)

Figure 4.11 Examples of types of carbon-nitrogen-oxygen indicators: (a) the chloride form of rhodamine 6G, (b) phenoxazones, and (c) benzo[a]phenoxazine.

hydrophobic than the B compound. This large fluorimetric sensitivity of rhodamines has been employed in fiber-optic sensors to detect various metallic ions by fluorescence quenching or enhancement [29]. Thus, rhodamine immobilized on a Nafion membrane is very attractive because the dye is both electrostatically and hydrophobically bound to the membrane in aqueous solution. Quenching of the rhodamine fluorescence can be reversed by the addition of reverser ions (e.g., H^+, Li^+, Na^+). Also, rhodamine 6G is quenched by sulfur dioxide [30]. Several derivatives, such as esters, are cation-sensitive dyes usable for the measurement of membrane potential (see Section 4.5).

4.3.4.2 Oxazines and Phenoxazines

One of the first intrinsic indicator-based fiber-optic sensors was developed for the determination of ammonia vapor [31]. An oxazine perchlorate dye was coated on capillary tubes with a permeable silicone acrylate cladding, and NH_3 induced a change in the fluorescence and a strong optical transmission loss at 560 nm (λ_{max}). Whereas the oxazines and phenoxazines have oxido-reduction indicator properties, compounds such as phenoxazones (Figure 4.11(b)) and benzo[a]phenoxazines (Figure 4.11(c)) are strongly colored in solution with acid-base properties. In general, the phenoxazones have two pK_a's. For instance, celestine blue with $R_1 = CONH_2$, $R_2 = H$,

R_3 = OH, and R_4 = C_2H_5 has pK_1' = 3.32 and pK_2' = 8.51 for λ_{max} = 547 nm (in the form RH^{2+}), 646 nm (in the form RH), and 530 nm (in the form R^-) with ε between 2 and 4.10^4 $cm^{-1} \cdot M^{-1}$.

Among the benzo[a]phenoxazines, the best known are Meldola blue with R_1 = R_2 = R_3 = H and R_4 = CH_3, and Nile blue with R_1 = R_2 = H, R_3 = NH_2, and R_4 = C_2H_5. They are used for measuring metallic anionic complexes having a single charge. An advantage of these compounds is their maximum absorption wavelength (around 570 to 660 nm), which is convenient for instrumentation with a light-emitting diode (LED) light source and *pin* photodetector (see Section 9.2.3). The synthesis of Meldola blue with 4-aminophenol and 4-aminobenzoic acid has been achieved by nucleophilic substitution with aromatic amines, producing pK_as around 6 to 8 and λ_{max} between 600 and 650 nm after immobilization on an optical fiber for a pH sensor [32].

4.3.4.3 Nitro, Nitroso, and Oxime Compounds

The nitro indicators, such as nitrophenol ($OH-C_6H_4-NO_2$), are acids whose absorption in the alkaline form depends on the relative position of OH and nitro groups; for example, the pK_a is 7.5 and 8.35 for p- and m-nitrophenols, respectively. P-nitrophenol (λ_{max} = 404 nm, ε = 19,000 $cm^{-1} \cdot M^{-1}$ at pH 10.4) has been used to demonstrate the basic concepts of fiber-optic enzymatic sensors, in which the transformation of p-nitrophenylphosphate into phosphate and p-nitrophenol is catalyzed by the alkaline phosphatase enzyme immobilized on a nylon membrane [33]. The dinitrophenols with two NO_2 groups are acid-base indicators (pK_a = 3.7, 4.1, 4.26, 5.2, and 5.35 for the 2,6-, 2,4-, 2,3-, 2,5-, and 3,4- forms, respectively). The dihydroxy compounds have two pK_as in general, such as 3.54 and 10.3 for 3,5-dinitropyrocathechol. Also, phenylhydrazine derivatives are weak bases of amphoteric character (having a base or anhydride acid part) having two pK_as (e.g., 2.8 and 11.2 for p-nitrobenzhydrazide).

The nitroso compounds (=C(NO)-C(OH)=) can form chelates by coordination with metals (cobalt, nickel) onto the nitrogen and oxygen with loss of hydrogen from the OH group.

The monoximes have a single (=NOH) group and α-dioximes have two neighboring (=NOH) groups; both can form colored metal complexes either through binding to the two nitrogens or through the nitrogen and oxygen. However, these chelates are only slightly soluble in water and must be extracted.

4.3.4.4 Other Carbon-Oxygen-Nitrogen Compounds

By oxidation, diphenylcarbazide (Figure 4.12(a)) gives diphenyl carbazone (Figure 4.12(b)) and diphenylcarbadiazone ($OC(N=N-C_6H_5)_2$). Diphenylcarbazide is an acid-base indicator and results in colored chelates (e.g., with Cr^{6+}). Diphenylcar-

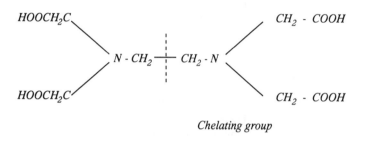

$$O = C \Big\langle{}^{NH - NH - C_6 H_5}_{NH - NH - C_6 H_5}$$

(a)

$$O = C \Big\langle{}^{N = N - C_6 H_5}_{NH - NH - C_6 H_5}$$

(b)

$$HOOCH_2C \diagdown \\ \diagup N - CH_2 \dashv CH_2 - N \diagup {}^{CH_2 - COOH} \\ HOOCH_2C \diagup \qquad\qquad \diagdown {}_{CH_2 - COOH}$$

Chelating group

(c)

Figure 4.12 Further examples of carbon-nitrogen-oxygen indicators: (a) symmetric diphenylcarbazide, (b) diphenylcarbazone, and (c) complexing agent with a chelating group: EDTA.

bazone reacts with many heavy metals such as mercury, rhodium, and copper to form chelates where the metal is bonded to the nitrogen and oxygen sites. Some derivatives with -SO_3H, nitroso or azo groups are applied to mercurimetric titrations of bromides and iodides.

The colorimetric complexation of ethylenediaminetetracetic (EDTA) (Figure 4.12(c)) with metals is well known. This reagent with symmetric iminodiacetic acid groups can be introduced into a chromophoric framework. Xylenol orange is a compound of this type that forms red complexes with palladium, zirconium, and hafnium. It has been used as a lead sensor [34]. However it is not very stable in solution.

Others types of indicators include carboxyindoles such as Indo 1 and aminobenzofuran such as Fura 2 immobilized on agarose gel for calcium determination [35].

Some indicators produce luminescence while undergoing a chemical reaction, known as chemiluminescence. Good examples of strong chemiluminescent compounds are luminol (Figure 4.13(a)) and lucigenin (Figure 4.13(b)) when the oxygen or hydroxyl bonds in the molecule react with hydrogen peroxide under alkaline conditions in the presence of a catalyst immobilized on a nylon membrane [36].

$$\text{Luminol} + H_2O_2 \rightleftharpoons \text{aminophthalate} + h\nu \ (\lambda = 430 \text{ nm}) \qquad (4.6)$$

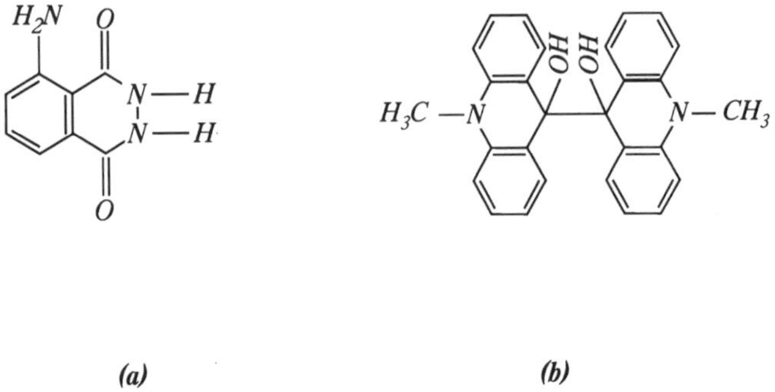

(a)   *(b)*

Figure 4.13   Structure of chemiluminescent compounds: (a) luminol and (b) lucigenin (carbinol base).

4.3.5 Compounds Incorporating Sulfur

Organic compounds incorporating sulfur and nitrogen are of interest for optical sensors, particularly in forming FITC and derivatives, which are suitable as reagents and labels (see Section 4.8). In addition, for inorganic media, many complexes formed from metals (iron, cobalt, and mercury) with thiocyanates are colored and are used for SCN^- determination. Thiocyanates are employed as reduction agents for iodates and chromates or as standard reagents for molybdenum, tungsten, and niobium.

Thiocarbazides (Figure 4.14(a)) and thiocarbazones are organic compounds forming chelates with many heavy metals such as platinum and ruthenium by bonding to sulfur and nitrogen, but in alkaline solutions a decomposition to S^{2-} can occur.

By binding with the sulfur and nitrogen atoms, dithizone (Figure 4.14(b)) forms dithizonates with many metals (e.g., with silver and mercury). After dithizone immobilization on XAD-type resin for a lead sensor [37], a maximum change in the reflectance spectrum has been found around 650 nm. At this wavelength, the photochromism, which is a reversible effect of light on the color, is partly reduced. However, dithizone is rather unstable when dissolved or immobilized, and good selectivity for metal dithizonates is difficult to obtain in practice in sensors because it is necessary to adjust the pH, alter the oxidation state of interfering species, and add a complex-forming agent to the solution to extract the measurand.

Thio-oxine (8-mercaptoquinoline) in Figure 4.14(c) can exist in aqueous solutions giving chelates of various forms, such as the $(HT^{\pm})$ zwitterion (contradictory ion with opposite charges). The isoelectric point occurs at about pH 5.2. Some mercaptoquinolates are strongly colored and can fluoresce in inorganic solutions; however, the stability and absorptivity of these chelates are less than those of dithizonates.

Thiazines are similar compounds to phenoxazines, in which a sulfur atom replaces a binding oxygen or nitrogen atom as shown in Figure 4.14(d). These dyes are oxido-reduction indicators, usually having two pK_a's. The most common are

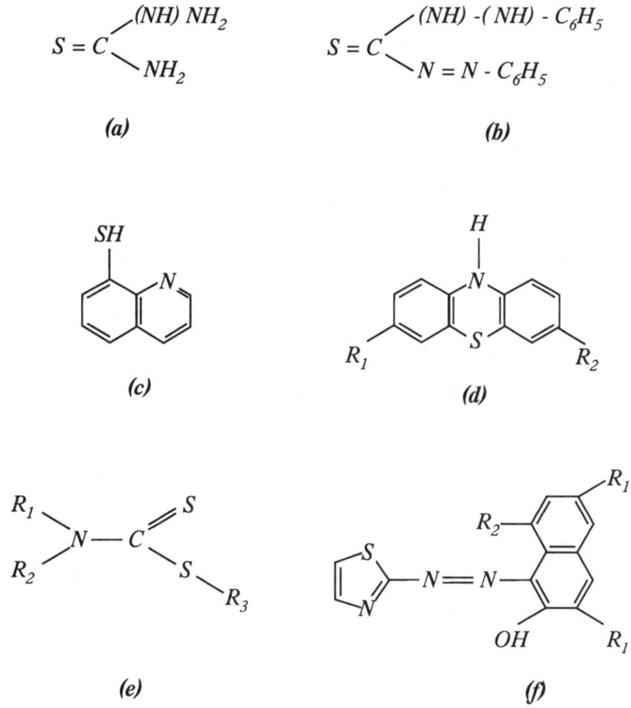

Figure 4.14 Some indicators containing sulfur: (a) thiosemicarbazide, (b) dithizone (thione form), (c) thio-oxine, (d) phenothiazines, (e) dithiocarbamates, and (f) thiazolyl-2-naphthol.

methylene blue with $R_1 = R_2 = N(CH_3)_2$ and $pK'_a = 4.52$ and 5.85, thionine with $R_1 = R_2 = NH_2$ and $pK'_a = 4.38$ and 5.3, and thiazine blue with $R_1 = N(CH_3)_2$ and $R_2 = N(C_2H_5)_2$. Their main applications are for end titrations of metals. As with phenoxazines, immobilization of thiazines requires new molecular arrangements.

Dithiocarbamates (Figure 4.14(e)) form chelates with at least 35 metals that are soluble in the usual organic solvents and exhibit a large molar absorptivity in the blue end of the spectrum. Without hydrophilic groups in the molecule, they are not selective and are therefore of only limited interest for sensors.

Thiazolylazo compounds of phenol or naphthol types (Figure 4.14(f)) are chelating reagents in which the hydrogen of the OH group can be replaced by most of the transition metals bonded onto the nitrogen of the azo group. The best known are TAR, 4-(2'-thiazolylazo)-resorcinol with two -OH groups in the phenol group and three pK'_as (1.0, 6.2, and 9.4), and TAN, 1-(2'-thiazolylazo)-2-naphthol with two pK'_as (2.4 and 8.7). However, these compounds are less important than the azo compounds PAR and PAN (see Section 4.3.3). Some derivatives have sulfonic groups in R_1 and another OH group in R_2. Another thiazolyl compound, luciferin, is well-known for its properties as a specific substrate in bioluminescence, particularly for the determination of adenosine triphosphate (ATP) [38].

4.3.6 Macrocyclic Compounds—Crown Ethers

Pedersen's discovery [39] of selective complexation in a biological medium of alkali and alkaline earth metal cations by crown ethers is the basis for coordination chemistry of the related host compounds [40]. A valuable review of these compounds in sensors is given by Beer [41].

Three basic configurations for monocyclic compounds (Figure 4.15) show the large diversity of these compounds. In Figure 4.15(a), the X element of the chain, repeated n times, is normally an oxygen atom, but can be replaced by a nitrogen or sulfur atom. In Figure 4.15(b), X can be an oxygen, nitrogen, or carbon atom with various radicals (e.g., -OH, =O). In Figure 4.15(c), the phenyl group is generally symmetrically attached. The chromophoric groups can possess dissociable protons or can be nonionic.

In ionic compounds, the color change is due to ion exchange between the proton and the metal cation. All these compounds, in lipophilic media, are in the form of a ring with an endohydrophilic and electronegative cavity able to accept metal guest

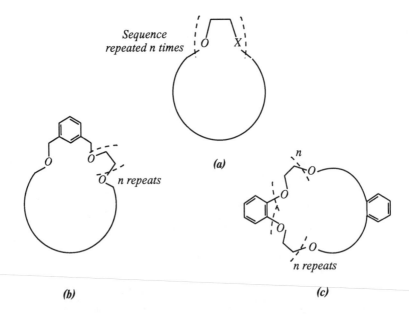

Figure 4.15 Basic configurations of macrocyclic compounds (crown ethers) that can be used as indicators: (a) without a phenyl group, (b) with one phenyl group, and (c) with two phenyl groups.

cations. The size of the guest cation depends on the number of ring atoms that act as a selectivity factor in the coordination chemistry. This selectivity (s) for a monovalent ion M_1^+ relative to another ion M_2^+ is expressed by the ratio of the ligand complexes (L):

$$s = K_{e1}\ (LM_1)\ /\ K_{e2}\ (LM_2) \tag{4.7}$$

the equilibrium constant value is normally expressed in logarithmic terms (pK_e). An important example of a crown ether used in an optical sensor has been mentioned previously (see Figure 4.8(b)) for K^+ determination [42].

Other classes are formed by cryptands, spherands, and hemispherands. The cryptands have an additional oligoether chain (X-R-X) of different oxygen donor atoms with varying lengths bridging two monocyclic crown ethers in a bicyclic configuration (Figure 4.16(a)). Generally, X is a CH_2, CONH, or CH_2OCO group, and R an aliphatic or benzenic group, but the chain can also be a -N=N- chain. The cryptate configuration, with its three-dimensional and spheroid-shaped host cavity, produces higher stability constants than for the configuration of monocyclic compounds. Some heterocyclic structures can be added in the place of the R radical, giving complicated ligand forms.

The spherands (Figure 4.16(b)) have more rigid structures, and complexation of a given ionic diameter is in relation to the size of the cavity. The semirigid hemispherands have the structure of half-spherand, half-monocyclic crown ether or half-spherand, half-cryptand types. These macrocyclic compounds not only allow recognition of cations (e.g., NH_4^+, Na^+, K^+, Li^+), but also the determination of anions of variable sizes and geometries by anion coordination chemistry. In this case, the cavity is positive and the anion binds onto the hydrogen of the N^+H group (e.g., N^+H . . X . . N^+H).

In nonionic compounds, the bonding of the metal ion to a donor or acceptor dye molecule induces a charge-transfer change of the dye. Complexation of noncharged molecules by hydrophilic interactions, which occurs in polar aqueous media, can be organized as an integral stoichiometry between the neutral host and the guest species in an intramolecular inclusion cavity, as in cyclodextrins.

The calix(n)arenes are cyclic compounds in which the basic sequence is repeated four to eight times in the cycle (Figure 4.16(c)). They can form complexes with metals, organic cations, and neutral organic molecules. In solution they can develop various geometrical configurations (cone, partial cone, and alternative forms). The properties of these compounds (homo-, oxo-, and heterocalixarenes) can be found in the specialized literature [43]. The R group is generally a t-butyl group. The Y group is an OH in calixarenes, and is an O-R' group in modified immobile calixarenes with R' = CH_3, C_2H_5, $COCH_3$, or $Si(CH_3)_2$. Other compounds, such as cavitands (with reinforced or rigid cavities suitable for simple molecules and ions), cryptophanes (pseudocavitands), and water-soluble cyclophanes (with a silicon frame), have considerable potential [41], but have not yet been exploited in optical sensors.

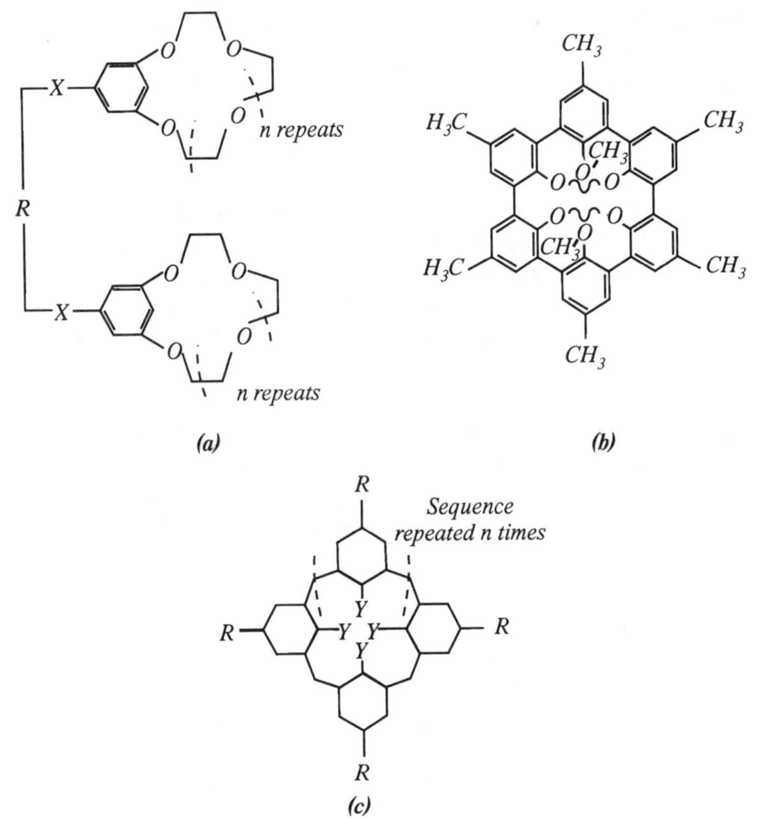

Figure 4.16 Macrocyclic compounds used as indicators: (a) the structure of cryptands, (b) an example of a spherand, and (c) calix(n)arenes, where n is the number of group sequences.

The bridged calix(4)arene and hemispherands appear to be attractive as ionophoric components in optical sensors [44]. High-selectivity coefficients (e.g., 2,800 with calixarene and 330 with hemispherand for K^+/Na^+) have been found for a simple configuration in which the OH function becomes an O*Me* bond with a metal cation (*Me*). A new compound of a calixarene type has been synthesized with an azo group (Figure 4.17(a,b)). The aryl (Ar) group is a phenyl group containing one or two NO_2 groups. The formation of azophenol salts with amines induces a typical shift in the absorption spectrum, and similar azo dyes have been suggested for fiber-optic sensors to measure amines [45].

In hemispherands, the X group can be an -O-, -O-phenyl-O-, or -OCH_2CH_2O- type. The advantages of the rigid structure of calixarenes are demonstrated by better K^+/Na^+ selectivity coefficients. For calixarenes the pK_e is −6.5 for K^+ and −9.7 for

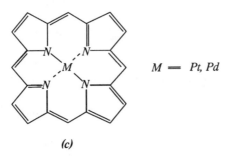

Figure 4.17 Examples of macrocyclic organic dyes for optical-fiber sensors: (a) azo-calix(4)arene, (b) azo-hemispherand (*from* [41]), and (c) chemical structure of the porphyrin tetrapyrrolic macrocycle with a coordinated metal (M) atom.

Na^+; for hemispherands pK_e is around -7 for K^+ and varies from -7 to -9.5 for Na^+. The most selective calixarene shows a 300:1 extraction ratio for K^+/Na^+ and does not extract Ca^{2+} or Mg^{2+}.

The porphyrins, another large class of macrocyclic compounds (Figure 4.17(c)), are used as highly efficient luminescent oxygen probes and enzyme biosensors [46]. The porphyrin molecules have a tetrapyrrolic macrocycle with substitutes attached to the side. The hydrophobic octaethylporphine, tetraphenylporphine, water-soluble coproporphyrin and tetra-(p-sulfophenyl)porphine derivatives are the most common indicators, coordinated with a metal atom (platinum or palladium). Advantages of these compounds are their intense Soret band (absorption band at short wavelength corresponding to a strong dipole moment) at about 400 nm, excitation bands at 450 to 650 nm, and emission in the 550- to 750-nm region.

4.4 REDOX INDICATORS

The previous section described indicators on the basis of chemical bonds involved in particle exchange systems, especially molecules and protons in acid-base and chelating reactions. A few redox indicators were mentioned; however, they form a separate class of indicators. They act by means of the oxido-reduction properties of these compounds, in which the chemical reaction is an exchange of electrons between an oxidizing (Ox) and a reducing agent (Red):

$$Ox + n \cdot e^- \rightleftharpoons Red \tag{4.8}$$

where n is the number of electrons (e) transferred. The electrode potential of this redox couple is given by the well-known Nernst equation:

$$E = E^{\otimes} + (RT/nF) \cdot \log_e(a_{ox}/a_{red}) = E_0 + (RT/nF) \cdot \log_e([Ox]/[Red]) \tag{4.9}$$

where a are the activities and the brackets [ ] correspond to the concentrations of oxidized and reduced forms, respectively. $E^{\otimes}$ is the standard potential and E_0 the normal potential (or formal potential) of this system. These potential values depend on the experimental conditions such as the concentration of Ox and Red species and complexation with ligands.

For an indicator (I) in which a color change occurs from the oxidized (I_{ox}) to the reduced form (I_{red}), two types of reactions can be considered [47]:

$$(I_{ox})^{z+} + n \cdot e^- \rightleftharpoons (I_{red})^{(z-n)+} \tag{4.10}$$

such as for the Fe^{3+}/Fe^{2+} pair, and

$$(I_{ox})^{z+} + n \cdot e^- + x \cdot H^+ \rightleftharpoons (I_{red})^{(z+x-n)+} \tag{4.11}$$

where n is the number of charges and x the number of protons in the reaction. Sometimes (I_{ox}) is neutral, as in Sb_2O_3, or negatively charged, as in MnO^{4-}. This equation shows the influence of the pH on E_0, the formal potential of the indicator. The measured potential E_m is

$$E_m = E_0' + (x/n) \, (RT/nF) \cdot \log_e[a_{H+}] + (RT/nF) \cdot \log_e([I_{ox}]/[I_{red}]) \qquad (4.12)$$

The potential E_0', related to the unit hydrogen activity, can be measured by photometric methods from the absorbance ratio $(S - S_{min})/(S_{max} - S)$ in (2.63) in Chapter 2. Thus, the relationship of E_m and E_0' to pH and the redox system of the indicator is

$$E_m = E_0' + (2.3 \, RT/nF) \cdot \{\log_{10}([I_{ox}]/[I_{red}]) - (x/n) \cdot pH\} \qquad (4.13)$$

Numerous redox indicators, listed with E_m at a given pH and E_0' at pH = 0, can be found in the literature [5].

Some applications have been examined for fiber-optic sensors, particularly for titrations [48]. As an example, the end-point determination of cerium (IV) with tin (II) was measured using N-N' diphenylbenzidine immobilized on a polyethylene terephthalate membrane as the redox indicator [49]. The effects have been studied of immobilization of some azoquinoid indicators on XAD-2 resin [50], and of pH on the equilibrium redox of immobilized dichlorophenol [51]. A fiber-optic oxygen sensor has been developed based on the absorbance variation of viologen indicators [52]. Cytochemical investigations of oxidation reactions of the metabolism by reflection photometry is an interesting application of continuously and noninvasively monitoring the tissue oxygen supply [53]. The variation of the fluorescence intensity is also a method for monitoring radical/redox systems [54].

4.5 POTENTIAL-SENSITIVE DYES

In a solution-membrane system, a potential can be developed in different ways [55]:

1. By the difference of mobilities of ions, either crossing the membrane and producing a diffusion potential or in a flowing sample solution (streaming potential);
2. From the difference of solubilities of ions in the membrane (distribution potential);
3. From the boundary conditions at the membrane-solution interface in the presence of charged functional groups (double-layer potential) or dipolar molecules (χ potential) on the membrane surface;
4. By adding an impermeable charged species to one side of the membrane as shown in Figure 4.18 (static Donnan potential). In electrochemical sensors, particularly ion-selective electrodes, the electrochemical equilibrium of transferred ions i

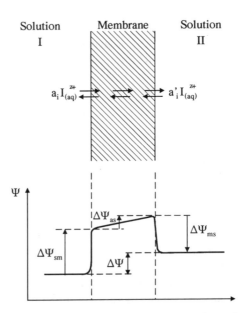

Figure 4.18 The variation of the potential across an ion-selective membrane for an I^+ ion with a_i and a'_i activities in the I and II solutions. $\Delta\Psi$ is the measured total potential. $\Delta\Psi_{sm}$, the Donnan potential, is positive in the direction of solution-membrane and negative, $\Delta\Psi_{ms}$, in the direction of membrane-solution. $\Delta\Psi_{as}$ is the asymmetric potential of the system. The thickness of the membrane is not to scale. (*After:* [56] with permission.)

through a membrane solution interface is equal inside the membrane and on both its sides [56]. The total potential difference ($\Delta\Psi$) between the two sides of the membrane is the sum of the two electric potentials (Donnan potentials) at the electrolyte-membrane interfaces ($\Delta\Psi_{sm}$ and $\Delta\Psi_{ms}$) and the potential coming from the possible asymmetric difference ($\Delta\Psi_{as}$) of chemical composition between the two interfaces :

$$\Delta\Psi = \Delta\Psi_{sm} + \Delta\Psi_{as} + \Delta\Psi_{ms} = \Delta\Psi_{as} + (RT/z_iF) \log_e(a_i/a'_i) \qquad (4.14)$$

5. By passing an electric current through the membrane system.

An applied external electric field ($E_e > 10^5$ V $\cdot$ cm^{-1}) shifts the wave number ($\Delta\nu$) of the absorption and emission spectra of ionophores with the Stark effect if a large difference in the permanent dipole moments ($\Delta\mu$) exists between the ground and excited states [57].

$$h \cdot c \cdot \Delta\nu = E_e \cdot \Delta\mu \qquad (4.15)$$

where h is Planck's constant and c is the velocity of the light. However, the Stark effect is normally very small, and for a detectable signal the dye in the optical sensor must have a large charge in the dipole moment in the excited state, and its transition moment must be parallel to the electric field.

Dye-based sensors are one approach to measuring the potential differences across biological cell membranes [58–60]. Some dyes, called PSDs or *electrochromic dyes*, are sensitive to the electric field potential created at the interface of a waveguide. Ion-active compounds called *ion carriers* with incorporated dyes in ion-sensitive membranes undergo exchange of ions between the membrane (m) and the aqueous phase (aq), with the equilibrium

$$(IS)_m^{z+} + J_{(aq)}^z \rightleftharpoons I_{(aq)}^{z+} + (JS)_m^{z+} \tag{4.16}$$

where IS is the ion-sensitive compound, J is the measurand, I is the active ion, and JS is the result of the reaction.

The charged ions are cations or anions, depending on the positive or negative charge of the dye. Some early experiments with positively charged carbocyanine dyes incorporated in optical probes demonstrated a fluorimetric response to membrane potentials of red blood cells [58]. However, these compounds (Figure 4.19(a)), which have a strong fluorescence in organic solvents when bound to membranes, show rapid photodecomposition and are unsuitable for long-term measurements. A large dynamic range has been obtained in a liposome-based potassium sensor using valinomycin as an ionophore and merocyanine 540 as the PSD [61]. Oxonol compounds have similar properties [62] but carry a negative charge (Figure 4.19(b)). Oxonol VI leads to a larger spectral shift [63]. Styryl dyes have an (aminostyryl)pyridinium chromophore (Figure 4.19(c)). Some styryl dyes are neutral zwitterionic PSDs with a faster response due to the change in the electronic distribution in the molecule [64].

In an apparently alternative approach, lipophilized fluorescent acridinium dyes have been used for K^+ determination [65]. These dyes, which are not proton donors, are highly sensitive to the polarity of the environment. In fact, they act as electrochromic and solvatochromic (sensitive to a solvatation effect) PSDs [66].

Rhodamine dyes, which are cationic PSDs, have been tried in optodes with a high fluorescence efficiency. For example, rhodamine B was incorporated in a Langmuir-Blodgett lipid membrane with arachidic acid or valinomycin and was used as a neutral ion carrier for K^+ determination by fluorescence [67]. The general relationship of the relative change of the fluorescent signal (ΔF) with metal concentration [Me] is

$$\Delta F/F = \Delta F_0/F + k \cdot \log [Me] \tag{4.17}$$

where k is a proportionality coefficient.

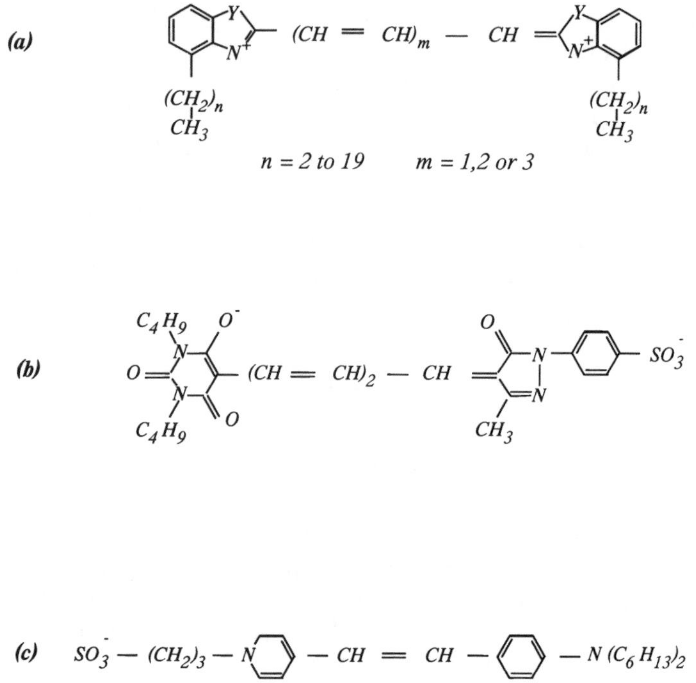

Figure 4.19 Some examples of PSDs: (a) carbocyanine in which Y is oxygen, sulfur, or >C(CH$_3$)$_2$; m is 1 to 3; and n is 2 to 19; (b) oxonol; and (c) an example of a styryl dye.

4.6 NEUTRAL ION CARRIERS AND OTHER RECEPTOR CARRIERS

Another approach, applied to optodes, originates from work on liquid electrolyte sensors and liquid membranes [68] and on two optical K$^+$ detection schemes [69,70]. It is based on ion exchange or a coextraction mechanism with complete mass transfer into and out of a very thin membrane (2 to 4 μm). The active compounds are trapped in the bulk of this sensing layer. Thermodynamic equilibrium of all the components in the sample and membrane phases causes this mass transfer, and the reaction rate obeys Fick's diffusion laws. Selective association to a ligand and desolvatation take place on the membrane acting as the solvent. Ion carriers (ionophores) combined with neutral lipophilic chromo-ionophores (dyes) within this polymeric membrane in a pH buffered solution lead to new possibilities for reversible optical sensing. A proton exchange mechanism is generally associated with the transport of ions by neutral carriers. In this case, the proton carrier is a dye that changes its optical properties by losing a proton. A spectral shift or color change is determined by absorbance or luminescence measurements.

The concepts concerning bulk membranes for optical sensors have been generalized theoretically in terms of three systems: ion exchange, ion extraction, and neutral analytes [71]. These three systems have been investigated experimentally in optodes [72] using an indicator (I) and a ligand (L) for the recognition and detection of charged analytes (cations C^+, anions A^-) and noncharged (neutral S species). This is represented schematically in Figures 4.20 to 4.22 for monovalent species. Electroneutrality in the membrane is controlled by adding lipophilic anions (R^- or R^+). The various exchange reactions between the different ions and species present in the sample and in the membrane are also shown; these determine the relationship of the

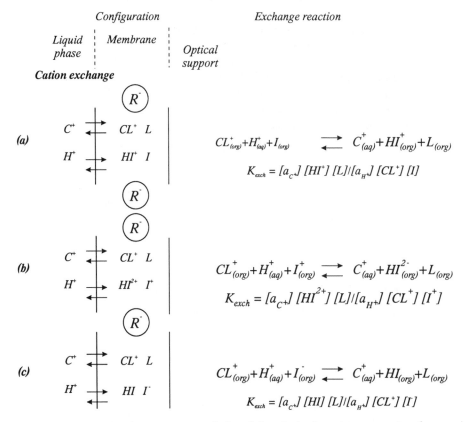

Figure 4.20 Cation (C^+) exchange reactions in the liquid phase (aq): schematic representation of an optode using neutral carriers in a lipophilic (organic phase) membrane. I is the indicator, L the ligand, and R^- the lipophilic anionic counterions. The exchange process between L and C^+ is accompanied by a parallel ion-exchange process for the indicator/pH system: (a,b) with counterions and (c) without counterions. The exchange reactions and K_{exch} coefficients correspond to these different situations. The term a represents the activities of the cations and hydrogen ions. (*After:* [73] with permission.)

analyte/optical signal to the exchange coefficients (K_{exch}) and the total indicator concentration.

In cation exchange reactions, cation extraction from the sample by the ligand (ionophore) is associated with the H^+ variation acting here as an internal reference. The chromo-ionophore in the membrane can be either charged (Figure 4.20(b,c)) or neutral (Figure 4.20(a)).

The second system is based on the coextraction of the analyte anion (A^-) and of a cation (e.g., hydrogen ions) in the membrane. This occurs in the presence of the ligand (L) or charged ligand (e.g., as L^+ for nitrate determination), the indicator (I), and the counterion (R^+). Two typical examples are shown in Figure 4.21.

The third system concerns neutral analytes and gas determination (see the three examples in Figure 4.22). In Figure 4.22(a), the ligand (L) acts as a chromophore or fluorophore. If L is not a chromophore, a label can be used. The exchange reaction is without a counterion in Figure 4.22(b) and with a counterion in Figure 4.22(c). Other examples include the formation of $(SOH)L^-$ compounds [73]. A permeable membrane is normally added for measurement of gases.

The chemical equilibria of these systems are expressed by the following equilibrium relationships given as typical examples.

1. For ion exchange with two different cation-selective ionophores L and P (one of

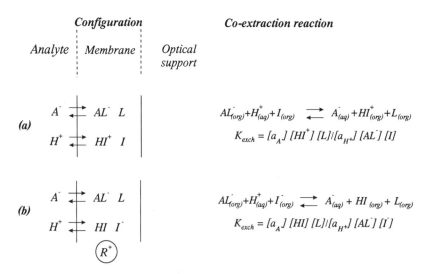

Figure 4.21 Coextraction reactions for anions (A^-) and cations (or H^+) in the presence of a selective ligand L or charged ligand L^+: (a) without a counterion and with an uncharged chromo-ionophore and (b) with a counterion R^+ and negatively charged chromo-ionophore. The exchange reactions and K_{exch} correspond to these different situations. The term a represents the activities of the anions and hydrogen ions. (*After:* [73] with permission.)

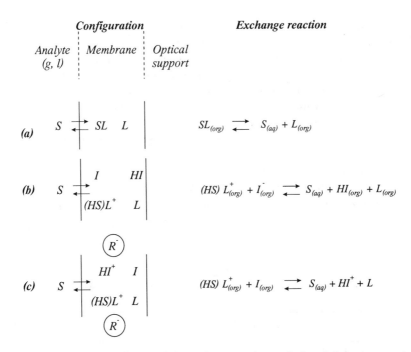

Figure 4.22 Exchange reactions for neutral chemical compounds (*S*) in the liquid phase (*l*) or gas phase (*g*): schematic representation of an optode using neutral carriers, where *I* is the indicator and *L* the ligand: the reaction in which the ligand acts as a chromophore (a) without a counterion, (b) with an indicator and without a counterion, and (c) with an indicator and counterions (*R⁻*) in the organic phase (membrane). (*After:* [73] with permission.)

which is the ionochromophore) producing the complexes CL_n^{z+} and P_pM^{r+} with cations C^{z+} and M^{r+} in the aqueous phase, and R^- the lipophilic compounds, the equilibrium is

$$pzP + zM^{r+}_{(aq)} + vCL_n^{z+} + vzR^- \rightleftharpoons nvL + vC^{z+}_{(aq)} + zP_pM^{r+} + vzR^- \quad (4.18)$$

2. For a coextraction system with the anion A^-, the cation C^{z+}, and the neutral ionophores X and L (one of which is the chromo-ionophore) giving AX_p^{v-} and CL_n^{z+} complexes, the equilibrium is

$$vnL + zpX + vC^{z+}_{(aq)} + zA^{v-}_{(aq)} \rightleftharpoons zAX_p^{v-} + vCL_n^{z+} \quad (4.19)$$

3. For neutral analytes (*S*) in a liquid (*l*) or gas (*g*) medium, the equilibrium is

$$S_{(l,g)} + mL \rightleftharpoons SL_m \quad (4.20)$$

In measurements using enzymes, the enzyme is immobilized on an additional membrane placed at the surface of the exchange membrane. Recent work has incorporated the enzyme within the membrane by means of reversed micelles having amphiphilic properties [74]. Some important examples of optical sensors using these systems are given in Table 4.6 for a plasticized high-molecular-weight polyvinyl chloride (PVC) membrane. Table 4.6 lists the chromo-ionophore, the ionophore or an electroactive compound (EAC) acting as the ligand, and the plasticizer.

The majority of dyes employed for cation determination are lipophilized phenoxazines or azo compounds. The structure of some typical chromo-ionophores and ionophores are presented in Figures 4.23 and 4.24.

New ion sensitive optodes (ISO) are very attractive for biological and pharmaceutical measurements with natural or artificial carriers and receptors [84]. The optode membranes are generally reversible and can sometimes achieve lifetimes of

Table 4.6
Characteristics of Some Optode Systems Using Lipophilic PVC Membranes

Measurand	CI	Ligand	P	R^-	λ_{ex}/λ_m	References
K^+	ETH 5294	Valinomycin	DOS	KT_pClPB	—/660	[75]
Na^+	ETH 2439	ETH 4120	BBPA	$NaT_m(CF_3)_2PB$	—/650	[76]
Ca^{2+}	ETH 5294	ETH 1001	DOS	$NaT_m(CF_3)_2PB$	—/660	[77]
Cl^-	ETH 7075	TOTC	BBPA	—	—/530	[78]
NH^{4+}	Nile Blue	Nonactin	DOS	KT_pClPB	550/630	[79]
NO^{3-}	ETH 2412	TDMACl	oNPOE	—	—/560	[80]
Salicylic acid	Nile Blue	BNAL	DOP	TDMACl	520/650	[81]
Bio-amines	DZ 49	tartrates	DOP	KT_pClPB	520/600	[82]
Thiamine	Eosine	DBVT	DOS	KT_pClPB	520/555	[83]
NH_3	TBPE	ETH 157	DOS	$NaT_m(CF_3)_2PB$	—/616	[84]

Note: Abbreviations are as follows: CI = chromo-ionophore; P = plasticizer; R^- = anionic site compound; λ_{ex} = excitation wavelength; λ_m = wavelength of absorption measurement or emission; ETH 5294 = 9-(diethylamino)5-octadecanoylimino-5H-benzo[a]phenoxazine; ETH 2439 = 9-(diethylamino)5-(4-(16-butyl-2,14-dioxo-3,15-dioxaeicosyl)phenylamino)-5H-benzo[a]phenoxazine; ETH 7075 = 4′,5′-dibromofluorescein octadecyl ester; ETH 2412 = 3-hydroxy-4-(4-nitrophenylazo)phenyl octadecanoate; 5-octadecanoyloxy-2-(4-nitrophenylazo)phenol; DZ 49 = lipophilic pH indicator (P. Czerney, Univ. Jena, Germany); TBPE = tetrabromophenolphthalein ethyl ester; ETH 4120 = 4-octadecanoyloxymethyl-N,N,N′,N′-tetracyclohexyl-1,2,phenylene dioxidi-acetamide; ETH 1001 = -(R,R′)-N,N′-[bis(11-ethoxy-carbonyl)undecyl]-N,N′,4,5-tetramethyl-3,6-dioxaoctanediamide; TOTC = trioctyltin chloride; TDMACl = tridodecylmethylammonium chloride; BNAL = n-Butyl-o-(1-naphthylaminocarbonyl) lactate; DBVT: di-ter.-butyl-o-o′-bis-(2,2-dichlorovinyl)-tartate; ETH 157 = N,N′-dibenzyl-N-N′-diphenyl-1,2-(phenylenedioxy) diacetamide; DOS = bis(2-ethylhexyl) sebacate; BBPA = bis(1-butylpentyl)adipate; oNPOE = 2-nitrophenyl octyl ether; DOP = bis(2-ethylhexyl) phthalate; KT_pClPB = potassium tetrakis (4-chlorophenyl) borate; (K) or (Na)$Tm(CF_3)_2PB$ = (Potassium) or (sodium) tetrakis (3,5-bis(trifluoromethyl)phenyl) borate; ETH (Swiss Federal Institute of Technology, Zurich).

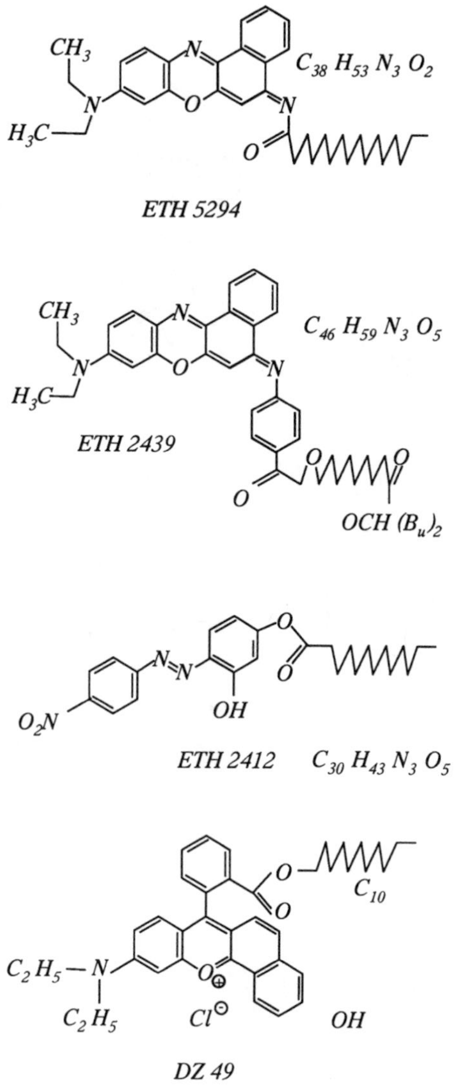

Figure 4.23 Chemical formulas of typical chromo-ionophores used in optode membranes (see Table 4.6).

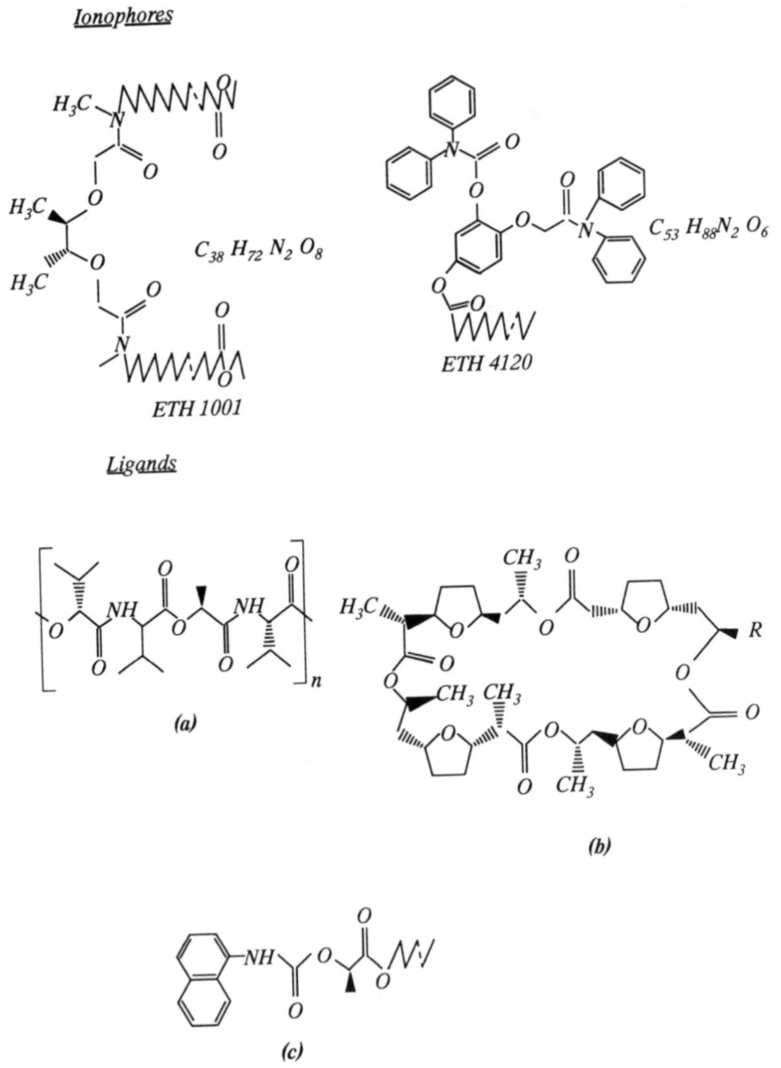

Ionophores

$C_{38} H_{72} N_2 O_8$

ETH 1001

$C_{53} H_{88} N_2 O_6$

ETH 4120

Ligands

(a)

(b)

(c)

Figure 4.24 Chemical formulas of two typical ionophores, ETH 1001 and 4120, developed at ETH. Formulas of various ligands: (a) valinomycin, (b) nonactin (R=CH₃) and monactin (R=C₂H₅), and (c) optically active BNAL (see Table 4.6).

several months [85]. Furthermore, the interaction of the receptor (host) with the substrate (guest), one of which is an enantiomer (optical isomer), introduces a specific selectivity (enantioselectivity). The technique can be extended to enzyme-based sensing of amino acids by transducers sensitive to organic ammonium ions. This can also provide an additional enzymatic selectivity and is well adapted for optical sensing techniques.

4.7 OPTICAL OPERATING REQUIREMENTS FOR INDICATORS

The practical requirements for the fabrication and operation of sensors destined for laboratory or industrial monitoring are very different; the latter have stricter requirements for long-term stability, ruggedness, and cost constraints. Additional constraints are imposed by the limited choice of optical fibers (which have a poor transmission in UV), the immobilization conditions for the indicator dye, and the choice of a low-cost light source (usually an LED) and photodetector (usually a PIN photodiode) with a suitable operating spectral range. Hence, from the large potential choice of absorbance-based indicators only a few are usable. This section outlines the criteria for selecting suitable candidates.

The choice of indicator is restricted by the LED as light source and the *pin* photodiode as detector (see Chapter 9). A large part of the absorption spectrum of the dye cannot be analyzed, since the LED and *pin* photodiode pair only covers a limited wavelength range, as shown in Figure 4.25. In addition, use of a blue LED requires a photomultiplier or an avalanche photodiode to produce an acceptable signal-to-noise ratio. These are more costly and have problems of long-term drift and temperature stability. Thus, indicators with a maximum absorbance at short wavelengths below 500 nm, such as *p*-nitrophenol, methyl red, and neutral red, are only suitable for laboratory experiments.

The technique of using an isobestic point as a reference, which often occurs at short wavelengths, can create problems in balancing the intensity in both measurement and reference channels. Dyes with an absorption between 550 and 650 nm, where LEDs have a higher light output, are generally preferred for industrial sensors. The reference wavelength is often chosen in the 830-nm region, where the dye does not absorb, and is measured with an infrared (IR) LED.

Fluorescent sensors impose different operating requirements than absorbance-based sensors. The choice of fluorophores is determined by immobilization conditions and the quantum yield of the fluorescence, or, in the case of quenchable fluorophores, the quenching efficiency. Fluorescent measurements also require an exciting source with a high output (a laser) and a very sensitive detector. With the present state of the art, these are not considered to be limiting factors. Current research is developing a range of fluorescent dyes in which the excitation and fluorescence occurs in the red and near IR spectrum (600 to 1,000 nm) [86–88]. This allows a solid-state LED or semiconductor laser to be used as the exciting source [89].

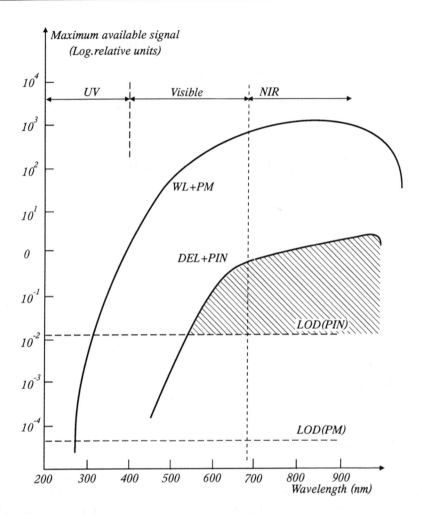

Figure 4.25 The maximum available signal, equivalent to the optical power margin, as a function of wavelength for two different light sources and detectors. The upper curve is for a combination of a white light source (WL), interference filters, and a photomultiplier (PM). The lower curve is for a combination of an LED source and *pin* photodiode. These are connected to a fiber-optic probe. The total loss for the optical fiber and connectors is assumed to be 3 dB. The available signal, shown in relative logarithmic units (similar to absorbance measurements), is calculated from the difference between the maximum power emitted by the source and the minimum detectable power for the detector at each wavelength. Extending the operating range of dyes into the near IR increases the available signal and power margin. The lower limit of detection (LOD) for the detector determines the possible wavelength range of operation. For example, the hatched area is for a *pin* and an LED, which could operate from 550 nm to near IR.

4.8 LABELS AND LABELING PROTOCOLS

4.8.1 Labels

A chromophore or a luminophore may be covalently attached to other molecules to act as a "tag" in a biochemical reaction. This is called a *label* and can be either chromogenic, fluorescent, chemiluminescent, bioluminescent, or phosphorescent. As opposed to indicators and noncovalently bound dyes, it does not form ionic, hydrophobic, or dipole-dipole interactions with other compounds or with substrates.

Several types of labels can be used, especially in immunosensors [90]. In addition, labels based on enzymes and coenzymes are becoming increasingly important because of the specificity of enzymatic reactions.

4.8.1.1 Fluorescent Labels

In immunoassays, the fluorescent label technique employing labeled antibodies or antigens is often preferred to radioactive markers because of the difficulties in handling radioactive products and the decay of the radioactive signal with time. For optical sensors, the protein can be conjugated (reacted) with a mixture of immunoglobulins and one or several dyes from their reactive functional groups. After labeling is completed, the unreacted dyes are removed (in homogeneous assays) or separated (in heterogeneous assays) and the labeled protein is purified to increase the specificity of the measurement.

Important examples of fluorescent labels with emission wavelengths longer than 420 nm are given in Table 4.7 [1,86,92,93]. These examples include the more common labels as well as ovalbumin [91]. The nature of the substrate or reagent determines the choice of available reactive groups that can be used in labeling [92]. These groups are mainly the amines ($-NH_2$), thiols (-SH), carboxylic acids (-COOH), and hydroxyls (-OH). Biomolecules and polymeric supports can be labeled on the surface by chemically modifying these groups.

Different aspects of the modification of these groups for adaptation to sensors have been well described in the literature [1,93]:

1. In amine modification, sulfonic acid chloride forms stable sulfonamide derivatives with primary and secondary amines and amino groups on proteins and solid supports used for label immobilization. Dansyl chloride and Texas red, a sulforhodamine, are examples of labels with emission wavelengths longer than 500 nm. Texas red, conjugated to avidin or streptavin as the acceptor, has been investigated experimentally with a phenytoin-B-phycoerythrin couple in a reversible fiber-optic immunosensor [94]. Isothiocyanates, which form thioureas with amines, include very well-known labels such as FITC, rhodamine-B-isothiocyanate (RBITC) or eosine-5-isothiocyanate (EITC), and their derivatives such as tetramethylrhodamine isothiocyanate (TRITC) are available as labels for large molecules [95].

Table 4.7
Common Fluorescent Labels and Reactive Systems for Immunosensing

Functional Group	Reagents	Examples of Labels	λ_{ex} (nm)	λ_{em} (nm)
Amines	Sulfonyl chloride	Dansyl chloride	350	535
		Texas red	514	577
	Isothiocyanates	FITC	490	525
		Rhodamine B-ITC	547	585
		Eosin-5-ITC	524	551
	o-Dialdehyde/thiol	OPA	326	424
	Benzo-diazole	N-BD-Cl	470	554
Amide	Sulfonic acid	Dansyl aziridine	346	511
Thiols	Maleimides	5-MSA	314	434
	Haloacetyl	Fluorescein-5-IA	490	522
		Eosin-5-IA	529	556
	Haloalkyl	Monobromobimane	386	463
	Amino-benzo-diazole	ASBD-F	378	502
Carboxylic acid	Diazo-alkyls	4-BMM-coumarin	365	450
Hydroxyl	Isocyanates	Anthracene IC	361	466
Carbonyl compounds	Amines	7-AM-4-coumarin	365	450
	Hydrazides	Lucifer Yellow	419	535

Note: Abbreviations are as follows: Texas red = sulforhodamine 101 sulfonyl chloride; ITC = isothiocyanate; FITC = fluorescein-5-isothiocyanate; OPA = o-pththalaldehyde; N-BD-Cl = 4-chloro-7-nitrodibenzo-2-oxa-1,3-diazole; 5-MSA = 5-maleimidyl salicylic acid; IA = iodoacetamide; ASBD-F = 4-(aminosulfonyl)-7-fluoro-2,1,3-benzooxadiazole; BMM = 4-bromomethyl-7-methoxy-; IC = isocyanate; 7-AM-4- = 7-amino-4-methyl-.

Umbelliferones are labels commonly employed in condensation reaction with amines and proteins [96]. Many examples of these labels in fiber-optic sensors can be found in the literature, especially for immunosensing.

2. In the presence of a thiol group, o-dialdehyde produces a fluorescent isoindole derivative (such as o-phthalaldehyde (OPA)) with amines. Also, benzodiazole (BD) derivatives such as nitro-BD-chloride are fluorescent reaction products of amines or thiols.

After amide modification with sulfonic acid, dansyl aziridine (DAZ) is stable in a basic medium. An example of an immunosensing system using the Langmuir-Blodgett technique has applied DAZ as a fluorescent thiol reagent for direct antigen detection, which modifies the hinge region of cystine residues in rabbit IgG [97].

After conjugation of the thiol (sulfydric) group, maleimide reagents become fluorescent across their double bonds, but these compounds can be hydrolyzed above pH 8. The thiol modifications with haloacetyl and haloalkyl derivatives give fluorescent compounds for probes, but only the covalent conjugates have an acceptable photostability. A fluorimetric reagent of the thiol group, 4-(aminosulfonyl)-7-fluoro-2,1,3-benzooxadiazole (ASBD-F) has been employed in the detection and isolation of proteins and peptides. Labeled egg albumin containing thiol groups is used as a standard for measuring proteins [98].

3. The modification of carboxylic acid by diazo-alkyls and carbonyl compounds by amines or hydrazides has been demonstrated with fluorescence emission wavelengths greater than 400 nm [1]. A major drawback of these compounds is that the excitation wavelengths are in the UV or violet part of the spectrum. In addition, their fluorescence properties are not easy to predict. For instance, a comparison of 7-amino-4-methyl coumarin-3-acetic acid (AMCA) with cascade blue (CB) conjugates of rabbit IgG showed that the fluorescence intensity increased linearly with increasing CB substitution, whereas the fluorescence was quenched with increasing AMCA substitution [93].

4. A number of workers have applied hydroxyl modification of isocyanates, carbonyl azides, carbonyl nitrides, carboxylic acid chlorides, and diazomethanes for the recognition of alcohol groups [1].

New activities in this field concerning sensor research concentrate on matching these fluorescent compounds with laser diodes coupled to an optical fiber or a planar waveguide for fluorescent excitation, particularly for immunosensors. For semiconductor laser diodes, the wavelength range is 670 to 850 nm or the frequency-doubled harmonic at around 420 nm [99]. Rhodamine 800 (λ_{ex} = 685 nm, λ_{em} = 700 nm), oxazine 750 (λ_{ex} = 673 nm, λ_{em} = 691 nm), phenoxazines such as Nile blue (λ_{ex} = 640 nm, λ_{em} = 672 nm), and thiazines dyes such as thionine (λ_{ex} = 600 nm, λ_{em} = 623 nm) are good candidate fluorophores for labeling biological molecules.

Rare-earth chelating compounds have a long photoluminescent lifetime (>200 μs) and a large Stokes shift around 300 nm [100]. In fluorescent immunoassays, europium and terbium chelates complexed with EDTA derivatives or β-diketones have received the most attention with a time-resolved measurement technique [101]. As an example, temporal rejection of backscattered light from fiber optics gives a better sensitivity for Eu-2-naphtholtrifluoroacetonate (λ_{ex} = 337 nm, λ_{em} = 613 nm) than for FITC [102]. Time-resolved methods can be divided into three main categories: fluorescence enhancement assays, fluorescent ligand stabilization methods, and assays relying on stable fluorescent chelates [103]. Narrowband emission of different lanthanides can also be used for simultaneous measurement of double, triple, or quadruple chelate label assays with an enhanced cofluorescence effect. This enhancement is based on an energy transfer between nonemitting and emitting chelate ions producing an increased emission efficiency.

4.8.1.2 Other Labels

At present, few labels have been used for absorption or scattering measurements because of the poor detection sensitivity compared to luminescent techniques. As an example, methylene blue, whose absorption peak at $\lambda = 664$ nm coincides approximately with He-Ne laser light output, has been tried as a dye label in solution for detecting small quantities of antigenic species with dye-labeled antibodies and enzymes [104]. However, the dye is unstable. It is also fluorescent ($\lambda_{em} = 683$ nm).

The advantage of phosphorescent labels over fluorescent ones is that the emission can be monitored after the excitation radiation has been removed, hence decreasing the light scattering typical of fiber-optic systems. Also, phosphorescent compounds are easily quenched by oxygen. The main disadvantage is the weaker light levels from phosphorescence, which can be partially compensated for by integrating the output signal.

Applications of bioluminescence, chemiluminescence and labeled enzymes in immunoassays are of high interest. By transferring macromolecules, such as proteins or DNA, from an electrophoretic gel to blotting paper (membrane) incubated with a primary antibody, it is possible to analyze a pattern copy on the membrane with a specific antisera. The second antibody labeled with an enzyme is detected by the luminescence generated by the enzymatic reaction.

Two classes of labels based on the kinetics of their light output have been defined by Jansen [105]: transient labels, such as luminol and its derivatives and acridinium compounds, and enzyme labels, including horseradish peroxidase (HRP), the alkaline phosphatase, and the xanthine oxidase (XOD). This last compound, using hypoxanthine as a substrate and an iron-EDTA complex, has a higher sensitivity and better kinetics. The chelate splits H_2O_2 into hydroxyl radicals and the reactive species reacts with luminol to yield chemiluminescence with a half-life decay time of about 30 hours.

4.8.2 Protein Labeling Protocols

The protocol or sequence of processing steps for the labeling of proteins consists of a description of the reaction conditions for protein-dye conjugation. The processing steps differ for each dye label, the nature of the protein and substrate, and the type of reaction that occurs in the immunosensor. In precipitation reactions, the antigen and antibody in equal proportions combine to form a complex network or lattice of Ag-Ab aggregates and precipitates. In agglutination, aggregates are also formed, but the antigen is absorbed or covalently linked to particles (proteins) or cells. For enzyme immunoassays (see Section 3.3.2), immobilization is achieved with the enzyme and either Ag or Ab on the surface after washing to remove extraneous molecules.

An outline of a typical sequence of process steps is described in Figure 4.26. The final protein concentration is normally determined by the weight of proteins equal to 1% of the final volume in milliliters. The steps use centrifuging, filtration, separation on columns, and dialysis to improve the specificity and sensitivity. These steps are also necessary to remove the unconjugated free dye, filter out insoluble species, or separate labeled compounds from interfering products. The absorbance and fluorescence properties of the final sample are determined from aliquots.

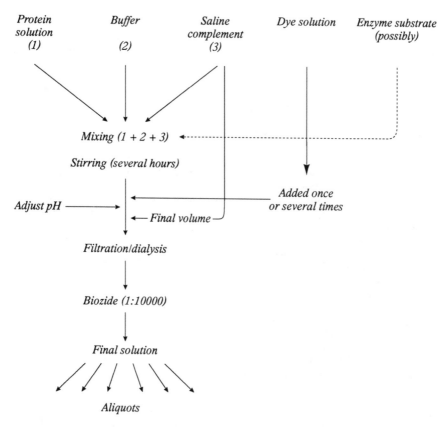

Figure 4.26 Typical method, showing the sequence of process steps, for labeling a protein used in immunosensing.

For fiber-optic immunosensors, the processing protocols must take into account the specific conditions for immobilization on the optical or polymer surfaces (see Chapter 5) and the perturbations due to labeling. Indeed, the labeled *reporter* group of proteins indicates the state of bound ligands in the whole molecular substrate. Nevertheless, the binding of the ligand can perturb the reporter group either by direct influence of the optical properties (e.g., by quenching), by a change in the structural arrangement of the molecule, or by modification of the aggregate state of the molecular receptor (e.g., antibodies) and thus the environment of the reporter group. Examination of these modifications on periplasmic binding proteins (PBP) of gram negative bacteria and on heparin and riboflavin binding proteins has been suggested as a method for fluorescent immunosensing without a reagent [106].

References

[1] Koller, E., and O. S. Wolfbeis, "Sensor Chemistry," Chap. 7 in *Fiber Optic Chemical Sensors and Biosensors, Vol. I*, O. S. Wolfbeis, ed., Boca Raton: CRC Press, 1991, pp. 303–353.

[2] Christian, G. D., *Analytical Chemistry*, 4th ed., New York: Wiley, 1986.

[3] Sandell, E. B., and H. Onishi, *Photometric Determination of Traces of Metals: General Aspects, Vol. 3*, Part I, New York: Wiley, 1978.

[4] Seitz, W. R., "Optical Ion Sensing," Chap. 9 in *Fiber Optic Chemical Sensors and Biosensors, Vol. II*, O. S., Wolfbeis, ed., Boca Raton: CRC Press, 1991, pp. 1–17.

[5] Bishop, E., ed., *Indicators*, Oxford: Pergamon Press, 1972.

[6] Wolfbeis, O. S., "Optical and Fiber Optic Fluorosensors in Analytical and Clinical Chemistry," Chap. 3 in *Molecular Luminescence Spectroscopy. Methods and Applications, Vol. 2*, S. G. Schulman, ed., New York: Wiley, 1988, pp. 129–281.

[7] Camara, C., M. C. Moreno, and G. Orellana, "Chemical Sensing With Fiberoptic Devices," in *Biosensors With Fiberoptics*, D. L. Wise and L. B. Wingard, eds., Clifton, NJ: Humana Press, 1991, pp. 29–84.

[8] Vo-Dinh,T., B. J. Tromberg, G. D. Griffin, K. R. Ambrose, M. J. Sepaniak, and E. M. Gardenshire, "Antibody-Based Fiberoptics Biosensor for Carcinogen Benzo(a)pyrene," *Appl. Spectr.*, Vol. 41, 1987, pp. 735–738.

[9] Kolthoff, I. M., P. J. Elving, and E. B. Sandell, *Treatise on Analytical Chemistry: Theory and Practice, Vol. 11*, Part I, New York: Wiley, 1975.

[10] Wolfbeis, O. S., N. V. Rodriguez, and T. Werner, "LED-Compatible Fluorosensor for Measurement of Near-Neutral pH Values," *Mikrochim. Acta*, Vol. 108, 1992, pp. 133–141.

[11] Atwater, B. W., and O. Laksin, "Emission Based Fiber Optic Sensors for pH and Carbon Dioxide Analysis," U.S. Pat., 5,280,548, 1994.

[12] Lee, L. G., G. M. Berry, and C. H. Chen, "Vita Blue: A New 633-nm Excitable Fluorescent Dye for Cell Analysis," *Cytometry*, Vol. 10, 1989, pp. 151–164.

[13] Vogel, A. I., *A Textbook of Quantitative Inorganic Analysis*, 3rd ed., London: Longmans, 1961, p. 56.

[14] Boisdé, G., and B. Sebille, "Co-immobilization of Several Dyes on Optodes for pH-Measurements," *Proc. SPIE–Int. Soc. Opt. Eng.*, Vol. 1510, 1991, pp. 80–94.

[15] Wolfbeis, O. S., E. Fürlinger, H. Kroneis, and H. Marsoner, "Fluorimetric Analysis: 1. A Study on Fluorescent Indicators for Measuring Near Neutral (Physiological) pH Values," *Fresenius Z. Anal. Chem.*, Vol. 314, 1983, pp. 119–124.

[16] Saari, L. A., and W. R. Seitz, "Optical Sensor for Beryllium Based on Immobilized Morin Fluorescence," *Analyst*, Vol. 109, 1984, pp. 655–657.

[17] Lübbers, D.W., and N. Opitz, "Die pCO₂/pO₂ Optode," *Z. Naturforschung*, Vol. 30C, 1975, pp. 532–533.

[18] Offenbacher, H., O. S. Wolfbeis, and E. Fürlinger, "Fluorescence Optical Sensors for Continuous Determination of Near Neutral pH Values," *Sensors and Actuators*, Vol. 9, 1986, pp. 73–84.

[19] Wolfbeis, O. S., and J. H. Baustert, "Synthesis and Spectral Properties of 7-(N-Arylsulfonyl)aminocoumarins: A New Class of Fluorescent pH Indicators," *J. Heterocyclic Chem.*, Vol. 22, 1985, pp. 1215–1218.

[20] Wolfbeis, O. S., and H. Marhold, "A New Group of Fluorescent pH-Indicators for an Extended pH-Range" *Fresenius Z. Anal. Chem.*, Vol. 327, 1987, pp. 347–350.

[21] Delisser-Matthews, L. A., and J. M. Kauffman, "3-Arylcoumarins as Fluorescent Indicators," *Analyst*, Vol. 109, 1984, pp. 1009–1011.

[22] Mohr, G. J., and O. S. Wolfbeis, "Optical Sensors for a Wide pH Range Based on Azo Dyes Immobilized on a Novel Support," *Anal. Chim. Acta*, Vol. 292, 1994, pp. 41–48.

[23] Al-Amir, S. M. S., D. C. Ashworth, and R. Narayanaswamy, "Synthesis and Characterisation of Some

Chromogenic Crown Ethers as Optical Potassium Ion Sensors," *Talanta*, Vol. 36, 1989, pp. 645–650.

[24] Wolfbeis, O. S., and W. Trettnak, "Fluorescence Quenching of 6-Methoxyquinoline by Pb, Hg, Cu, Ag and HS$^-$," *Spectrochim. Acta*, Vol. 43 A, 1987, pp. 405–408.

[25] Urbano, E., H. Offenbacher, and O. S. Wolfbeis, "Optical Sensor for Continuous Determination of Halides," *Anal. Chem.*, Vol. 56, 1983, pp. 427–429.

[26] Narayanaswamy, R., and F. Sevilla, "Flow Cell Studies With Immobilised Reagents for the Development of an Optical Fibre Sulphide Sensor," *Analyst*, Vol. 111, 1986, pp. 1085–1088.

[27] Roe, J. N., F. C. Szoka, and A. S. Verkman, "Fiber Optic Sensor for the Detection of Potassium Using Fluorescence Energy Transfer," *Analyst*, Vol. 115, 1990, pp. 353–358.

[28] Hirschfeld, T., T. Deaton, F. Milanovich, and S. Klainer, "Feasibility of Using Fiber Optics for Monitoring Groundwater Contaminants," *Opt. Eng.*, Vol. 22, 1983, pp. 527–531.

[29] Bright, F. V., G. E. Poirier, and G. M. Hieftje, "A New Ion Sensor Based on Fiber Optics," *Talanta*, Vol. 35, 1988, pp. 113–118.

[30] Sharma, A., and O. S. Wolfbeis, "The Quenching of the Fluorescence of Polycyclic Aromatic Hydrocarbons and Rhodamine 6G by Sulfur Dioxide," *Spectrochim. Acta*, Vol. 43A, 1987, pp. 1417–1421.

[31] Giuliani, J. F., H. Wohltjen, and N. L. Jarvis, "Reversible Optical Waveguide Sensor for Ammonia Vapors," *Opt. Lett.*, Vol. 8, 1983, pp. 54–56.

[32] Biatry, B., "Capteur chimique à fibre optique pour la mesure du pH en milieu médical et industriel," Thesis, Université Paris XII, 1993.

[33] Arnold, M. A., "Enzyme-Based Fiber Optic Sensor," *Anal. Chem.*, Vol. 57, 1985, pp. 565–566.

[34] Klimant, I., and M. Otto, "Optical Fiber Sensor for Heavy Metals Ions Based on Immobilized Xylenol Orange," *Mikrochim. Acta*, Vol. 108, 1992, pp. 11–17.

[35] Vo-Dinh, T., P. Viallet, L. Ramirez, and A. Pal, "Gel Based Indo-1 Probe for Monitoring Calcium (II) Ions," *Anal. Chem.*, Vol. 66, 1994, pp. 813–817.

[36] Freeman, T. M., and W. R. Seitz, "Chemiluminescence Fiber Optic Probe for Hydrogen Peroxide Based on the Luminol Reaction," *Anal. Chem.*, Vol. 50, 1978, pp. 1242–1246.

[37] De Oliveira, W. A., and R. Narayanaswamy, "A Flow Cell Optosensor for Lead Based on Immobilized Dithizone," *Talanta*, Vol. 39, 1992, pp. 1499–1503.

[38] Blum, L. J., and P. R. Coulet, "Chemiluminescence and Bioluminescence Based Optical Probes," Chap. 20 in *Fiber Optic Chemical Sensors and Biosensors, Vol. II*, O. S. Wolfbeis, ed., Boca Raton: CRC Press, 1991, pp. 301–313.

[39] Pedersen, C. J., "Cyclic Polyethers and Their Complexes With Metal Salts," *J. Amer. Chem. Soc.*, Vol. 89, 1967, pp. 7017–7036.

[40] Lehn, J. M., B. Dietrich, and P. Viout, *Aspects de la Chimie des Composés Macromoléculaires*, Meudon: Paris Inter. Editions, Ed. CNRS, 1991.

[41] Beer, P. D., "Molecular and Ionic Recognitions by Chemical Methods," Chap. 2 in *Chemical Sensors*, T. E. Edmonds, ed., Glascow: Blackie and Sons, 1991, pp. 17–70.

[42] Alder, J. F., D. C. Ashworth, R. Narayanaswamy, R. E. Moss, and I. O. Sutherland, "An Optical Potassium Ion Sensor," *Analyst*, Vol. 112, 1987, pp. 1191–1192.

[43] Gutsche, C. D., *Calixarenes*, London: Royal Society of Chemistry, 1989.

[44] Sandanayake, K. R. A. S., and I. O. Sutherland, "Organic Dyes for Optical Sensors," *Sensors and Actuators*, Vol. B 11, 1993, pp. 331–340.

[45] King, A. M., "Chromogenic Reagents for Amines and Cations," Ph.D. Thesis, Univ. Liverpool, 1991.

[46] Papkovsky, D. B., "Luminescent Porphyrins as Probes for Optical (Bio)sensors," *Sensors and Actuators*, Vol. B 11, 1993, pp. 293–300.

[47] Oehme, F., "Liquid Electrolyte Sensors: Potentiometry, Amperometry and Conductometry," Chap. 7 in *Sensors: A Comprehensive Survey, Vol. 2*, W. Göpel, J. Hesse, and J. N. Zemel, eds., 1991, pp. 239–339.

[48] Wolfbeis, O. S., "Optical Fibers in Titrimetry," Chap. 13 in *Fiber Optic Chemical Sensors and Biosensors*, O. S. Wolfbeis, ed., Boca Raton: CRC Press, Vol. 2, 1991, pp. 121–133.

[49] Dybko, A., J. Maciejewski, Z. Brzozka, and W. Wroblewski, "Application of Optical Fibres in Oxidation-Reduction Titrations," *Sensors and Actuators*, Vol. B 29, 1995, pp. 374–377.

[50] Narayanaswamy, R., and F. Sevilla, "Reflectometric Study of Oxidation-Reduction Equilibria of Indicators Immobilized on a Non-Ionic Polymer," *Mikrochim. Acta*, Vol. I, 1989, pp. 293–301.

[51] Goodlet, G., and R. Narayanaswamy, "Effect of the pH on the Redox Equilibria of Immobilised 2,6-Dichloroindophenol," *Anal. Chim. Acta*, Vol. 279, 1993, pp. 335–340.

[52] Mitchell, G. L., J. C. Hartl, D. McCrae, R. A. Wolthuis, E. W. Saaski, K. C. Garcin, and H. R. Williard, "Viologen-Based Fiber Optic Oxygen Sensors: Optics Developments," *Proc. SPIE–Int. Soc. Opt. Eng.*, Vol. 1587, 1992, pp. 16–20.

[53] Heinrich, U., J. Hoffmann, and D. W. Lübbers, "Quantitative Evaluation of Optical Spectra of Blood-Free Perfused Guinea Pig Brain Using a Nonlinear Multicomponent Analysis," *Pflügers Arch.*, Vol. 409, 1987, pp. 152–157.

[54] Blough, N. V., D. and J. Simpson, "Chemically Mediated Fluorescence Yield Switching in Nitroxide-Fluorophore Adducts: Optical Sensors of Radical/Redox Reactions," *J. Am. Chem. Soc.*, Vol. 110, 1988, pp. 1915–1917.

[55] Lakshminarayanaiah, N., "Transport Processes in Membranes: A Consideration of Membrane Potential Across Thick and Thin Membranes," Chap. 7 in *Subcellular Biochemistry*, *Vol. 6*, R. B. Roodyn, ed., New York: Plenum Press, 1979, pp. 401–494.

[56] Wiemhöfer, H. D., and K. Cammann, "Specific Features of Electrochemical Sensors," Chap. 5 in *Sensors: A Comprehensive survey*, *Vol. 2*, Göpel, W., Hesse, J., Zemel, J.N., eds., Weinheim, VCH, 1991, pp. 160–189.

[57] Opitz, N., and D. W. Lübbers, "Electrochromic Dyes, Enzyme Reactions and Hormone-Protein Interactions in Fluorescence Optic Sensor (Optode) Technology," *Talanta*, Vol. 35, 1988, pp. 123–127.

[58] Sims, P. J., A. S. Waggoner, C. H. Wang, and J. F. Hoffman, "Studies on the Mechanism by Which Cyanine Dyes Measure Membrane Potential in Red Blood Cells and Phosphatidylcholine Vesicles," *Biochem.* Vol. 13, 1974, pp. 3315–3330.

[59] Waggoner, A. S., "Optical Probes of Membrane Potential," *J. Membrane Biol.*, Vol. 27, 1976, pp. 317–334.

[60] Szöllösia, J., S. Damjanovich, S. A. Mulhern, and L. Tron, "Fluorescence Energy Transfer and Membrane Potential Measurements Monitor Dynamic Properties of Cell Membranes: A Critical Review," *Prog. Biophys. Molec. Biol.*, Vol. 49, 1987, pp. 65–87.

[61] Zhujun, Z., and W. R. Seitz, "Ion Selective Sensing Based on Potential Sensitive Dyes," *Proc. SPIE–Int. Soc. Opt. Eng.*, Vol. 906, 1988, pp. 74–78.

[62] Chused, T. M., A. H. Wilson, B. E. Seligmann, and R. Y. Tsien, "Probes for Use in the Study of Leukocyte Physiology by Flow Cytometry," in *Application of Fluorescence in the Biomedical Sciences*, D. L. Taylor, A. S. Waggoner, F. Lanni, R. F. Murphy, R. R. Birge, eds., New York: A.R. Liss, 1986, pp. 531–544.

[63] Smith, J. C., and B. Chance, "Kinetics of the Potential Sensitive Extrinsic Probe Oxonol VI in Beef Heart Submitochondrial Particles," *J. Membr. Biol.*, Vol. 46, 1979, pp. 255–282.

[64] Grinvald, A., R. Hildesheim, I. Farber, and C. Anglister, "Improved Fluorescence Probes for the Measurement of Rapid Changes in Membrane Potential," *Biophys. J.*, Vol. 39, 1982, pp. 301–308.

[65] Kawabata, Y., T. Yamamoto, and T. Imasaka, "Theoretical Evaluation of Optical Response to Cations and Cationic Surfactant for Optrode Using Hexadecyl-Acridine Orange Attached on Plasticized Poly(Vinyl Chloride) Membrane," *Sensors and Actuators*, Vol. B 11, 1993, pp. 341–346.

[66] Wolfbeis, O. S., "Fluorescence Based Ion Sensing Using Potential Sensitive Dyes," *Sensors and Actuators*, B 29, 1995, pp. 140–147, and references herein.

[67] Schaffar, B. P. H., O. S. Wolfbeis, and A. Leitner, "Effect of Langmuir-Blodgett Layer Composition

on the Response of Ion-Selective Optrodes for Potassium, Based on the Fluorimetric Measurement of Membrane Potential," *Analyst*, Vol. 113, 1988, pp. 693–697.

[68] Morf, W. E., P. Wuhrmann, and W. Simon, "Transport Properties of Neutral Carriers Ion Selective Membranes," *Anal. Chem.*, Vol. 48, 1976, pp. 1031–1039.

[69] Charlton, S. C., R. L. Fleming, and A. Zipp, "Solid Phase Colorimetric Determination of Potassium," *Clin. Chem.*, Vol. 28, 1982, pp. 1857–1861.

[70] Charlton, S. C., R. L. Fleming, P. Hemmes, and A. L. Y. Lau, "Unified Tests Means for Ion Determination," US. Pat., 4,645,744, 1987.

[71] Seiler, K., and W. Simon, "Theoretical Aspects of Bulk Optode Membranes," *Anal. Chim. Acta*, Vol. 266, 1992, pp. 73–87.

[72] Spichiger, U., W. Simon, E. Bakker, M. Lerchi, P. Bühlmann, J. P. Haug, M. Kuratli, S. Ozawa, and S. West, "Optical Sensors Based on Neutral Carriers," *Sensors and Actuators*, B, Vol. 11, 1993, pp. 1–8, and references herein.

[73] Spichiger, U., "Chemical Sensors and Biosensors for Medical and Biological Applications: A Developing Field in Analytical Chemistry," Thesis, ETH Zurich, 1995.

[74] Vaillo, E., P. Walde, and U. E. Spichiger, "Development of a Micellar Biooptode Membrane for an Urea Sensor," AMI, 1995, in press.

[75] Wang, K., K. Seiler, W. E. Morf, U. E. Spichiger, W. Simon, E. Lindner, and E. Pungor, "Characterisation of Potassium-Selective Optode Membranes Based on Neutral Ionophores and Application in Human Blood Plasma," *Anal. Sciences*, Vol. 6, 1990, pp. 715–720.

[76] Seiler, K., K. Wang, E. Bakker, W. E. Morf, B. Rusterholz, U. E. Spichiger, and W. Simon, "Characterization of Sodium Selective Optode Membranes Based on Neutral Ionophores and Assay of Sodium in Plasma," *Clin. Chem.*, Vol. 37, 1991, pp. 1350–1355.

[77] Morf, W. E., K. Seiler, B. Rusterholz, and W. Simon, "Design of a Calcium Selective Optode Membrane Based on Neutral Ionophores," *Anal. Chem.*, Vol. 62, 1990, pp. 738–742.

[78] Tan, S. S. S., P. C. Hauser, K. Wang, K. Fluri, K. Seiler, B. Rusterholz, G. Suter, M. Kuratli, U. E. Spichiger, and W. Simon, "Reversible Optical Sensing Membrane for The Determination of Chloride in Serum," *Anal. Chim. Acta*, Vol. 255, 1991, pp. 35–44.

[79] Wolfbeis, O. S., and H. Li, "LED-Compatible Fluorosensor for Ammonium Ion and Its Application to Biosensing," *Proc. SPIE–Int. Soc. Opt. Eng.*, Vol. 1587, 1991, pp. 48–58.

[80] Hauser, P. C., P. M. J. Perisset, S. S. S. Tan, and W. Simon, "Optode for Bulk-Response Membrane," *Anal. Chem.*, Vol. 62, 1990, pp. 1919–1923.

[81] He, H., G. Uray, and O. S. Wolfbeis, "Optical Sensor for Salicylic Acid and Aspirin Based on a New Lipophilic Carrier for Aromatic Carboxylic Acids," *Fresenius, J. Anal. Chem.*, Vol. 343, 1992, pp. 313–318.

[82] He, H., G. Uray, and O. S. Wolfbeis, "Eniantoselective Optodes," *Anal. Chim. Acta*, Vol. 246, 1991, pp. 251–257.

[83] He, H., G. Uray, and O. S. Wolfbeis, "A Thiamine-Selective Optical Sensor Based on Molecular Recognition," *Anal. Lett.*, Vol. 25, 1992, pp. 405–415.

[84] Wolfbeis, O. S., "Optical Sensing Based on Analyte Recognition by Enzymes Carriers and Molecular Interactions," *Anal. Chim. Acta*, Vol. 250, 1991, pp. 181–201.

[85] West, S.J., S. Ozawa, K. Seiler, S. S. S. Tan, and W. Simon, "Selective Ionophore Based Optical Sensors for Ammonia Measurement in Air," *Anal. Chem.*, Vol. 64, 1992, pp. 533–540.

[86] Akiyama, S., "Near Infrared Luminescence Spectroscopy," in *Molecular Luminescence Spectroscopy, Vol. 3*, S. G. Schulman, ed., New York: Wiley, 1990, p. 229

[87] Casay, G. A., T. Czuppon, J. Lipowski, and G. Patonay, "Near-Infrared Fluorescence Probes," *Proc. SPIE–Int. Soc. Opt. Eng.*, Vol. 1885, 1993, pp. 324–336.

[88] Shriver-Lake, L. C., J. P. Golden, G. Patonay, N. Narayanan, and F. S. Ligler, "Use of Three Longer-Wavelength Fluorophores With the Fiber Optic Biosensor," *Sensors and Actuators*, Vol. B 29, 1995, pp. 25–30.

[89] Kawabata, Y., K. Yasunaga, T. Imasaka, and N. Ishibashi, "Fiber Optic Chemical Sensors Using Semiconductors Lasers and Plasticized Poly(Vinyl Chloride) Membrane," *Anal. Sci.*, Vol. 7, 1991, pp. 1465–1468.

[90] Vo-Dinh, T., G. D. Griffin, and M. J. Sepaniak, "Fiberoptics Immunosensors," Chap. 17 in *Fiber Optics Chemical Sensors and Biosensors, Vol. II*, O. S. Wolfbeis, ed., Boca Raton: CRC Press, 1991, pp. 217–257.

[91] Chen, R. F., and C. H. Scott, "Atlas of Fluorescence Spectra and Lifetimes of Dyes Attached to Protein," *Anal. Lett.*, Vol. 18, 1985, pp. 393–421.

[92] Haugland, R. P., "Covalent Fluorescent Probes," Chap. 2 in *Excited States of Biopolymers*, R. F. Steiner, ed., New York: Plenum Press, 1983, pp. 29–58.

[93] Haugland, R. P., "Fluorescent Labels," in *Biosensors With Fiberoptics*, D. L. Wise, and L. B. Wingard, eds., Clifton, NJ: Humana Press, 1991, pp. 85–110.

[94] Astles, J. R., and W. G. Miller, "Reversible Fiber Optic Immunosensor Measurements," *Sensors and Actuators*, Vol. B 11, 1993, pp. 73–78.

[95] Shriver-Lake, L. C., R. A. Ogert, and F. S. Ligler, "A Fiber-Optic Evanescent-Wave Immunosensor for Large Molecules," *Sensors and Actuators*, Vol. B 11, 1993, pp. 239–243.

[96] Khalfan, H., R. Abuknesha, and D. Robinson, "Fluorigenic Method for the Assay of Proteinase Activity With the Use of 4-Methylumbelliferyl-Casein," *Biochem. J.*, Vol. 209, 1983, pp. 265–267.

[97] Turko, I. V., I. A. Pikuleva, I. S. Yurkevich, and V. L. Chashchin, "Langmuir-Blodgett Fims of Immunoglobulin G and Direct Immunochemical Sensing," *Proc. SPIE–Int. Soc. Opt. Eng.*, Vol. 1510, 1991, pp. 53–56.

[98] Toyo'oka, T., and K. Imai, "Isolation and Characterization of Cysteine Containing Regions of Proteins Using 4-(Aminosulfonyl)-7-Fluoro-2,1,3-Benzoxadiazole and High Performance Liquid Chromatography," *Anal. Chem.*, Vol. 57, 1985, pp. 1931–1937.

[99] Imasaka, T., and N. Ishibashi, "Diode Lasers," *Anal. Chem.*, Vol. 62, 1990, pp. 363A–371A.

[100] Hemmilä, I., *Applications of Fluorescence in Immunoassays*, New York: Wiley-Interscience, 1991.

[101] Hemmilä, I., S. Dakubu, V. Mukkala, H. Siitari, and S. Lövgren, "Europium as a Label in Time Resolved Immunofluorometric Assays," *Anal. Biochem.*, Vol. 137, 1984, pp. 335–343.

[102] Petrea, R. D., and M. J. Sepaniak, "Fiber Optic Time Resolved Fluorimetry for Immunoassays," *Talanta*, Vol. 35, 1988, pp. 139–144.

[103] Hemmilä, I., "Progress in Delayed Fluorescence Immunoassay," in *Fluorescence Spectoscopy. New Methods and Applications*, O. S. Wolfbeis, ed., Berlin: Springer-Verlag, 1992, pp. 259–265.

[104] Villarruel, C. A., D. D. Dominguez, and A. Dandridge, "Evanescent Wave Fiber Optic Chemical Sensor," *Proc. SPIE–Int. Soc. Opt. Eng.*, Vol. 798, 1987, pp. 225–229.

[105] Jansen, E. H. J. M., "Chemiluminescence Detection in Immunochemical Techniques: Applications to Environmental Monitoring," in *Fluorescence Spectoscopy: New Methods and Applications*, O. S. Wolfbeis, ed., Berlin: Springer-Verlag, 1992, pp. 267–278.

[106] Sohanpal, K., T. Watsuji, L. Q. Zhou, and A. E. G., Cass, "Reagentless Fluorescence Sensors Based Upon Specific Binding Proteins," *Sensors and Actuators*, Vol. B 11, 1993, pp. 547–552.

Chapter 5

Supports and Immobilization Techniques

5.1 SUPPORTS

5.1.1 Materials

In heterogeneous systems, indicators are normally immobilized on a solid support, usually a polymer, copolymer, or various glasses. The support can be the waveguide itself, the fiber cladding, or an optical element coupled to optical fibers. The support must perform two functions:

1. It must serve as the liquid-solid or gas-solid interface and sometimes as a double interface between liquid, gas, and solid, as in some enzymatic sensors.
2. It must be optically transparent to allow transmission of the light signal (the variation of the light) from the interface.

Additional requirements are that the support remains inert to the chemical reaction being analyzed, couples efficiently to the optical fibers or waveguides that conduct light to the probe, and has an ability to act as a selective element, such as a membrane for gases or neutral carriers.

The main types of support are presented in Table 5.1, which shows the wide range of materials available for the construction of optical sensors. Figures 5.1 and 5.2 give the basic chemical formulas for these organic compounds. Details of their general properties, the method of fabrication, and the use of these materials are not given here, but may be found in a number of specialized books covering polymers [1–4], membranes [5,6], and films [7].

In the construction of the sensor, it is often necessary to distinguish between the optical part of the support, the mechanical support structure, and the membrane whose main function is to enhance selectivity and protect the active part of the sensor.

Table 5.1
Materials Used as Supports for Fiber-Optic and Waveguide Sensors

Materials	Example	Figure	Main Use	Main Properties
Polymers				
Cellulose	Agarose	5.1(a)	Membrane	Biodegradable
Hydrogels	PHEMA, PVP		Membrane	Biodegradable
Polyamides				
• Synthetic	Nylon	5.1(b)	Membrane	Flexibility
• Natural	Proteins	5.1(c)	Membrane	Biodegradable
Polyethylenes (PE)	High-pressure form	5.1(d)	Packaging	Chemically resistant
	PE-glycol		Film	Easy swelling
	Copolymers		Membrane	Large varieties
Polypropylene	High-pressure form	5.1(e)	Packaging	Chemical-proof
PVC	Solid form	5.1(f)	Packaging	Low cost
• Plasticized	PVC film		Membrane	Easy to use
• Copolymers	PVC/acrylic		Film	Many varieties
Polyacrylic	PMMA	5.1(g)	Fiber	Transparent
Polystyrene (PS)	Solid form	5.1(h)	Fiber	Transparent
• Copolymers	Resins		Ion exchange	Large surface area
Polyurethanes	Resins	5.2(a)	Packaging	
Polycarbonates (PC)	Solid form	5.2(b)	Packaging, Fiber	
Epoxy	Resins		Packaging	
Polyethylsulfones (PES)	Films		Membrane	Hydrophobic
Polyimides (PI)	Films	5.2(c)	Membrane	Hydrophilic
• Copolymers	PVI	5.2(d)	Membrane	Biodegradable
Polyfluorocarbons	PTFE™	5.2(e)	Membrane	Chemically resistant
• Ionomers	Nafion™	5.2(f)	Ion exchange	Chemically resistant
Inorganic				
Silicones		5.2(g)	Cladding	Permeability
• Siloxanes (SX)		5.2(h)	Rubbers	Absorbs gas
• Fluorinated	Films		Membrane	Chemically resistant
• Copolymers	SX-PC		Membrane	Permeability
Glasses				
• Silica gel			Dialysis	Easy to use
• Porous glasses			Beads, Fiber	Transparent
• Quartz			Waveguide	Transparent
• Sapphire			Waveguide	High temperature
• Silica			Fiber	Transparent
• Fluoride glasses			Fiber	NIR
• Chalcogenides			Fiber	Mid-IR
• Halogenides			Fiber	IR
• Tellurides			Fiber	IR

Note: Abbreviations are as follows: PHEMA = poly(hydroxyethyl methacrylate); PVP = poly(vinylpyrrolidone); PMMA = polymethylmethacrylate; PTFE = poly(tetrafluoroethylene); PVI = polyvinylimidazole.

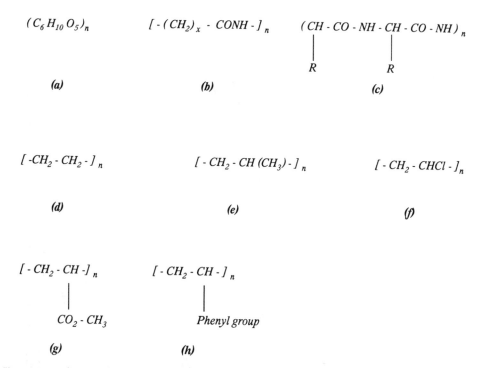

$(C_6 H_{10} O_5)_n$

(a)

$[- (CH_2)_x - CONH -]_n$

(b)

$(CH - CO - NH - CH - CO - NH)_n$
$\qquad\quad |\qquad\qquad\qquad |$
$\qquad\quad R\qquad\qquad\qquad R$

(c)

$[-CH_2 - CH_2 -]_n$

(d)

$[- CH_2 - CH (CH_3) -]_n$

(e)

$[- CH_2 - CHCl -]_n$

(f)

$[- CH_2 - CH -]_n$
$\qquad\quad |$
$\qquad CO_2 - CH_3$

(g)

$[- CH_2 - CH -]_n$
$\qquad\quad |$
$\qquad$ *Phenyl group*

(h)

Figure 5.1 Chemical formulas of main polymeric supports for optical sensors: (a) cellulose, (b) nylon, (c) protein segments (R is a functional side chain), (d) polyethylene, (e) polypropylene, (f) PVC, (g) PMMA, and (h) polystyrene. n is the number of repetitive sequences.

Extra opaque layers or films may also be added to block the ambient light. In addition, plasticizers may be incorporated; these are nonvolatile solvents, such as bis(2-ethyl-hexyl)sebacate, referred to as DOS (dioctyl sebacate) or 2-nitrophenyloctylether (NPOE), which interact with the polymer molecules to make the materials more flexible. These compounds are generally used for ion carrier membranes and lipophilic systems. A variety of pH buffer compounds (e.g., sodium acetate) and additives such as sodium tetrakis 3-bis-(tri-fluoromethyl) phenyl borate, $NaTm(CF_3)_2PB$, can also be incorporated into membranes and act as anionic sites (see Table 4.6). Other compounds used, such as tridodecylmethylammonium chloride (TDMACl), can act as carriers for anions such as chloride and nitrate ions.

Since indicators and labels are organic compounds, the support is generally an organic polymer, preferably a protein in the case of biosensors. The choice of the polymer is governed by its suitability for dye immobilization, its stability, and its permeability to the analyte, for example, in gas measurements. In an intrinsic sensor, the cladding or external coating of the optical fiber determines the type of polymer.

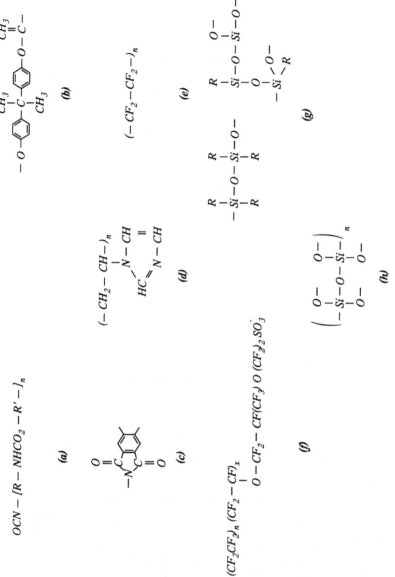

Figure 5.2 Chemical formulas of polymeric supports for optical sensors: (a) polyurethane, (b) sequence of a typical polycarbonate, (c) sequence of polyimide, (d) PVI, (e) PTFE, (f) Nafion™, (g) silicone, and (h) siloxane. n and x are numbers of repetitive sequences.

The choice and type of polymer strongly influence the response time of the sensor, because the liquid-solid and gas-liquid interfaces are subject to the diffusion laws and control the surface reactions of the liquids, ions, and gases.

Other types of support, such as glasses and transparent polymers, are rigid and play an essential role in the optical transmission, either as the waveguide or as beads on the end of the optical fiber. In this case the indicator is grafted (covalently attached) or entrapped, and the reaction takes place only on the surface in contact with the analyte; the response time is then a function of the equilibrium rate at the surface of the sensor.

The majority of supports (e.g., glasses, polystyrenes, and polyethylenes) and plasticizers may exhibit an intrinsic fluorescence, especially with UV excitation. This fluorescence, which produces background noise, is a disadvantage. However, it can sometimes be turned to advantage as an intrinsic reference in the sensor.

5.1.2 Theoretical Aspects of Membranes

5.1.2.1 Diffusion Through Polymers

The permeability (P_m) of a membrane or membrane transfer coefficient and its inverse, the mass transfer resistance (R_m), are defined by the equation:

$$P_m = b \cdot D_m/L = D_{eff}/L = 1/R_m \qquad (5.1)$$

where D_m is the diffusion coefficient that depends on the permeand (generally a gas) and the type of polymer used for the membrane. D_{eff} is the effective diffusion and L the thickness of the membrane. The partition coefficient b is defined as a ratio between the membrane and the external solution concentration equilibrium:

$$b = p_m/t_m \qquad (5.2)$$

where p_m is the porosity and t_m the tortuosity coefficients of the membrane.

For small gas molecules (i), the common expression of the permeation coefficient (P_i) is the relation

$$P_i = D_i \cdot S_i \qquad (5.3)$$

where D_i and S_i are the diffusion and solubility coefficients, respectively.

5.1.2.2 Hydrophobic and Hydrophilic Characteristics

The hydrophobic and hydrophilic characteristics of a membrane are important for controlling the charge transfer and selectivity in sensors. For instance, protons cannot cross hydrophobic membranes and interact with an immobilized indicator. Hence, they do not interfere with gases, as in a CO_2 gas sensor. Conversely, hydrophilic membranes are permeable to low molecular-weight analytes, but not to proteins.

The characteristics of the basic polymer and copolymer can be modified by a number of techniques to improve selectivity, immobilization, and the realization of a specialized *mixed polarity* composition with positive and negative sites:

1. By adding charged functions to hydrophobic compounds, such as Nafion™—a perfluorinated compound with a SO_3H group;
2. By introducing a lipophilic compound, such as long-chain fatty acids or apolar groups, which modifies the surface of hydrophilic polymers;
3. By introducing functional groups to convert the polymer into an ion exchange membrane.

Hydrophilic compounds, such as cellulose, polyacrylamides, poly(vinyl alcohol), polyglycols, polyacrylates, and polyimides, possess several OH, NH_2, CO_2, or SO_3 groups. They generally swell in water. They have limited compatibility with hydrophobic compounds, such as PVCs, polysulfones, polyethylenes, polypropylenes, polystyrenes, and polyfluorinated compounds (e.g., PTFE, PTFCE, and PVDF).

For a composite flat membrane having a part consisting of hydrophilic pores in the aqueous phase (w) and another part of hydrophobic pores in the organic phase (o), the relation between the overall mass transfer coefficients K_o for the organic phase and K_w for the aqueous phase is given by [8]:

$$1/K_o = 1/k_o + 1/k_{mo} + m/k_{mw} + m/k_w \qquad (5.4)$$

and

$$1/K_w = 1/k_w + 1/k_{mw} + 1/(m \cdot k_o) + 1/(m \cdot k_{mo}) \qquad (5.5)$$

where k_w, k_o, k_{mw}, and k_{mo} are the mass transfer coefficients for the aqueous and organic phases in solution and in the membrane (subscript m), respectively. The concentration profile of the analyte can be studied from these general functions for hydrophilic and hydrophobic composite membranes. One way could use the mass transfer in a chemical reaction. In this case, the reaction zone can be inside the membrane or in the liquid film outside the membrane.

5.2 CONFIGURATIONS FOR SENSORS

Different configurations of optical guides, polymers, and membranes acting as supports for indicators or for immobilization of labels are shown schematically in Figures 5.3 and 5.4. These have been generalized to ten basic structures, but

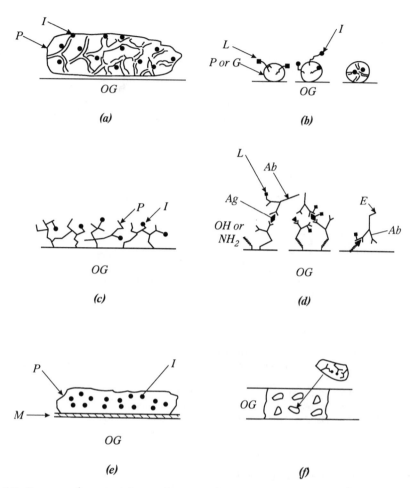

Figure 5.3 Structures of optode surfaces without a membrane showing the optical guide (OG), indicator (I), label (L), antigen (Ag), antibody (Ab), enzyme (E), polymer (P), glasses (G), and metal (M) components. (a) The OG is separate from the polymer containing the entrapped indicator; (b) beads or a gel are placed on the OG with three configurations, a label, with a grafted indicator, or with an entrapped indicator; (c) the polymer-indicator system is grafted directly onto the OG; (d) the polymer-antibody system is grafted indirectly onto the OG in various ways; (e) the polymer and metallic layer are bound to the OG; and (f) the indicator is included with, or grafted onto, a porous guide. The dimensions of the layers are not to scale.

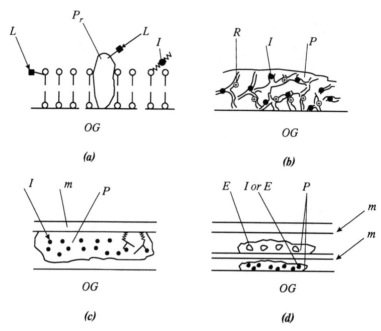

Figure 5.4 Structures of optode surfaces with membranes showing the optical guide (OG), indicator (I), label (L), membranes (m), charged sites (R), proteins (Pr), and polymer (P) components. (a) Configuration of an LB bilayer acting as the membrane; (b) a polymeric lipophilic membrane with indicator and charged sites; (c) configuration with a polymer-indicator or polymer-antibody system grafted onto a membrane and put on the OG; and (d) configuration with two membranes and enzyme(s) (E). The dimensions of the layers are not to scale.

combinations of these arrangements are also possible. Figure 5.3 shows structures without a membrane. In Figure 5.3(a), the polymer or gel, acting as the polymeric support of the indicator, is not bonded to the optical guide, nor are the beads (Figure 5.3(b)). In Figure 5.3(c,d), the polymer is bonded onto the optical guide by grafting, either with an indicator or an antibody-antigen, respectively. In Figure 5.3(e), the polymer and metal layer are bound to the optical guide (e.g., for surface plasmon resonance measurements). In Figure 5.3(f), the indicator is included in the porous structure of the optical guide.

Systems with one or two membranes are shown in Figure 5.4: a bilayer membrane (e.g., made by the Langmuir-Blodgett (LB) technique) (Figure 5.4(a)), a lipophilic system (Figure 5.4(b)), a semirigid or rigid membrane and a polymer holding the indicator (Figure 5.4(c)), and two membranes as found in some enzymatic sensors (Figure 5.4(d)). This last configuration is useful for sensors based on

Table 5.2

Types of Binding for Different Configurations of the Sensing Part of Optical Sensors

Configuration (Figure)	Immobilization Support	Optical Guide	Type of Binding, Indicator-to-Support	Type of Binding, Support-to-Optical Guide
5.3(a)	G, P	G, P	Dispersion[a]	Mechanical, deposit
	G	G	Dispersion	Sol-gel technique
	P	G	Entrapment	Deposit
5.3(b)	G, P	G, P	Entrapment	Deposit
	G, P	G, P	Grafting on beads	Deposit
5.3(c)	G, P	G, P	Grafting	Grafting[b]
	G, P	G, P	Ionic	Grafting
5.3(d)	P	G, P	Grafting	Grafting
5.3(e)	P and M	G	Grafting, entrapment	P-M-G layers[c]
5.3(f)	G	G	Grafting, entrapment	Grafting
5.4(a)	P	G, P	Thin-layer film	LB technique
5.4(b)	P	P, G	Lipophilic sites	Deposit
5.4(c)	P	P, G	Grafting, Entrapment	Mechanical
5.4(d)	P	P, G	Grafting, Entrapment	Mechanical

Note: Abbreviations are as follows: G = glass; P = polymer; M = metal; LB = Langmuir-Blodgett.
[a]Dispersion = the indicator is distributed at random.
[b]Grafting = covalent attachment.
[c]P-M-G = polymer-metal-glass layers for surface plasmon resonance (SPR) technique.

a double enzyme response, because the two enzymatic layers are in separate compartments [9].

The nature of the binding between the indicator, the support, and the optical guide is given in Table 5.2 for the examples in Figures 5.3 and 5.4, in the first case between the indicator and the support, and in the second case between the support and the optical guide.

5.3 IMMOBILIZATION TECHNIQUES

The wide variety of supports and analytes gives rise to a range of immobilization techniques for the fabrication of optodes. A number of works describe surface modification of polymers [10,11], chemical modification of membranes and films [12], and immobilization on components used in sensors such as ion sensitive field effect transistors (ISFETs) or chemical sensitive field effect transistors (CHEMFETs) [13]. Three immobilization methods have been considered for optical sensors by Koller and Wolfbeis [14]: mechanical (adsorption and entrapment), electrostatic (or ionic), and chemical (covalent) methods.

Surface modification for immobilization includes mechanical polishing and sometimes electrochemical modification or chemical etching of the surface as a preliminary step, but these techniques are not confined exclusively to optical sensors and are not described here.

The most widely used immobilization methods for optical sensors are physical entrapment, physicochemical techniques, including adsorption, sol-gel, the use of lipophilic membranes, and chemical (both electrostatic and covalent) methods. As noted in the previous section, the different steps in the immobilization and fabrication of the sensor must take into account both the interaction between the indicator and the support, and the binding of the polymer to the optical waveguide. These two requirements are combined in a single step only when the polymer is the optical guide itself.

5.3.1 Physical Methods

Mixing is the most common physical method. Chelating agents, enzymes, and indicators can be mixed directly in the polymeric support in a silica gel or in a solution that is subsequently polymerized. The molecules are incorporated in the interior of the polymer. A common technique is to trap an indicator in microspheres with diameters of less than 1 μm. Sometimes this mixing can be realized at the same time as the polymerization by emulsion polymerization [15] or copolymerization [16]. This technique is simple and effective. However, the reproducibility from mixing is generally poor compared to the sol-gel technique (see Section 5.3.2.2).

5.3.2 Physicochemical Methods

5.3.2.1 Adsorption

Adsorption is a physicochemical immobilization due to weak electrostatic and van der Waals forces of interaction between the dye and the support. Dye immobilization can be done by polymerization or adsorption on a preactivated surface. It is a simple technique. Thus, to fabricate optodes, the easiest method is to dissolve a dye in methanol or another alcohol, to immerse a polar cross-linked polymer or copolymer (e.g., Amberlite™) placed on the end of the fiber into the dye solution, and then to wash off the unadsorbed dye.

A number of pH sensors have been obtained with this method [17]. Resin microspheres, in which the adsorbed dye is active on the surface, are widely used for reflectance measurements. However, there are many different parameters that need to be controlled in the fabrication, such as particle size, the quantity of the microspheres (the matrix), the concentration of the immobilized dye (resulting from the time of contact between the dye solution and the polymer), and the porosity of the matrix. This leads to difficulties in controlling the repeatability of the optode. The main problem of this kind of sensor is ionic-strength-dependent swelling and the change in the work function over time.

Furthermore, when the microspheres are small, an additional porous membrane is needed to contain the spheres, leading to other drawbacks (such as a longer response time). Thus the immobilization procedures and the physical characteristics must to be strictly controlled. The standard deviation of the signal for several optodes is rarely better than 10% [18]. Adsorbed dye layers made by an automated LB technique usually have better reproducibility.

Adsorption is achieved following preactivation of the support surface (see Section 5.3.4.2). For instance, for internal reflection spectroscopy (IRS), a hydrophobic surface can be generated by washing the reflection element in a silane that allows protein adsorption [19]. However, the preparation and the resulting reproducibility of antibody-coated waveguides remains empirical [20].

5.3.2.2 Sol-Gel Method

The sol-gel method is a chemical process in which a solution (a sol) of metal oxides (normally metal alkoxides in alcohol) undergoes polymerization and gelation to form a rigid matrix of a cross-linked metal-oxide network. After heating to remove the products of the polycondensation (alcohols and water), a solid matrix with small interconnecting pores is produced. The chromophore or fluorophore can be immobilized in a coating on the surface of the pores, or inside a monolithic gel. This technique has been tried for fiber-optic sensors [21]. There are three distinct steps in the preparation of the support:

- Mixing of the precursors in solution;
- Polymerization and gelling of the solution, usually catalyzed by water;
- Heating to eliminate condensation products and transformation into a solid glass.

Alkoxy compounds, such as tetraorthoethylsilicate (TEOS), of the $R_n Si(OR')_{4-n}$ type are generally used as starting materials. Figure 5.5 shows the process of dye immobilization adapted for optical sensors [22]. The different methods include coatings, incorporation of the dye within the gel, or impregnation with the indicator after silanization. The application of porous sol-gel glass coatings has also been proposed to enhance the performance of evanescent wave sensors [23] and realize lost-cost portable devices [24]. Fairly reproducible results of sensors are obtained with this technique [25], but leaching of the reagent from the inorganic support is currently the main drawback of this technology. However, enzymes have been permanently immobilized by the sol-gel technique without observable leaching even after extensive washing [26].

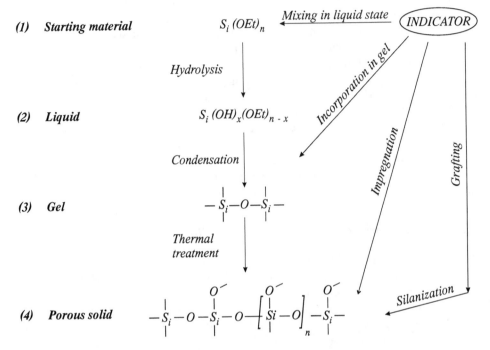

Figure 5.5   Dye immobilization with the sol-gel technique for optical sensors. The indicator can be mixed in the liquid state, incorporated into the gel, or impregnated into the solid porous matrix with possible grafting after silanization. The sol-gel technique is one of the more promising methods being developed.

5.3.2.3 Lipophilic Membranes

Lipophilic membranes can be in the form of lamellar vesicles, monolayers or multi-layers, and other surfaces modified by lipophilic compounds. In practice, lipophilic indicators are immobilized either by incorporation in a lipophilic polymer [27] as vesicles [28] or by simply dissolving them in lipid monolayers or bilayers with the LB technique [29], which has become widely used for optical sensors [28]. For a biomatrix labeled with a fluorophore, the dye is generally incorporated before the introduction of the protein. Immobilization can be assisted by adding alkyl chains onto the proteins or by electrostatic or covalent attachment of the proteins to the membrane. These membranes consist of fatty acid layers and can be deposited on inorganic supports, such as glasses and quartz, or immobilized covalently [12]. The distribution of the indicator within the membrane depends on the structure of the membrane.

5.3.3 Electrostatic (Ionic) Binding

A polymer containing charged groups can be bound to oppositely charged ions by electrostatic forces. Negatively charged sulphonic groups or positively charged quaternized (creating a positive charge between two nitrogen atoms in a cyclic compound) ammonium groups are the most commonly used. These functional groups either compete with cations and anions, as for weak or strong ion-exchange resins, or may be bound to indicators that interact with strong acids or bases.

The main advantage of this technique is the simplicity of the procedure by immersion of charged polymers in an alcohol or methanol solution. The loading conditions for the indicator, such as the immersion time, must be closely controlled to obtain reproducible optodes. This method of immobilization is widely applied, especially for pH measurements with various indicators immobilized on quaternized amino-styrene [30], on anion or cation resins, and on quaternized PVI grafted onto optical fibers [31]. This technique is also very effective with ion carrier membranes (see Sections 4.5 and 4.6 and Figure 5.4).

5.3.4 Covalent Immobilization

Covalent binding has the advantage of being more stable, and it is commonly used in three ways:

- Immobilization of polymers by grafting on a surface (surface modification). If an indicator is employed, it can be incorporated in the polymer by physical, physicochemical, or electrostatic binding techniques (support-polymer binding plus indicator).
- Immobilization of the indicator by grafting onto the polymer. The polymer is attached noncovalently to the support (support plus polymer-indicator binding).
- Immobilization of the indicator by grafting onto the polymer, which is grafted onto the support (support-polymer-indicator double binding). In this case, it is necessary to maintain the functional activity of the indicator, and the chemistry of dual immobilization is more complex.

Thus, several immobilization methods can be present in the same sensor for incorporating the indicator and for attaching the polymer to the support; for instance, physical-covalent binding, covalent-physicochemical, covalent-electrostatic, ionic-covalent, and covalent-covalent binding. In covalent-covalent binding, the polymer is covalently grafted onto both the support and the indicator; this is the most complicated type of binding to make but is the most stable.

5.3.4.1 Silanization

Silica, with its surface hydroxy groups (-SiOH), is the basic material for glass optical fibers and waveguides, and silanization is an obvious choice for immobilization in optical-fiber sensors [32]. Silanization consists of substituting a hydrogen atom (e.g., in R-SiOH) bound to a hetero-atom (the oxygen atom) of an inorganic compound, with a silyl group ($R_y SiX$) to form a silico-hetero-atom without changing the basic molecule. The simplified reaction is

$$R-SiOH + R_ySiX \Rightarrow R-Si-O-Si-R_y + XH \qquad (5.6)$$

where X is a chlorine or a hydrolyzable group such as a methoxy (-OCH_3) or ethoxy (-OC_2H_5) groups, normally in an organosilane $R_ySiX_{(4-y)}$.

The functional group (part of the R_y) of the silane can be an acrylic compound for acrylamide film polymerization, or an epoxy (R-O-R) function oxidized in aldehyde for antibody covalent grafting by amino groups of the protein in a Schiff base reaction [33]. Dimethylamine silanes are more reactive compared to chlorosilanes and alkoxysilanes. Principal silanization agents are given in Table 5.3.

Table 5.3
Principal Silanes Used for Grafting on Waveguides
in Chemical Sensing

Silanization Agents	Common Abbreviation
Aminopropyltriethoxysilane	APTES
n-Decyltrichlorosilane	DTCS
Dimethyldichlorosilane	DMDCS
n-Trimethylsilyldimethylamine	TMSDMA
Diphenyldichlorosilane	DPDCS
3-Glycidoxypropyltrimethoxysilane	GOPTS
γ-Methacryloxypropyltrimethoxysilane	MAPS
Octadecyltrichlorosilane	OTCS
Octadecyltriethoxysilane	OTES
Triacetoxyvinylsilane	TAVS
Tributylchlorosilane	TBCS

Notes: The silanes listed in the table lend themselves to modification of carbon (graphite). They can also be used to modify metal surfaces. Typical metals that can be functionalized include titanium, aluminum, gold, and tin.

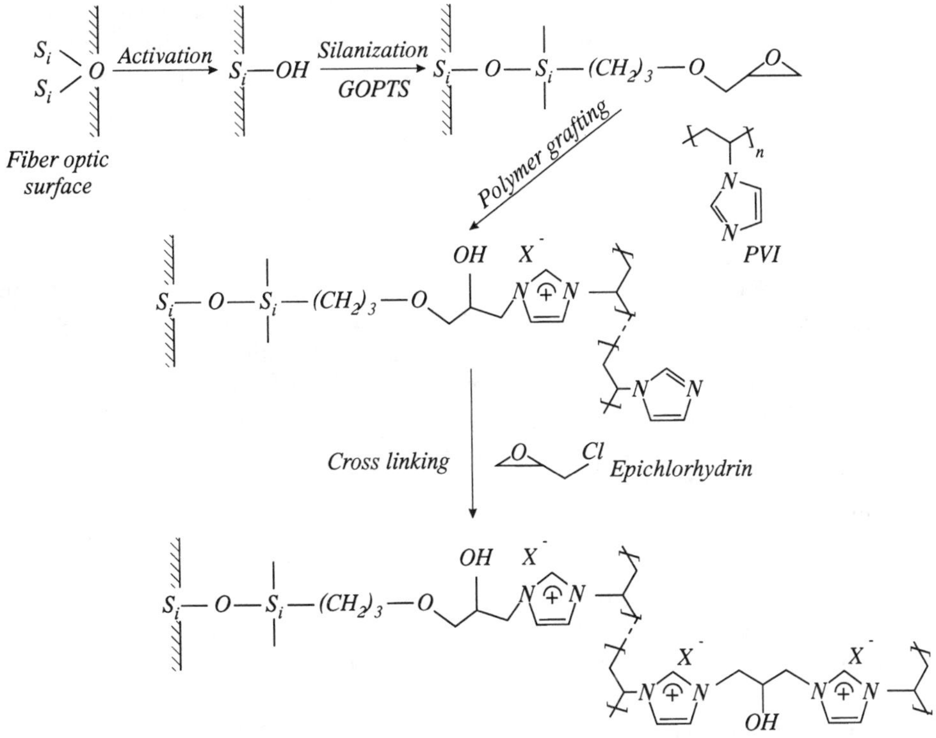

Figure 5.6 Steps for ionic-covalent grafting of PVI on the silica surface of an optical fiber and ionic binding of the dye on polymer. X can be OH^- or Cl^-. GOPTS is a silane (see Table 5.3).

An example of silanization of PVI, a compatible compound in biosensing, is given in Figure 5.6 [31]. The different steps of surface preactivation, silanization (in this case with GOPTS possessing an epoxy function), and covalent grafting of the polymer onto the fiber are generally followed by a cross-linking (e.g., by a bifunctional agent such as epichlorhydrin). This particular example allows subsequent electrostatic binding between the polymer and the indicator.

5.3.4.2 Surface Preactivation

Some membranes, resins, beads, or copolymers, which are preactivated for immobilization, are commercially available [14]. Munkholm et al. [34] have developed another activation technique for the surface of a silica fiber, employing consecutive hexane and ammonia plasmas for making an alkylamine surface. Preactivation of polymer surfaces by hydrophilic conversion (modification of -COOH) followed by coupling to an amino group or cross-linking is a suitable technique for polystyrene optical fibers.

5.3.4.3 Covalent Binding on Organic Polymers

Cellulosic and polyacrylamide compounds and carboxylic-modified PVC and polystyrenes are suitable polymers for binding amines and proteins after undergoing surface treatment. Polystyrenes, which have no active functional groups, can be chloromethylated, sulfonated and halogenated to allow binding of the chromophore via the OH group. A summary of commonly used methods for surface modification of different polymers, the coupling reagents, and the type of the molecule that can be attached to the surface is given by Wolfbeis [35].

5.3.4.4 Covalent-Covalent Attachment

Experimental techniques have been investigated for covalent-covalent attachment between the support and the polymer and between the polymer and the indicator [36]. Typical examples are given here for silica, membranes attached to silica, and fluoropolymers.

The example step consists of grafting a polymer (polyvinylchloroformate chloride (PVCFC)) or a copolymer (vinylchloroformiate and N-vinylpyrrolidone (NVP)) onto a silica fiber, and then grafting a modified phenoxazine (with an absorption maximum of $600 \, nm < \lambda < 650 \, nm$) onto the polymer-silica support. Figure 5.7 shows schematically the different steps of the process [36]:

- Activation and modification of the dye by nucleophilic substitution with an aromatic amine without significantly changing the spectral characteristics of the dye;
- Preparation of the polymer or copolymer;
- Grafting of the polymer by reaction between silane amine functions and PVCFC;
- Grafting of the dye onto free chloroformiate groups in the presence of ethanolamine.

The second example uses a lipophilic membrane grafted onto silica followed by covalent immobilization of an enzyme (urease) onto fatty acid monolayers [12]. The urease is immobilized by activating the carboxylic surface with either

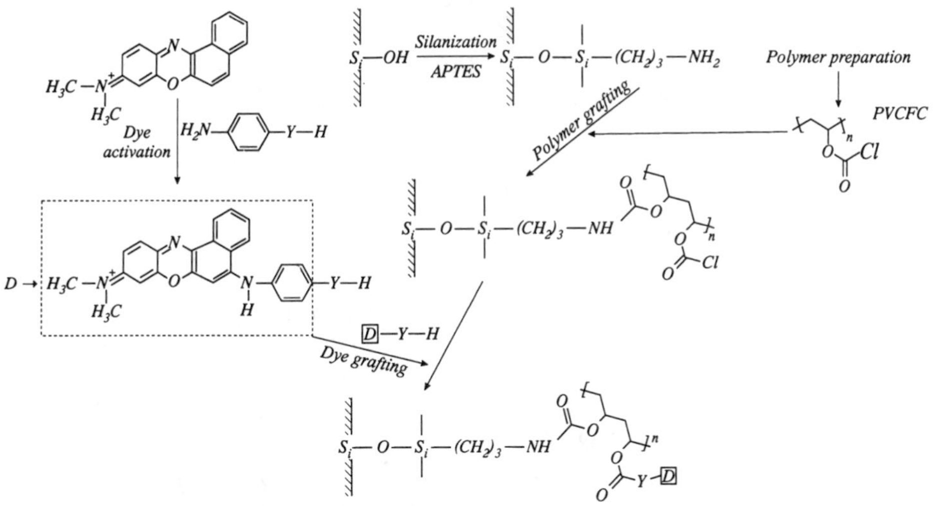

Figure 5.7 Steps for covalent-covalent grafting of PVCFC polymer on the optical fiber with activation and covalent immobilization of a phenoxazine-type indicator (see Chapter 4). D is the dye, Y-H is the group -COOH or -OH. APTES is a silane (see Table 5.3).

N-hydroxysuc-cinimide (NHS) or carbonyldiimidazole (CDI). Figure 5.8 demonstrates the different steps of this covalent-covalent immobilization using the NHS activation method and dicyclohexyl carbodiimide (DCC) as a coupling agent.

The third example is devoted to covalent attachment of proteins on a planar-waveguide [37]. APTES is used for the silanization resulting in pendant amino groups on the silica surface. An active aldehyde is then attached to these groups by reaction with glutaraldehyde. In the final step, the active aldehyde reacts with a primary amine of the protein to form the final bond, such as silica -CH=N-antibody.

A similar experiment is shown in Figure 5.9 for immunosensing with covalent-covalent grafting of a fluorescent-labeled antibody or F_{ab} fragments onto a fiber-optic probe [38]. Modified fluoropolymer materials (MePTFE and fluoroethylene propylene (FEP)) are used for the covalent immobilization of the antibody. In this scheme, the APTES-modified compounds are activated by glutaric dialdehyde (GDA). The long-term stability of the biomolecule was significantly improved and the performance was equally good for both quartz and fluoropolymer probes.

A general method to immobilize enzymes uses the interaction between the glycoprotein avidin and the vitamin biotin [39,40]. Experiments to immobilize enzymes on fluoride and chalcogenide glasses show the possibilities of extending these techniques to other sensor materials [41].

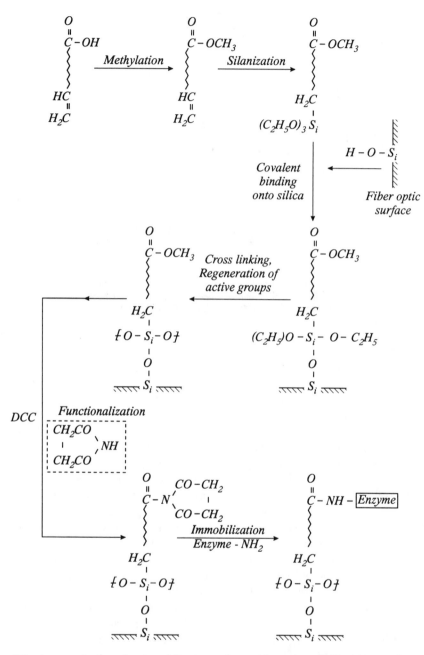

Figure 5.8 An example of covalent immobilization on fatty acid monolayers followed by covalent immobilization of the enzyme (urease) onto the monolayers. DCC is a coupling agent. (*After:* [12], with permission.)

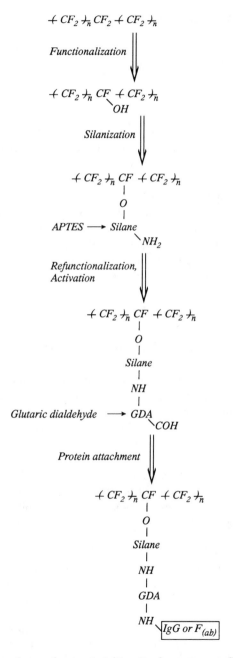

Figure 5.9 Steps for the attachment of a protein (IgG) or F_{ab} fragments on a fluorinated polymeric membrane. (*After:* [38], with permission.)

5.4 PROPERTIES OF IMMOBILIZED COMPOUNDS

This section is devoted to the principal effects of immobilization and the change in optical properties produced by the immobilization.

5.4.1 Characterization

Many researchers only evaluate an immobilized indicator in terms of the optical performance of the sensor. However, a more detailed knowledge is often necessary in order to control the immobilization technique, to characterize its performance, and to compare the relative value of several techniques. The amount of bound indicator can be evaluated by chemical analysis, such as using potentiometric titration after soaking [42]. Attempts to define an immobilization parameter by optical absorbance measurements have not been successful [43]. However, an evaluation of activity losses of sulfophthaleins, electrostatically bound inside a polymer grafted onto a fiber, has shown the importance of halogenides in the dye molecules for stabilizing optodes [31]. The desorption kinetics in a solvent were estimated to be first-order.

$$\mathrm{Ln}[(A_t - A_\infty) / (A_0 - A_\infty)] = -k \cdot t \tag{5.7}$$

where A_0 is the initial absorbance, A_∞ the residual absorbance, and A_t the absorbance at a time t (Figure 5.10).

In immunosensing, the quantity of immobilized molecules can be determined by the amount bound to the surface (e.g., micrograms of labeled IgG per square centimeter of support surface) measured by fluorescence after binding, compared to a standard curve [32]. The immobilization technique for an antibody has been evaluated with fiber-optic sensors that were regenerable for four reagents linked by different bonding methods [44]. The sensitivity of a sensor is directly proportional to the amount of antibody present. The amount immobilized (loaded) and the ability of the immobilized antibody to maintain antibody recognition (activity) are the two main criteria affecting the sensitivity of the sensor. The immobilization procedures were evaluated by the following:

1. The amount of antibody that can be loaded onto beads was determined by fluorescence measurements of the antibody solution before and after incubation.
2. The retention of antibody activity (i.e., the ability of an immobilized antibody to recognize a specific antigen) was determined by the ratio of the bound antigen to immobilized antibody.

These tests, investigated on two antibody-antigen systems, show that the ideal immobilization reagents and conditions will undoubtedly change from system to system.

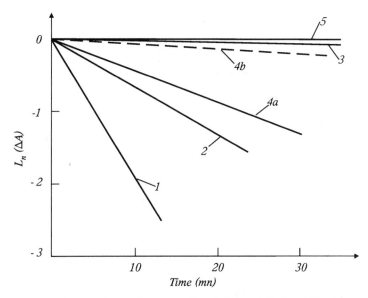

Figure 5.10   Desorption kinetics of some electrostatically immobilized sulfophenolphthaleins on polyvinylimidazole as the polymer. The measurements are made at the maximum absorption wavelengths: (1) cresol red (CR) at 590 nm; (2) phenol red (PR) at 560 nm; (3) chlorophenol red (CPR) at 586 nm; (4) bromothymol blue (BTB); at 620 nm; (4a) after an initial immobilization (2 minutes); (4b) after a second immobilization (2 minutes), shown dotted; (5) -3,4,5,6-tetrabromophenolsulfonephthalein (TBPSP) at 592 nm. The absorbance variation ΔA is given on a Naperian logarithmic scale (Ln) (see (5.7)).

5.4.2 Optical Characteristics in pH Sensors

A loss of dye from the immobilized layers produces an undesirable variation in the optical properties of an optode. Other effects resulting from the immobilization procedure can also change the optical characteristics.

For example, the acid base properties of pH indicators (hence the pK_a') are affected by the microenvironment with a possible broadening of the pH response [45] by the indicator concentration, which can decrease the pK_a', and by the electronic properties of the matrix [46,47], which normally increase the pK_a' compared to the values in solution. This increase due to the adsorption of a sulfophthalein dye on a hydrophobic styrene/divinylbenzene copolymer can be explained in terms of the formation of a π-π bond by electron transfer from a donor (matrix) to an acceptor (the bromine of the indicator) complex [46,48]. These characteristics can also be examined by the relationship between the pK_a' and the structure of the sulfophthaleins. A summary of the results, taken from references cited in this section, is presented in Table 5.4 and Figure 5.11.

Table 5.4

The pK'_a Values of the Main Sulfophthalein Dyes at the Maximum Absorption Wavelength in Solution (Ionic Strength $\mu = 0$) and Immobilized on Various Supports

Indicator[a]	Solution	AG1X4 Anion Exchange Resin	AG1X8 Anion Exchange Resin	IRA400 Anion Exchange Resin	PVI	IRA45 Anion Exchange Resin	Dowex 4 Anion Echange Resin	XAD7 Resin	XAD4 Resin	XAD2 Resin
Thymol blue (acid) (TB)	1.65								2.1	
Tetrabromothymol blue (TBTB)	3.8								3.8	
Bromophenol blue (BPB)	4.1	1.2	1.5		2.2				5.0	4.65
Bromocresol green (BCG)	4.9	2.3	2.7		3.35				6.5	
Chlorophenol red (CPR)	6.25				4.6				6.5	
Bromocresol purple	6.4	1.3	3.5		3				8.5	
Bromophenol red (BPR)	6.8	1.9	1.5		4.25				7.3	
Bromothymol blue	7.3	5.9	6.3		7.05				11.5	9.95
Tetrabromophenol-sulfophthalein	7.9			6.17	6.73	7.01	7.24	7.9	8.66	8.86
Phenol red	8.0	6.4	6.3		7.14				9.6	
Cresol red	8.0	6.2	6.3		7.4				10.0	
m-Cresol purple	8.4	6.2	6.6						11.4	
Thymol blue (basic) (TB)	9.2	7.3	8.8		9				12.1	

Notes: IRA45 and 400 are Amberlite™; XAD2, 4, 7 are styrene-divinylbenzene copolymers.
[a]Data taken from [31,36,43,46,47,49,50].

Immobilization can modify the absorption spectrum of a dye, particularly the wavelengths of the absorption maxima for the acid and basic forms. The broadening of absorption bands can be used for extending the pH range [51]. Table 5.5 shows the observed wavelength shifts for various sulfophthaleins and Meldola blue derivatives immobilized on XAD4™, PVI [43], and PVCFC [36] polymers compared to the wavelengths in solution.

A bathochromic effect (an increase in λ) of about 2 to 16 nm is observed with the basic form of sulfonephthaleins on PVI, and XAD4 and a strong hypsochromic effect (a decrease in λ) with the Meldola blue derivatives on PVCFC. For the acid form, these effects are less marked. Results from different authors apparently differ, supposedly because the experimental procedures are not identical [17].

The variation of the absorption spectra due to immobilization also leads to a

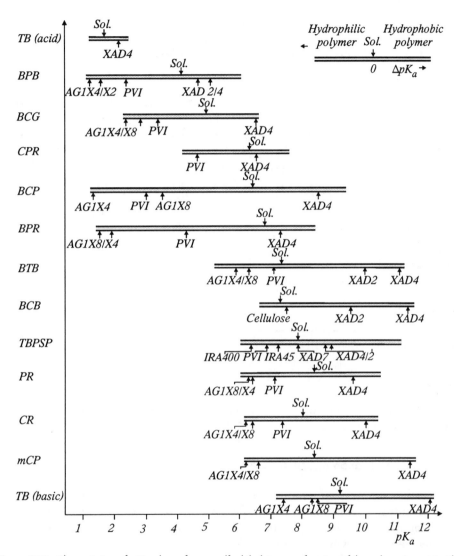

Figure 5.11 The variation of pK'_a values of some sulfophthaleins as a function of the polymeric matrix. The pK'_a in solution (sol), taken arbitrarily as zero, shows the degree of relative hydrophilic or hydrophobic characteristics of the polymer and its interaction with the dye. BCB = bromocresol blue, mCP = meta-cresol purple (see other chemical formulas in Table 4.2). AG1X4 and AG1X8 are XAD modified with anion exchange functional groups [-N(CH$_3$)$_3^+$].

Table 5.5
Wavelength λ (in nanometers) of the Maximum Absorption for Dyes (in Their Basic Form) Either in Solution (ionic strength μ = 0)[b] and Immobilized on Supports

Indicator	λ in solution (at μ = 0) (1)	λ for PVI (2)	λ for XAD4 (3)	λ for PVCFC (4)	Δλ PVI (2–1)	Δλ XAD4 (3–1)	Δλ PVCFC (4–1)
Phenol red	558	574	570		16	12	
Bromophenol red	572	586			14		
Cresol red	572	586			14		
Chlorophenol red	573	586	600		13	27	
Tetrabromophenol sulfophthalein	578	592	602		14	24	
Bromocresol purple	586	602			16		
Bromophenol blue	592	604	604		12	12	
Thymol blue	596	600	600		4	4	
Bromothymol blue	616	618	620		2	4	
Bromocresol green	618	620	625		2	7	
4-Aminophenol Meldola blue (APMB)	656			576			−80[a]
4-Aminobenzoic Meldola blue (ABMB)	664			538			−126[a]

[a]The large shift is unexplained.
[b]μ = ionic strength at ambient temperature.

change in the fluorescence emission spectra. This variation results in a wavelength shift as well as a change in intensity.

5.4.3 Coimmobilization of Several Dyes

Simultaneous immobilization of several indicators (coimmobilization) is a relatively novel procedure that has been tried experimentally to extend the range of a single optode for pH and medical applications [52]. Stable and reversible pH indicators have been coimmobilized with the sol-gel technique [53].

A pseudolinear absorbance-pH function can be obtained when several indicators are used simultaneously [54]. The condition to be fulfilled by two consecutive dyes, with concentrations C_1 and C_2, is determined by

$$d_1 C_1 / d_2 C_2 = \{(1 - R_2) \cdot \varepsilon_{I2}\} / \{(1 - R_1) \cdot \varepsilon_{I1}\} \tag{5.8}$$

where d_1 and d_2 are the optical path lengths (taken as $d_1 = d_2$ to a first approximation), ε_{I1} and ε_{I2} the extinction molar coefficients of dissociated forms (I^-) of the indicators

1 and 2, respectively, and R_1 and R_2 are the ratio of the extinction molar coefficients of undissociated and dissociated (HI) forms:

$$R_1 = \varepsilon_{HI1}/\varepsilon_{I1} \text{ and } R_2 = \varepsilon_{HI2}/\varepsilon_{I2} \quad (5.9)$$

The difference of pK_a' of two consecutive indicators $(\Delta pK_a' = pK_{a2}' - pK_{a1}')$ must also be less than 1.737.

Figure 5.12 shows the results obtained with an ionic-covalent system in solution and after coimmobilization from a mixture of bromophenol red, bromothymol blue, and thymol blue of C_1, C_2, and C_3 concentrations, respectively. The shift of pK_a's and the change of final concentrations C_1', C_2', and C_3' of immobilized indicators are particularly evident.

In this method, the effects of competitive binding have also been observed, either due to selective adsorption of one of the indicators or to displacement of an already

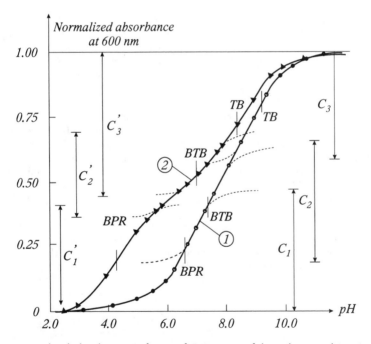

Figure 5.12 Normalized absorbance-pH function for a mixture of three dyes in solution (curve 1) and coimmobilized on PVI polymer grafted on an optical fiber (curve 2). The respective pK_a's are specified by vertical lines. The initial concentrations of BPR ($C_1 = 2.9 \times 10^{-5}$ M), BTB ($C_2 = 2.9 \times 10^{-5}$ M), and TB ($C_3 = 3.7 \times 10^{-5}$ M) in solution and coimmobilization are calculated for a pseudolinear function (see (5.8) and (5.9)). The concentrations (C_1', C_2', and C_3', respectively, for the three dyes) are different after immobilization on the polymer. The dotted lines represent the theoretical curves corresponding to each dye.

immobilized dye [31]. Consequently, the optical properties and performance of the coimmobilized phases are difficult to predict without a basic experimental study of the hydrophilic/hydrophobic characteristics of the indicator and support. Other aspects to be considered in multiple indicator systems are the response times and the lifetime. At present, no systematic experimental work has apparently been done on coimmobilization of covalent-covalent systems.

Another promising approach with a photodeposition method consists of making patterned sensor arrays on a polymer, which is coated on the distal tip of an imaging optical-fiber bundle 350 μm in diameter [55]. The fluorescent image of the array can contain eight pH-sensing regions for different dyes. Potentially, this technique could be valuable in extending the pH range of fiber-optic sensors.

References

[1] Bandrup, J., and E. H. Immergut, eds., *Polymer Handbook*, New York: Wiley, 1981.

[2] Feast, W. J. and H. S. Munro, eds., *Polymeric Surfaces and Interfaces*, New York: Wiley, 1987.

[3] Seymour, R. B., H. F. Mark, eds., *Applications of Polymers*, New York: Wiley, 1988.

[4] Nicholson, J. M., *The Chemistry of Polymers*, Cambridge: Royal Society of Chemistry, 1991.

[5] Kesting, R. E., *Synthetic Polymeric Membranes: A Structural Perspective*, 2nd ed., New York: Wiley and sons, 1985.

[6] Ho, W. S. W., and K. K. Sirkar, *Membrane Handbook*, New York: Van Nostrand-Reinhold, 1992.

[7] Roberts, G., ed., *Langmuir Blodgett Films*, New York: Plenum Press, 1990.

[8] Prasad, R., and K. K. Sirkar, "Membrane-Based Solvent Extraction," in *Membrane Handbook*, W. S. W, Ho, and K. K. Sirkar, eds., New York: Van Nostrand-Reinhold,1992, pp. 727–763.

[9] Berger, A., and L. J. Blum, "Enhancement of the Response of a Lactate Oxidase-Peroxidase Based Fiber-Optic Sensor by Compartmentalization of the Enzyme Layer," *Enzyme Microb. Technol.*, Vol. 16, 1994, pp. 979–984.

[10] Leyden, D. E., and W. T. Collins, eds., *Chemically Modified Surfaces, Vols. 2 and 3*, New York: Gordon and Breach, 1986; 1988.

[11] Mottola, H. A., and J. R. Steinmetz, eds., *Chemically Modified Surfaces*, Amsterdam: Elsevier, 1992.

[12] Brennan, J. D., R. S. Brown, Ghaemmaghami, V., Kallury, R.K., Thompson, M., and U. J. Krull, "Immobilization of Amphiphilic Membranes for the Development of Optical and Electrochemical Biosensors," in *Chemically Modified Surfaces*, H. A. Mottola and J. R. Steinmetz, eds., Amsterdam: Elsevier, 1992, pp. 275–305.

[13] Clechet, P., "Membranes for Chemical Sensors," *Sensors and Actuators*, Vol. B 4, 1991, pp. 53–63.

[14] Koller, E., and O. S. Wolfbeis, "Sensor Chemistry," Chap. 7 in *Fiber Optic Chemical Sensors and Biosensors, Vol. 1*, O. S. Wolfbeis, ed., Boca Raton: CRC Press, 1991, pp. 303–353.

[15] Peterson, J. I., R. V. Goldstein, R. and V. Fitzerald, D. K. Buckhold, "Fiber Optic pH Probe for Physiological Use," *Anal. Chem.* Vol. 52, 1980, pp. 864–869.

[16] Munkholm, C., D. R. Walt, and F. P. Milanovich, "A Fiber Optic Sensor for CO_2 Measurement," *Talanta*, Vol. 35, 1988, pp. 109–112.

[17] Kirkbright, G. F., R. Narayanaswamy, and N. A. Welti, "Fiber Optic pH Probe Based on the Use of an Immobilised Colorimetric Indicator," *Analyst*, Vol. 109, 1984, pp. 1025–1028.

[18] Alabbas, S. H., D. C. Ashworth, and R. Narayanaswamy, "Design and Characterisation Parameters of an Optical Fibre pH Sensor," *Proc. SPIE–Int. Soc. Opt. Eng.*, Vol. 1172, 1989, pp. 251–257.

[19] Elwing, H., and M. Stenberg, "Biospecific Bimolecular Binding Reactions: A New Ellipsometric Method for the Detection, Quantitation and Characterization," *J. Immunol. Methods*, Vol. 44, 1981, pp. 343–349.

[20] Lundström, I., "Surface Physics and Biological Phenomena," *Phys. Scrip.*, Vol. T4, 1983, pp. 5–13.

[21] Badini, G. E., K. T. V. Grattan, A. W. Palmer, A. C. C. Tseung, "Development of pH Sensitive Substrates for Optical Sensor Applications," in *Optical Fiber Sensors*, H. J. Arditty, J. P. Dakin, and R. T. Kersten, eds., Berlin: Springer-Verlag, Vol. 44, 1989, pp. 436–442.

[22] Grattan, K. T. V., G. E. Badini, A. W. Palmer, and A. C. C. Tseung, "Use of Sol-Gel Techniques for Fibre-Optic Sensor Applications," *Sensors and Actuators*, Vol. A 25–27, 1991, pp. 483–487.

[23] MacCraith, B. D., "Enhanced Evanescent Wave Sensors Based on Sol-Gel Derived Porous Glass Coatings," *Sensors and Actuators*, Vol. B11, 1993, pp. 29–34.

[24] MacCraith, B. D., G. O'Keeffe, A. K. McEvoy, C. M. McDonagh, J. F. McGilp, B. O'Kelly, J. D. O'Mahony, and M. Cavenagh, "Light-Emitting-Diode Based Oxygen Sensing Using Evanescent Wave Excitation of a Dye-Doped Sol-Gel Coating," *Opt. Eng.*, Vol. 33, 1994, pp. 3861–3866.

[25] Lev, O., "Diagnostic Applications of Organically Doped Sol-Gel Porous Glass," *Analusis*, Vol. 20, 1992, pp. 543–553.

[26] Braun, S., S. Rappoport, R. Zusman, D. Avnir, and M. Ottenlenghi, "Biochemically Active Sol Gel Glasses: The Trapping of Enzymes," *Mater. Lett.*, Vol. 10, 1990, pp. 1–5.

[27] Wolfbeis, O. S., H. E. Posch, and H. W. Kroneis, "Fiber Optic Fluorosensor for Determination of Halothane and/or Oxygen," *Anal. Chem.*, Vol. 57, 1985, pp. 2556–2561.

[28] Krull, U. J., R. S. Brown, and E. T. Vandeberg, "Fiber Optic Chemoreception," Chap. 21 in *Fiber Optic Chemical Sensors and Biosensors, Vol. 2*, O. S. Wolfbeis, ed., Boca Raton: CRC Press, 1991, pp. 315–340.

[29] Schaffar, B. P. H., O. S. Wolfbeis, and A. Leitner, "Effect of Langmuir-Blodgett Layer Composition on the Response of Ion-Selective Optrodes for Potassium, Based on the Fluorimetric Measurement of Membrane Potential," *Analyst*, Vol. 113, 1988, pp. 693–697.

[30] Zhujun, Z., and W. R. Seitz, "A Fluorescence Sensor for Quantiying pH in the Range From 6.5 to 8.5," *Anal. Chim. Acta*, Vol. 160, 1984, pp. 47–55.

[31] Boisdé, G., B. Biatry, B. Magny, B. Dureault, F. Blanc, and B. Sebille, "Comparisons Between Two Dye Immobilization Techniques on Optodes for the pH-Measurement by Absorption and Reflectance," *Proc. SPIE–Int. Soc. Opt. Eng.*, Vol. 1172, 1989, pp. 239–250.

[32] Baldini, F., and S. Bracci, "Optical-Fibre Sensors by Silylation Techniques," *Sensors and Actuators*, Vol. B 11, 1993, pp. 353–360.

[33] Petrea, R. D., and M. J. Sepaniak, "Fiber Optic Time Resolved Fluorimetry for Immunoassays," *Talanta*, Vol. 35, 1988, pp. 139–144.

[34] Munkholm, C., D. R. Walt, F. P. Milanovich, and S. M. Klainer, "Polymer Modification of Fiber Optic Chemical Sensors as a Method of Enhancing Fluorescence Signal for pH Measurement," *Anal. Chem.*, Vol. 58, 1986, pp. 1427–1430.

[35] Wolfbeis, O. S., "Fiber Optical Fluorosensors in Analytical and Clinical Chemistry," Chap. 3 in *Molecular Luminescence Spectroscopy: Methods and Applications*, Part II, S. J. Schulman, ed., New York: Wiley, 1988, pp. 129–281.

[36] Biatry, B., "Capteur chimique à fibre optique pour la mesure du pH en milieu médical et industriel," Thesis, Université Paris XII, 1993.

[37] Sutherland, R. M., C. Dähne, J. F. Place, and A. R. Ringrose, "Optical Detection of Antibody-Antigen Reactions at a Glass Liquid Interface," *Clin. Chem.*, Vol. 30, 1984, pp. 1533–1538.

[38] Litwiler, K. S., T. G. Vargo, D. J. Hook, J. Gardella, and F. V. Bright, "Novel Supports for the Development of High Stability Fiber Optic Based Immunoprobes," in *Chemically Modified Surfaces*, H. A. Mottola, and J. R. Steinmetz, eds., Amsterdam: Elsevier, 1992, pp. 307–317.

[39] Walt, D. R., C. Munkholm, P. Yuan, S. Luo, and S. Barnard, "Design, Preparation and Applications of Fiber Optic Chemical Sensors for Continuous Monitoring," Chap. 17 in *Chemical Sensors and*

Microinstrumentation, R. W. Murray, R. E. Dessy, W. R. Heineman, J. Janata, and W. R. Seitz, eds., Washington, D.C., ACS, *Vol. 403*, 1989, pp. 252–272.

[40] Wang, Y., and D. R. Bobitt, "Binding Characteristics of Avidin and Surface Immobilized Octylbiotin Implications for the Development of Dynamically Modified Optical Fiber Sensors," *Anal. Chim. Acta*, Vol. 298, 1994, pp. 105–112.

[41] Taga, K., S. Weger, R. Göbel, and R. Kellner, "Colorimetric Activity Assays of Enzyme-Modified MIR Fibers," *Sensors and Actuators*, Vol. B 11, 1993, pp. 553–559.

[42] Saari, L. A., and W. R. Seitz, "pH Sensor Based on Immobilized Fluoresceinamine," *Anal. Chem.*, Vol. 54, 1982, pp. 821–823.

[43] Boisdé, G., and B. Sebille, "Co-immobilization of Several Dyes on Optodes for pH-Measurements," *Proc. SPIE-Int. Soc. Opt. Eng.*, Vol. 1510, 1991, pp. 80–94.

[44] Alarie, J. P., M. J. Sepaniak, and T. Vo-Dinh, "Evaluation of Antibody Immobilization Techniques for Fiber Optic Based Fluoroimmunosensing," *Anal. Chim. Acta*, Vol. 229, 1990, pp. 169–176.

[45] Ashworth, D. C., and R. Narayanaswamy, "The Effects of Surfactants on the pH Transition Interval of Bromothymol Blue Immobilised on XAD-4," *Mikrochim. Acta*, Vol. 106, 1992, pp. 287–292.

[46] Motellier, S., and P. Toulhoat, "Modified Acid-Base Behaviour of Resin-Bound pH Indicators," *Anal. Chim. Acta*, Vol. 271, 1993, pp. 323–329.

[47] Baker, M. E. J., and R. Narayanaswamy, "The Modelling and Control of the pH Response of an Immobilised Indicator," *Sensors and Actuators, Vol. B 29*, 1995, pp. 368–373.

[48] Andres, R. T., and F. Sevilla, "Fibre Optic Reflectometric Study on Acid-Base Equilibria of Immobilised Indicators: Effect of the Nature of Immobilising Agents," *Anal. Chim. Acta*, Vol. 251, 1991, pp. 165–168.

[49] Narayanaswamy, R., and F. Sevilla, "Reflectometric Study of Oxidation-Reduction Equilibria of Indicators Immobilized on a Non-ionic Polymer," *Mikrochim. Acta*, Vol. I, 1989, pp. 293–301.

[50] Wolfbeis, O. S., and G. Boisdé, "Applications of Optochemical Sensors for Measuring Chemical Quantities," Chap. 17 in *Sensors: A Comprehensive Survey, Vol. 2*, W. Göpel, J. Hesse, and J. N. Zemel, eds., "Chemical and Biochemical Sensors," Part II, Weinheim: VCH, 1991, pp. 867–930.

[51] Baldini, F., S. Bracci, and F. Cosi, "An Extended Range Fibre Optic pH Sensor," *Sensors and Actuators*, Vol. A37–38, 1993, pp. 180–186.

[52] Posch, H. E., M. J. P. Leiner, and O. S. Wolfbeis, "Towards a Gastric pH Sensor: An Optode for the 0–7 Range," *Fresenius Z. Anal. Chem.*, Vol. 334, 1989, pp. 162–165.

[53] Ding, J. Y., M. R. Shariari, and G. H. Sigel, "Fiber Optic pH Sensors Prepared by Sol-Gel Immobilization Technique," *Electron. Lett.*, Vol. 27, 1991, pp. 1560–1562.

[54] Boisdé, G., F. Blanc, and X. Machuron-Mandard, "pH Measurements With Dyes Co-immobilization on Optodes: Principles and Associated Instrumentation," *Intern. J. of Optoelectronics*, Vol. 6, 1991, pp. 407–423.

[55] Bronk, K. S., and D. R. Walt, "Fabrication of Patterned Sensor Arrays With Aryl Azides on a Polymer-Coated Imaging Optical Fiber Bundle," *Anal. Chem.*, Vol. 66, 1994, pp. 3519–3520.

$P_{art\ III:}$

Optical Signal Processing and Optical Electrical Transduction

Part I considered different aspects of ionic and molecular recognition and Part II optical chemoreception. The third step in the general scheme of biochemical or chemical sensing (see Figure 2.1) is the conditioning of the optical signal into an electrical signal. This optical-electrical transduction is based on the properties of light described in Chapter 6 and is achieved by using optical waveguides (Chapter 7) and optoelectronic components (Chapter 9) with appropriate optical techniques (Chapter 8). The various concepts of optodes and measurement cells as chemical-optical components used in chemical and biochemical sensing are discussed in Chapter 10. These five chapters form Part III of this book.

The present summary of optical theory and techniques is intentionally limited, since it applies only to sensors and is intended as a condensed overview for the reader not familiar with the subject. For those requiring more detailed information, numerous books are available on specific topics: optics [1,2], waveguide concepts [3–8], and fiber optics [9–12]. In addition, optical techniques for physical measurements with sensors have been extensively reviewed [13–15].

Chapter 6

Essential Theory of Optics

6.1 LIGHT

Light may be considered in three different ways [16,17]; the appropriate viewpoint depends on the nature of the phenomena to be understood.

- Light is composed of photons, vectors of energy exchange, which can explain absorption, emission, and related phenomena.
- Light is an electromagnetic wave that can propagate and be diffracted in a nondispersive medium without energy exchange.
- Light is composed of luminous rays that obey the principles of geometrical optics.

6.1.1 Particle Nature

Planck showed that light as electromagnetic energy is discontinuous; that is, the energy occurs in discrete energy packets as photons or particles. The quantum energy (E) associated with a photon is proportional to the frequency (ν) of the light:

$$E = h \cdot \nu \qquad (6.1)$$

where h is Planck's constant $(6.626 \times 10^{-34} \text{ J} \cdot \text{s})$.

Atoms and molecules have discrete energy levels associated with different electron states. An atom that absorbs a photon can increase its energy by changing between two states. Conversely, when an atom changes its energy state from a higher to a lower level, a photon can be emitted (fluorescence) with a frequency (wavelength) determined by (6.1), corresponding to the energy difference between the two levels.

Other discrete energy levels due to vibrational states are associated with an electronic state. An atom normally exists in a ground state (S_0). The relative number (N_0) of atoms in this state and the number (N_1) in an excited energy level corresponding to an energy difference $\Delta E = (E_1 - E_0)$ is given, in thermal equilibrium at a temperature T, by the Boltzmann relation:

$$N_1/N_0 = \exp\left[-\Delta E/k_B T\right] \tag{6.2}$$

with Boltzmann's constant $k_B = 1.38 \times 10^{-23}$ J $\cdot$ T^{-1} (T is Kelvin temperature).

The characteristics of a photon are:

1. A zero rest mass and a relativistic mass related to its energy by

$$m = E/c^2 = h\nu/c^2 \tag{6.3}$$

with c being the velocity of light in a vacuum (2.9979246×10^8 m $\cdot$ s^{-1}). Inside solid matter, this speed (v) is reduced by

$$v = c/n \tag{6.4}$$

where n is the refractive index of the medium.

2. Photons do not transfer energy to a perfectly transparent medium, since it does not absorb light. However, solid matter is inhomogeneous, which causes scattering of the light. In an absorbing solid, optical energy is lost by absorption and is transformed into vibrational energy or heat. Hence, light interacts with matter through scattering, and absorption or emission of photons by individual atoms or molecules with a characteristic set of energy levels. A photon can interact with another photon only through matter, acting as an intermediary, and not directly with each other. Hence, photons are superimposable. This is very important for optical communications and sensors, because many light beams can intersect without affecting each other.
3. Photons have a zero electrical charge and a kinetic moment of spin $s = 1$ (in $h/2\pi$ units).

6.1.2 Electromagnetic Waves

6.1.2.1 Basic Equations

Light can also be considered as an electromagnetic wave. A wave is a vibratory phenomenon that propagates with a given velocity and is hence a varying function of time. An electromagnetic wave can be represented by an oscillation in a given plane that propagates in a direction normal to this plane. In free space, the velocity (c) of an electromagnetic wave or light in a vacuum is related to two constants, the electrical permittivity (ε_0) and the magnetic permeability (μ_0):

$$c = (\varepsilon_0 \cdot \mu_0)^{-1/2} \tag{6.5}$$

In a homogeneous medium, the wave velocity (v) is

$$\varepsilon \cdot \mu \cdot v^2 = 1 \qquad (6.6)$$

where ε and μ are the electric permittivity and the magnetic permeability of the medium, respectively.

The wavelength (λ) is related to this velocity and the frequency (ν) by

$$\lambda = v/\nu = c/n\nu \qquad (6.7)$$

hence $\lambda_0 = c/\nu$ in a vacuum ($n = 1$).

A complete mathematical description of electromagnetic waves and their propagation is provided by the well-known Maxwell's equations. In three-dimensional Cartesian axes, one typical sinusoidal solution of Maxwell's equations for the wave function of the electric field (E) along the z-axis as a function of time (t) is

$$E_x = E_0 \exp\{i(\omega t - kz)\} \qquad (6.8)$$

and for the magnetic field (H):

$$H_y = H_0 \exp\{i(\omega t - kz)\} \qquad (6.9)$$

where $\omega = 2\pi\nu$ is the angular frequency, k is the wave vector, and E_0 and H_0 are constants.

The two field waves oscillate sinusoidally in phase along the z-direction, and they are orthogonal to each other with the E-field in the x-direction and the H-field in the y-direction.

The phase velocity, that is the velocity of a point on the wave that travels with constant phase, is

$$v = \omega/k = (\varepsilon\mu)^{-1/2} \qquad (6.10)$$

and the wave number is

$$1/\lambda = \nu/v \qquad (6.11)$$

The frequency and wave number are related by

$$k = 2\pi/\lambda = \omega/v \qquad (6.12)$$

Figure 6.1 shows the different units of wavelength (e.g., angstroms, nanometers,

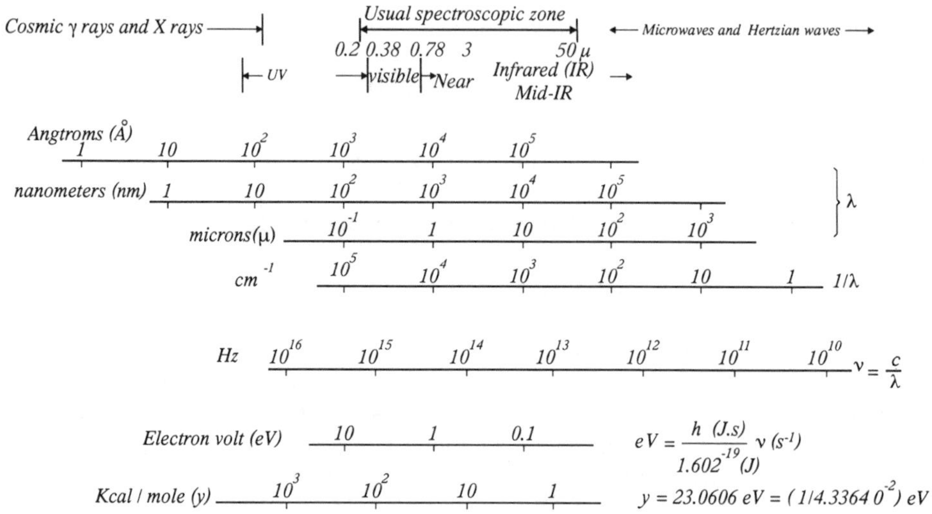

Figure 6.1 Optical conversion factors commonly used in optochemical sensing. The usual spectroscopic range (UV to IR) is the most interesting for optical measurements. Units are expressed in wavelength λ (in angstroms, nanometers, microns), wave number $1/\lambda$, frequency v, energy in electron volts (h is Planck's constant, 6.625×10^{-34} J · s) or in kilocalories/mole (y).

microns, cm^{-1}), frequency, and energy (electron volts, kilocalories/mole) that are commonly used in optical-chemical and biochemical sensing.

The intensity (I) of the wave, or the power per unit area of the flow of energy carried by an electromagnetic field in an isotropic dielectric medium, is proportional to the number of photons and to the square of the electric field [17]:

$$I = \varepsilon \cdot c \cdot E^2 \tag{6.13}$$

If ε is not a constant (in a dispersive medium), but is a function of the frequency $\varepsilon(v)$, then a group of waves is observed with a frequency modulation around v having a group velocity u on a carrier wave of velocity v and frequency v. The relation between the group and phase velocities is

$$1/u = 1/v[1 - (v/v)^2 \cdot dv/dv] \tag{6.14}$$

6.1.2.2 Polarization

The general solution of Maxwell's equations for a plane wave can be written in terms of two waves (one in the xz-plane and the other in the yz-plane):

$$E_x = |A_x| \cos(\omega t - kz + \varphi_x) \quad \text{(in the } yz\text{-plane)} \tag{6.15}$$

$$E_y = |A_y| \cos(\omega t - kz + \varphi_y) \quad \text{(in the } xz\text{-plane)}} \tag{6.16}$$

where φ_x and φ_y are arbitrary phase angles and $|A_x|$, $|A_y|$ are constants.

E_x and E_y are linearly polarized in the x- and y-directions, respectively (Figure 6.2).

The electric field vector is the sum of E_x and E_y and generally describes an ellipse (Figure 6.3(a)):

$$E_x{}^2/A_x{}^2 + E_y{}^2/A_y{}^2 + 2\, E_x \cdot E_y \cdot \cos \varphi / (A_x \cdot A_y) = \sin^2 \varphi \tag{6.17}$$

with $\varphi = (\varphi_y - \varphi_x)$.

Most crystals exhibit polarization anisotropy, depending on their internal symmetry. Cubic crystals do not, since they are symmetrical along all three cartesian axes, but all other crystals (e.g., hexagonal symmetry with a sixfold axis along the z-direction) have an asymmetry that leads to a different refractive index along different crystal axes. A crystal with two different refractive indexes produces a so-called *linear birefringence*. For light that propagates down the main axis (e.g., along the hexagonal axis in the z-direction), the refractive index is uniform in the plane of polarization (the xy-plane) and the linear birefringence vanishes. For light propagating in any other direction (e.g., along the x-axis), the state of polarization (the orientation of the polarization with respect to the crystal axes) experiences two different refractive indexes (in this case along the z- and y-axis) and a biaxial birefringence is apparent. Molecules that are anisotropic also exhibit polarization anisotropy.

The crystal exhibits circular birefringence if it causes the direction of polarization of a linearly polarized wave to rotate as it propagates. In the presence of both circular and linear birefringence, the light experiences an elliptical birefringence.

The total light intensity, I_0, remains constant

$$I_0 = I(0°, 0) + I(90°, 0) = A_x^2 + A_y^2 \tag{6.18}$$

The polarization of light as it propagates with elliptical birefringence, the direction of

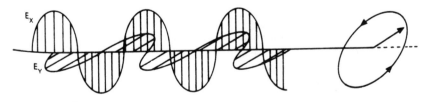

Figure 6.2 A polarized light wave showing the electric E-field components. The sum of the two field components can be represented by an elliptically polarized wave. (*After:* [17], with permission.)

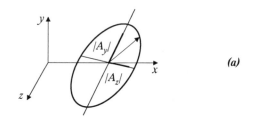

(a)

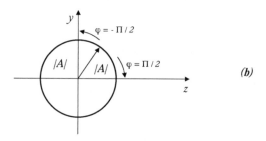

(b)

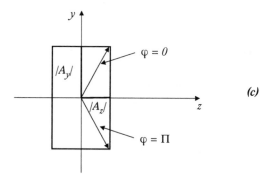

(c)

Figure 6.3 Schematic representation of polarized waves: (a) elliptical polarization; (b) circular polarization; and (c) linear polarization.

the polarization of the field components, and their relative intensities can be analyzed in a powerful way by the Poincaré sphere.

An elliptical polarization can become circular (Figure 6.3(b)) when $|A_x| = |A_y| = |A|$. If $\varphi = \pm\pi/2$, (6.17) becomes

$$E_x^2 + E_y^2 = |A|^2 \tag{6.19}$$

Also, the polarization can be linear or plane (Figure 6.3(c)), with

$$E_x/|A_x| = \pm E_y/|A_y|$$ (6.20)

and has a positive sign for $\varphi = 0$ and negative sign for $\varphi = \pi$.

The ellipse becomes a straight line if E_x or E_y is zero with $\varphi = m\pi$, where m is a positive integer for E_x and E_y in phase, and a negative integer for E_x and E_y in antiphase.

6.1.3 Geometrical Optics

In a homogeneous medium, with constant refractive index n, light propagates as a rectilinear ray (in a straight line) and changes direction only when it strikes an object. This is the basic principle of geometrical optics.

6.1.3.1 Light Flux

For luminous rays coming from a point source, the solution of Maxwell's equations describes the electric field E_r at a distance r from the source, for a wave that spreads out spherically:

$$E_r = (E_0/r) \cdot \exp\{i(\omega t - kr)\}$$ (6.21)

where (E_0/r) is an amplitude factor for $r \geq 1$ indicating the reduction in the electric field strength from the initial field E_0 at $(r = 1)$ as the light spreads out. The total area over which the energy flux occurs is simply the surface area of a sphere $4\pi r^2$ enclosing the source.

The flux (W) passing through a solid angle Ω, which is defined as the fraction of the area S of the light beam to the surface of a sphere with the radius r $(\Omega = S/r^2)$, is constant within this solid angle (Figure 6.4):

$$dW/dS \approx 1/r^2$$ (6.22)

The conservation of light energy implies that the product $n^2 \cdot dS \cdot d\Omega$ remains constant between, for example, an object plane (subscript 1) and the image plane (subscript 2) having the same refractive index (n) and cannot be increased by optical means:

$$S_1\Omega_1 = S_2\Omega_2$$ (6.23)

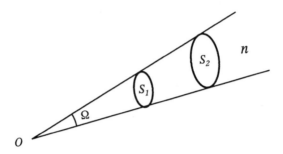

Figure 6.4   A cone of light from a point source, spreading out within a solid angle Ω in a constant refractive index medium. The energy remains constant over the object planes S_1 and S_2.

6.1.3.2 Reflection and Refraction

An incident wave arriving at an oblique angle at the interface (the surface P' in the xy-plane) between two homogeneous materials of refractive index n_1 and n_2 will be both reflected and refracted (Figure 6.5). The reflected ray lies in the plane yz (P) formed by the incident ray and the normal to the plane P'. The angle of reflection (θ_r) is equal to the angle of incidence $(\theta_r = \theta_i)$.

The refracted ray, after crossing the plane P', also lies in the same plane P. According to Snell's law, the sine of the angle at an interface multiplied by the refractive index is constant:

$$n_1 \cdot \sin \theta_i = n_2 \cdot \sin \theta_t \tag{6.24}$$

where θ_i and θ_t are the incident and refracted angles, respectively. Most bulk optic components (e.g., lens and prisms) can be described in terms of the laws of reflection and refraction.

6.1.3.3 Total Internal Reflection

If the medium containing the incident ray (refractive index n_1) is denser than the second medium (refractive index n_2), hence $n_1 > n_2$, no refracted ray exists for angles of incidence θ_i greater than a limiting value when $\sin \theta_t = 1$. This limiting angle (θ_c) is given from (6.24) by

$$\theta_c = \sin^{-1}(n_2/n_1) \tag{6.25}$$

For higher angles of incidence, $\theta_i > \theta_c$, the light is totally reflected at the boundary. Optical waveguides are based on this principle of total internal reflection (TIR); light propagates within the optical guide by multiple total reflections. Conversely, the

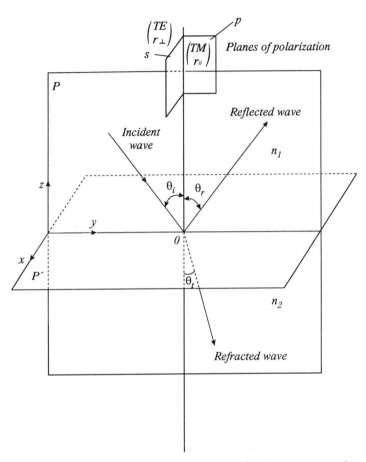

Figure 6.5 Reflection and refraction of a light beam at an interface between two media with different refractive indexes, $n_1 < n_2$. P is the incidence plane, and P' is the interface plane normal to P. The angles θ (i incident, r reflected, t refracted) are measured with respect to the normal of P'. The planes of polarization s (perpendicular) and p (parallel) are shown with the notations TE (transverse electric) and TM (transverse magnetic), with reflection coefficients r for s- and p-polarizations.

reflection is called external when the light comes from the rarer medium and is reflected at the more dense medium $(n_1 < n_2)$.

6.1.4 Light as Combined Photons and Waves

6.1.4.1 Fresnel Equations

Snell's law may be extended by electromagnetic wave theory to determine the relative intensities of the reflected (subscript r) and transmitted (subscript t) light rays crossing an interface, relative to the incident ray (subscript i). The four Fresnel equations, which are not derived here (see [17]), describe the relative wave amplitudes E_r/E_i, E_t/E_i, E_r'/E_i' and E_t'/E_i' of the field components, either with the electric field E in the plane of incidence and H normal to this plane (the form E), or with the orientations of E and H reversed (the form E').

For a light ray at normal incidence, $\theta_i = \theta_r = \theta_t = 0$, the ratios of the wave amplitudes $(E_t/E_i, E_r/E_i)$ are identical for both polarizations. The light intensity (I) is proportional to the square of the amplitude of the electric field, and thus the reflection (R) and transmission (T) coefficients in terms of optical power, relative to the refractive indices n_1 and n_2 of the two media (Figure 6.5), are

$$R = I_r/I_i = (n_1 - n_2)^2/(n_1 + n_2)^2 \tag{6.26}$$

$$T = I_t/I_i = 4n_1{}^2/(n_1 + n_2)^2 \tag{6.27}$$

The reflected wave, if it is polarized with E parallel to the incidence plane, is eliminated (i.e., E_r/E_i becomes zero) when

$$n_1 \cos \theta_t = n_2 \cos \theta_i \tag{6.28}$$

and $(\theta_i + \theta_t) = \pi/2$; hence,

$$\tan \theta_i = n_2/n_1 = \tan \theta_B \tag{6.29}$$

This particular angle of incidence when the reflected wave disappears is known as the Brewster angle θ_B. However, if the incident wave is polarized with E normal to the incidence plane, there is no Brewster angle and the intensity of the polarized reflected wave increases monotonically as the incident angle is increased.

A detailed theoretical examination of the condition of total internal reflection reveals that the incident light penetrates slightly, to a depth y_d, into the second medium in the form of an *evanescent wave*, and its intensity declines rapidly away from the surface of the interface (see Chapter 8 for a description of evanescent wave techniques). In addition, it is found that the reflected ray appears to come from a point that is displaced from the point of incidence (Figure 6.6). This shift or displacement, first demonstrated by Goos-Hänchen, produces a shift of $2z_d$ along the z-direction of propagation with a value of

$$z_d = y_d \cdot \tan \theta_i \tag{6.30}$$

The magnitude of the shift is small, but for multiple reflections the effect is magnified and can be detected by interferometric techniques.

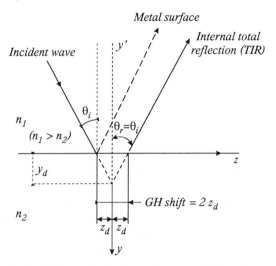

Figure 6.6 Goos-Hänchen (GH) shift, equal to $2z_d$ for total internal reflection of a wave. y_d is the penetration depth of the incident wave into the second medium $(n_1 > n_2)$. Note that a wave reflected from a metal surface is not displaced.

6.1.4.2 Phase Changes During Total Internal Reflection

Changes in the propagation of the light for TIR can be analyzed from the Fresnel equations. When a ray is reflected in a denser medium at the interface with a less dense medium (in this case $n_2 < n_1$ in Figure 6.5, as for an optical fiber in Figure 7.1), with an angle of incidence greater than the critical angle $(\theta_i > \theta_c)$, $\sin\theta_t > 1$ and no refracted ray exists. A phase change occurs that depends on both the incident angle and the polarization. The phase change in E, between the polarization parallel (δ_p) and perpendicular (δ_s) to the incident surface, is a function of θ_i:

$$\tan[(\delta_p - \delta_s)/2] = \cos\theta_i \cdot (n_1^2 \cdot \sin^2\theta_i - n_2^2)^{1/2}/n_1 \cdot \sin^2\theta_i \qquad (6.31)$$

Thus, the state of polarization of the light changes during TIR as a result of this differential phase change $(\delta_p - \delta_s)$.

6.1.4.3 Interference

Electric and magnetic fields may be conveniently represented by vectors, and the superposition of two sinusoidal light waves is easily expressed as the resulting sum of the two vectors representing each wave. For the same polarization, the electric fields of the two waves are

$$e_1 = E_1 \cos(\omega t + \varphi_1) \qquad (6.32)$$

$$e_2 = E_2 \cos(\omega t + \varphi_2) \tag{6.33}$$

and the resulting vectorial sum (denoted by the subscript T) is

$$e_T = E_T \cos(\omega t + \varphi_T) \tag{6.34}$$

If $E_1 = E_2 = E$,

$$\tan \varphi_T = \tan[(\varphi_1 + \varphi_2)/2] \tag{6.35}$$

In the well-known Young's slits experiment, light from a single source is passed through two slits and produces an interference pattern on a screen placed on the other side (Figure 6.7). The two slits, separated by a distance p, are illuminated by a plane wave and act as secondary sources of two cylindrical waves that leave the slits in phase. The difference (δ) in phase between these two waves, at a distance s off-axis from the midline of the screen placed at a distance d from the slits, is (when d is much greater than both s and p, $d \gg s, p$):

$$\delta = (2\pi/\lambda) \cdot (s \cdot p \ / \ d) \tag{6.36}$$

The interference pattern has a light intensity that varies across the screen (as a function of s) and is given by

$$E_T^2 = 4E^2 \cdot \cos^2[(\varphi_2 - \varphi_1) \ / \ 2] \tag{6.37}$$

This describes the well-known interference pattern of Young's fringes.

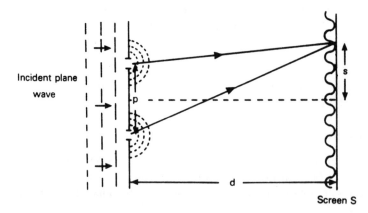

Figure 6.7 Interference pattern produced by a light beam diffracted by Young's slits. (*After:* [17], with permission.)

6.1.4.4 Diffraction

When a plane wave passes through a narrow slit, the light on the other side diverges; this is known as diffraction. If the observing screen is placed far away, the light arriving there will be a plane wave and is known as Fraunhofer diffraction. If the screen is close to the slit, it is referred to as Fresnel diffraction.

The amplitude of the wave as it spreads out after passing through the slit is a function of the angle θ (see Figure 6.8). For small angles, the angular distribution of the diffracted light corresponds to the Fourier transform of the amplitude distribution of the light passing through the aperture of the slit. For a uniformly illuminated slit, the light has a maximum intensity in the center (for $\theta = 0$) of the diffraction pattern with minima on either side. The first minima occur at

$$\sin \theta = \pm \lambda/s \qquad (6.38)$$

or for small angles $\theta = \pm\lambda/s$.

For nonuniform illumination for example, with a sinusoidal intensity distribution across the aperture, the diffraction pattern consists of two lines on each side of the central position, exactly the inverse of the Young's slits situation, as would be expected. If the intensity distribution across the aperture is a rectangular wave amplitude function, coming from a set of narrow slits equivalent to a diffraction grating

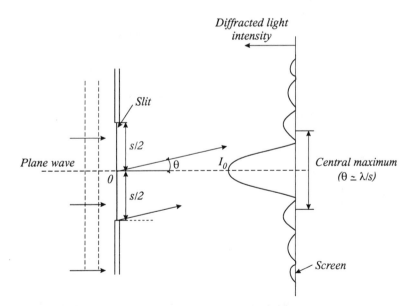

Figure 6.8 Diffraction through a narrow slit. (After: [17] with permission.)

(Figure 6.9), the Fourier transform and hence the Fraunhofer diffraction pattern gives a set of discrete lines. For different wavelengths, each wavelength produces its own separate diffraction pattern. This could be used to distinguish between the different wavelengths, and this type of diffraction grating is characterized by the resolving power (ρ)

$$\rho = \lambda/\delta\lambda \qquad (6.39)$$

where ($\delta\lambda$) is a measure of the smallest resolvable wavelength separation.

6.1.4.5 Coherence

The spatial coherence of a wave means that any point on the wave front has the same phase and is transverse to the direction of propagation, giving a high *transverse correlation*. A phase difference $\Delta\Phi$ between two different phase fronts is equivalent

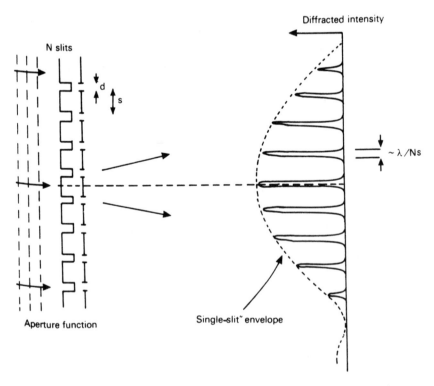

Figure 6.9 Diffraction of light from a grating consisting of a series of narrow slits. (*After:* [17], with permission.)

to a distance d ($\Delta\Phi = 2\pi nd/\lambda$) between the two phase fronts, which is called *temporal coherence* (see Figure 6.10).

A coherence length L_c is defined by

$$L_c = c \cdot \tau_c \tag{6.40}$$

where τ_c is the coherence time, which corresponds to the maximum temporal delay that preserves a good fringe visibility, and c is the velocity of light in a vacuum.

A self-correlation function $f(\tau_{sc})$ of the amplitude of a periodic disturbance at one place and/or time can also be defined [17]. It oscillates with a frequency ω and a constant amplitude $a^2/2$:

$$f(\tau_{sc}) = (a^2/2) \cdot \cos(\omega t) \tag{6.41}$$

Taking this measure as unity for a sine wave, this value can be normalized at $\tau_c = 0$; hence $f(0) = a^2/2$, giving a normalized function $\gamma(\tau_{sc})$, called the *coherence function* and represented in Figure 6.10:

$$\gamma(\tau_{sc}) = |f(\tau_{sc})|/|f(0)| \tag{6.42}$$

In a dual-beam interference setup, the fringe visibility is defined from the maximum (I_{max}) and minimum (I_{min}) light intensities of the interference pattern:

$$V = (I_{max} - I_{min}) / (I_{max} + I_{min}) \tag{6.43}$$

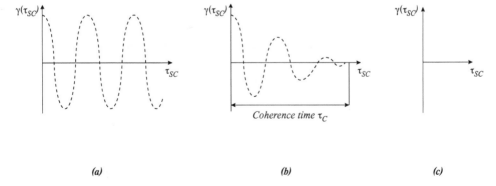

Figure 6.10   Relationship between the coherence function $\gamma(\tau_{SC})$ and the displacement time τ_{SC} of a periodic disturbance: (a) totally coherent wave; (b) partially coherent wave; and (c) incoherent wave.

and the visibility function $V(\tau_{sc})$ is related to the normalized coherence function $\gamma(\tau_{sc})$, coming from two sources of intensity I_1 and I_2:

$$V(\tau_{sc}) = 2(I_1 \cdot I_2)^{1/2} \cdot \gamma(\tau_{sc}) / (I_1 + I_2) \qquad (6.44)$$

The concept of coherence is extremely important for interferometric fiber-optic sensors, such as the fiber-optic gyroscope [15], but is not common for chemical and biochemical sensors. However, coherent sensing is attracting more attention in different applications such as coherent sensors with modulated laser sources [18], CARS [19], single-mode evanescent wave spectroscopy [20], spectral interferometry on thin multilayers [21], and sensor networks.

6.2 INTERACTION OF LIGHT WITH MATTER

The interaction of light with matter is important in chemical and biochemical sensors, since it forms the basis of sensing techniques. Electromagnetic radiation of frequency v interacting with charged matter creates a change in energy (hv), a change of linear momentum $(hv/c = h/\lambda)$, or a change of angular momentum $(h/2\pi)$. This results in a change of the optical properties of the atoms or molecules, or can lead to stimulation of atomic electrons, causing them to radiate as elementary electric dipoles. This interaction process results in optical changes in reflection, refraction, diffraction, interference, and polarization, which were examined in the previous section. The effects of scattering, dispersion, absorption, luminescence, energy transfer, and nonlinear or photochemical effects are summarized in this section. More complete discussions on interactions between light and matter can be found elsewhere [22].

6.2.1 Scattering

In a uniform dielectric medium, the refractive index n can be defined as a complex quantity that includes the attenuation and phase behavior of the wave:

$$n = n' - in'' \qquad (6.45)$$

Using this form, (6.4), (6.8), and (6.12) can be written

$$E = E_0 \exp(-\omega n'' z/c) \cdot \exp\{i\omega[t - (n'z/c)]\} \qquad (6.46)$$

The first exponential term is an attenuation factor and the second exponential term a propagation factor.

The distribution of the intensity of light propagating in matter depends on the dipole distribution in the matter, since interference occurs between the original wave and scattered radiation coming from these elementary dipoles. The refractive index function can be separated into two components [17]:

$$n = n_f + n_s \qquad (6.47)$$

one for forward scattering (n_f) of the secondary waves (the real part of the index $n_f = \varepsilon^{1/2}$), which can experience phase displacements, and the other for scattering (n_s) in all other directions (the imaginary component), which occurs with a random angular distribution of the intensity. The intensity of the elastically scattered light is greatest in the backward direction and certain preferential forward directions [23].

For small molecules with a size much less than the wavelength of the light (diameter $< \lambda/100$) the effects of n_s scattering are negligible. For larger molecules (diameter $> \lambda/10$ for visible light), this elastic scattering is measurable and called *Rayleigh scattering* (Figure 6.11(a)). The amount of scattered light is proportional to λ^{-4} and thus increases considerably in the UV region. If the diameter of the molecules is larger than the wavelength, the scattered waves interfere with one another from several centers of scattering. This is called *Mie scattering*.

If absorption occurs at the same time as the scattering, part of the energy of the incident light induces vibrations and/or rotations in the molecule and hence produces a change in the energy states of the molecule. This is nonelastic *Raman scattering*. When the molecule returns to its equilibrium ground state, energy can be emitted as radiation. The emission lines (Stokes lines) occur at lower energies or longer wavelengths than the unscattered radiation (Figure 6.11(b)). Emission can also occur at higher energies than the unscattered light (anti-Stokes lines). The probability of

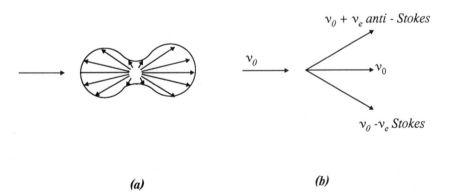

(a)   **(b)**

Figure 6.11   Representations of scattering phenomena: (a) in Rayleigh scattering, the excitation and scattered frequencies (wavelengths) are identical. (b) In nonelastic Raman scattering, the frequencies of excitation (ν_0) and emission (ν_e) are different, with $(\nu_0 - \nu_e)$ for Stokes lines and $(\nu_0 + \nu_e)$ for anti-Stokes lines.

transition, or the intensity, of these anti-Stokes emission lines is lower than for the Stokes lines. The relative emission intensities in terms of the excitation (v_0) and emission (v) frequencies are given by

$$I(\text{Stokes})/I(\text{Anti-Stokes}) = [(v_0 - v)/(v_0 + v)]^4 \cdot \exp(-hv/kT) \qquad (6.48)$$

It should be noted that the Raman effect is relatively weak. The intensity of the Raman emission is about 10^{-13} that of the incident light. In addition, if the excitation is in the UV region, the fluorescence emission (about 4 to 6 orders of magnitude larger) can mask the Raman spectrum of the compounds being measured.

6.2.2 Dispersion

The real part of the refractive index is related to the wave frequency, since $\omega/k = c/n$. Hence, in a medium where the refractive index changes with the frequency (e.g., one in which the polarizability varies), different wavelengths will travel at different velocities. This causes *dispersion* of the light energy within the dispersive medium. The group velocity c_g of a pulse of light with a given spectral distribution can be defined in terms of the group refractive index n_g:

$$n_g = c/c_g = n - \lambda \, (dn/d\lambda) \qquad (6.49)$$

The effect of this dispersion is that the width of a pulse of light increases as it propagates. This is extremely important in optical communications, as propagation over long distances in an optical fiber introduces significant pulse broadening, which limits the bandwidth of the transmission system. The propagation time (t) of the pulse over a given distance L is

$$t = L/c_g = (L/c) \cdot [n - \lambda \, (dn/d\lambda)] \qquad (6.50)$$

It is constant only if $\lambda(dn/d\lambda)$ is constant for all the wavelengths emitted by the source. If $(dn/d\lambda)$ varies, the shape of the pulse changes and the pulse spread in terms of time (τ_p) for a narrow range of wavelengths ($\delta\lambda$) is

$$\tau_p = (dt/d\lambda)\delta\lambda = (-L\lambda/c) \cdot (d^2n/d\lambda^2)\delta\lambda \qquad (6.51)$$

Minimum dispersion is obtained for $d^2n/d\lambda^2 = 0$, which in the case of silica occurs at around 1.3 μm, a wavelength chosen for fiber-optic communications.

The dispersion (Figure 6.12) due to variations of $dn/d\lambda$ is called *material dispersion*, since it depends on the properties of the material. In fiber-optic communications, other dispersion effects must be considered, such as modal dispersion, which derives from the difference in group velocities of different modes propagating in the

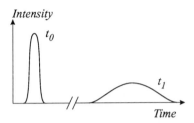

Figure 6.12 Dispersion of a pulse of light in an optical fiber. The initial pulse at time t_0 is broadened after propagating for a time t_1.

waveguide. This modal dispersion is usually expressed as an electrical bandwidth-distance product (e.g., MHz · km).

In general, the refractive index increases with increasing optical frequency, the case of *normal dispersion*; for example, blue light is diffracted more than red light in a prism made to separate different wavelengths. Molecules can exhibit resonant oscillations (eigenfrequencies), which are due to vibrations between neighboring atoms coupled together or to changes in the distribution of electrons relative to the atom nuclei. At the resonant frequency, the dielectric constant and the refractive index are drastically reduced as the optical frequency increases. This can occur in the UV, visible, and IR regions and is known as *anomalous dispersion*.

6.2.3 Attenuation and Absorption

Light passing through a material can be absorbed by interaction with the molecules and atoms of the host material, or by impurities and defects. At any given wavelength λ, the intensity of the light passing through the material is attenuated as an exponential function of its path length (x), as shown in Figure 6.13. The absorption coefficient (α_λ) is defined relative to the concentration $[M]$ and the cross section (s) of the absorbing molecules.

$$I_\lambda(x) = I_\lambda(0) \cdot \exp(-\alpha_\lambda \cdot x) = I_\lambda(0) \cdot \exp(-s_\lambda[M]x/N) \tag{6.52}$$

where $I_\lambda(x)$ is the light intensity at a distance x, $I_\lambda(0)$ is the incident light intensity at $x = 0$, and N is Avogadro's number $(6.022 \times 10^{23}\ \mathrm{mol}^{-1})$.

Changes in the chemical environment can alter the absorption coefficient α, and absorbance-based sensors measure these changes by the transmitted light intensity in terms of absorbance (A_λ) units:

$$A_\lambda = \log[I_\lambda(0)/I_\lambda(x)] \tag{6.53}$$

Attenuation (a_λ) is often used in characterizing optical fibers. It is defined in terms of

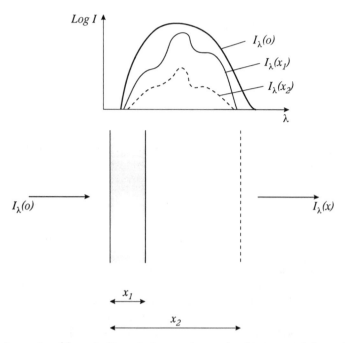

Figure 6.13 Attenuation of the optical intensity in traversing an absorbing material after a distance x_1 and x_2. The shape of the absorption spectrum remains unchanged. (See (6.52) and (6.53).)

optical power loss, expressed in decibels per unit length (L), and is related to absorbance by

$$a_\lambda = 10 \, A_\lambda/L \qquad (6.54)$$

6.2.4 Molecular Transitions

After a molecule has been excited by light, it may return to its equilibrium ground state in a number of ways, by transitions with the emission of radiation or by radiationless transitions. These transitions can be represented on a Jablonski energy level diagram (Figure 6.14). Each energy level corresponds to a characteristic arrangement of electron orbitals and vibrational states of the atom or molecule. Different transitions can occur with different probabilities and transition times, which determine the relative strengths of the absorption and emission.

As an example, an absorption (Figure 6.14(a)) is represented by a vertical line corresponding to a transition from a singlet ground state (S_0) to the first singlet excited state without a change in spin but with a new electronic configuration. The transition occurs in a short time (on the order of 10^{-15} sec or less). In this excited state, one of

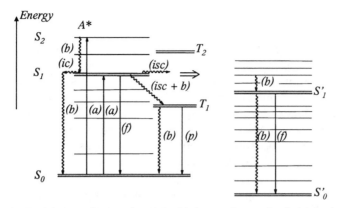

Figure 6.14 Energy level diagram showing electronic and vibrational energy levels: (a) An absorption occurs by a transition from the ground state S_0 to the first excited singlet state S_1 or a higher level. Relaxation occurs by thermal processes (b), by internal conversion (ic), and by intersystem crossing (isc). Deactivation of the molecule occurs by fluorescence (f) or phosphorescence (p) from the triplet state (T_1). By means of a photochemical reaction and/or energy transfer, another molecule with S'_0 and S'_1 states can be activated by intersystem crossing (isc) (on the right-hand side of the figure).

the two electrons is unstable in this configuration (A^*) and the molecule rapidly relaxes to a more stable configuration (Figure 6.14(b)) or may undergo a chemical reaction (i.e., decomposition).

In the relaxation process, energy dissipation occurs by radiationless transitions within the same molecule either by internal conversion, or by an intersystem crossing to a lower energy triplet state via a change of the multiplicity (m) of the quantum number of spins (s). *Multiplicity* is a term that means that one of the electronic spins in different orbitals is reversed (with $m = 2s + 1$). Through radiative transitions, the luminescence appears as fluorescence from the singlet state (spin $s = 0$) to the singlet ground state, or phosphorescence from the triplet state (a state in which the spins are parallel and $m = 3$) to the singlet ground state with a change in spin.

6.2.5 Photochemical Effects

An intersystem crossing can occur by photochemical processes, by the modification of the arrangement of the atoms in a molecule or a change of the molecule through the influence of photons (energy transfer). For example, this occurs for photoisomerization (the formation of new isomers under the influence of photons) by breaking bonds to form new molecules. The new configuration leads to other transitions from the excited state (seen on the right of the energy level diagram for the singlet state, Figure 6.14). This can also occur from a triplet state (not shown in the figure).

6.2.6 Luminescence

After a photon of energy ($h\nu$) is absorbed between two levels S_1 and S_2, luminescence can occur through the emission of a photon by a transition from the S_2 energy level or, more usually, from a lower energy level. In spontaneous emission, the polarization and propagation directions of the emitted photon are random with respect to the exciting photon (Figure 6.15). In the special case of stimulated emission, an emitted photon induces atoms to release further photons, creating optical amplification, and the polarization and propagation direction are the same as the incident photon. This is the basic principle of the *laser* (light amplification by stimulation emission of radiation).

Excitation and subsequent emission can occur not only by photoluminescence (excitation by photons), but also by electroluminescence (excitation by an electric field), chemiluminescence (excitation from a chemical reaction), and bioluminescence (excitation from a biological reaction). Many of these processes are important in optical-fiber sensors.

6.2.6.1 Fluorescence

Fluorescence occurs when an atom or molecule makes an allowed transition from a higher to a lower energy state in a short time (10^{-9} to 10^{-3} sec) emitting a photon, as shown in Figure 6.14 (line f).

In atomic fluorescence, emission gives rise to a series of discrete sharp spectral lines, since the atoms (e.g., in a gas) do not interact with one another and the electronic levels are not broadened by phonon interaction (change of quantized vibrational energy). In resonant fluorescence, absorption and emission take place between the same two energy levels, and the wavelengths of the incident and emitted light are

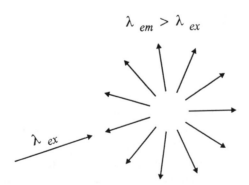

Figure 6.15 Fluorescence occurs at a longer wavelength (λ_{em}) than the exciting wavelength (λ_{ex}) and has a uniform spatial distribution.

identical because the fluorescent decay corresponds to a return to the equilibrium ground state (Figure 6.16(a)). The states may be slightly changed by phonon broadening, giving rise to thermally excited levels S_1 within the ground state S_0 (Figure 6.16(b)), and the emission wavelength is then not identical.

Nonresonant emission can occur as Stokes fluorescence in which the emission occurs at a longer wavelength (photon of lower energy) than the excitation (see Figure 6.16(c)), or as anti-Stokes fluorescence in which the emission occurs at shorter wavelengths with a higher photon energy than the excitation (see Figure 6.16(d)). In addition, collisional coupling can induce an increase or decrease of the emission energy from levels populated by collisions. This is called *fluorescence in cascade* and is shown in Figure 6.16(e).

In molecular fluorescence, generally used in optical sensors, emission is the result of transitions of one electron between the various electronic energy levels (E_e) in the molecule. These energy states are modified because of the interaction between several atomic nuclei, giving rise to vibrational energy (phonon) states (E_v) due to the relative

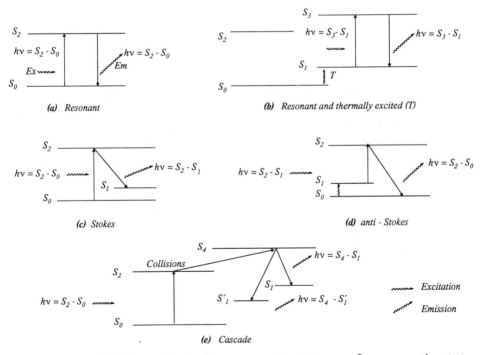

Figure 6.16 Energy level diagram showing fluorescent transitions: (a) resonant fluorescence with excitation and emission transitions between the same levels; (b) resonant fluorescence from thermally excited levels; (c) Stokes fluorescence; (d) anti-Stokes fluorescence; and (e) fluorescence in cascade from energy levels populated by collisions.

displacement of atoms with respect to each other, and rotational energy states (E_r) coming from the rotation of the molecule about specific axes. In the Born-Oppenheimer approximation, the total energy (E_t) is the sum of these different energies:

$$E_t = E_e + E_v + E_r \qquad (6.55)$$

The decay rate dN^*/dt of the fluorescence for a two-level system is

$$dN^*/dt = -k_t \cdot N^* \qquad (6.56)$$

where k_t is the total fluorescence rate (in $\sec^{-1}$) and N^* is proportional to the number of electrons excited to a fluorescent state in a time t. Hence,

$$N^* = N_0^* \cdot \exp(-k_t \cdot t) \qquad (6.57)$$

where N_0^* is the number of excited electrons at time t_0.

For a molecule, the overall lifetime $(\tau = 1/k_t)$ corresponds to a global decay constant:

$$\tau = 1/(k_{ic} + k_{st} + k_c + k_f) \qquad (6.58)$$

where k_{ic} is the internal conversion constant, k_{st} the constant for crossing from a singlet to a triplet state, k_c is the collisional deactivation constant (a constant that is a function of the environment, e.g., from the effect of quenching—see Chapter 8), and k_f the constant of radiative emission (fluorescence).

The decay time constant τ_0 is defined as the time for the steady-state fluorescence intensity to decay to $1/e$ of its original value (Figure 6.17). For $k_{st} = 0$, the decay time is equal to

$$\tau_0 = 1/(k_f + k_{ic}) \qquad (6.59)$$

The quantum yield for a given transition (η_λ) at a well-defined fluorescence wavelength λ is the ratio between the number of radiative transitions (N_r^*) and the total number (N^*) of radiative and nonradiative transitions:

$$\eta_\lambda = N_r^*/N^* \qquad (6.60)$$

The general relationship between η and τ is

$$\eta = k_f \cdot \tau \qquad (6.61)$$

An identical relationship exists between η_0 and τ_0.

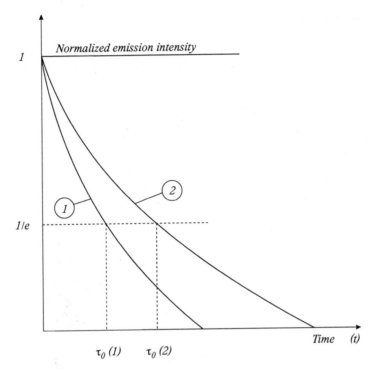

Figure 6.17 Time decay curve of fluorescence. Two molecules (1) and (2) can be distinguished by their different fluorescent decay time constants (τ_0).

Fluorescence techniques (see Chapter 8) are used widely in fiber-optic sensors, such as intensity, quenching, anisotropy, resonance energy transfer, polarization, and inhomogeneous broadening of electronic spectra of dye molecules [24].

6.2.6.2 Phosphorescence

After conversion from a singlet to a triplet state by intersystem crossing or energy transfer to another molecule, a metastable excited state (transition T) is observed with a relatively long lifetime τ of emission between 10^{-3} to 10 sec [25]. This is phosphorescence (see line p in Figure 6.14), which is frequently observed in molecules with heavy atoms. It is strongly influenced by temperature; the decay time of phosphorescence can be used to measure temperature and has been employed in some fiber-optic temperature and chemical sensors. Often the terms *fluorescence* and *phosphorescence* are arbitrarily interchanged when the decay times are similar.

6.2.7 Energy Transfer

A number of fiber-optic chemical sensors are based on energy transfer as the method of ion determination. Five different processes can occur for energy transfer between ions, atoms, and molecules [25].

1. Radiative transfer (radiation trapping). Whereby an atom A is excited and emits a photon, which is not released but is immediately captured by a second atom B. This excites atom B and produces fluorescence (see the band model illustrated in Figure 6.18).
2. Resonant exchange. A nonradiative energy transfer occurs between two molecules (e.g., from a donor to an acceptor) that can radiate at the same frequency (wavelength) and are separated by less than a critical distance, typically 20 Å to 60 Å.

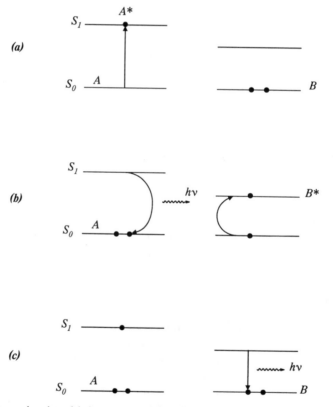

Figure 6.18 Energy band model showing energy transfer between two molecules: (a) excitation by an absorbed photon; (b) deactivation of molecule A, and energy transfer and excitation of molecule B; and (c) deactivation of B. The fundamental state S_0 of B can be different from that of A.

It is governed by long-range dipole-dipole interaction with the condition of conservation of total energy [26]. The probability of energy transfer is a direct function of the overlap of the energy states between atoms A and B. This overlap occurs between two spectral bands, as shown in Figure 6.19. Important examples have been demonstrated for quenching fluorescence in an oxygen sensor [27] and for the measurement of spectral shifts accompanying labeled polymer-polymer binding [28]. Broadening (by phonon interaction) of the spectral emission band of A or the absorption band of B increases the overlap and the probability of energy transfer.

3. Spatial process. The excited energy is carried by a phonon to a receiving site. The probability of energy transfer depends on the type of coupling involved. In addition, energy transfer does not occur beyond a critical distance.

4. Spin coupling. This occurs only between nearest neighbors, illustrated by direct energy exchange in the band model of Figure 6.20. Both ions are initially in an excited state, A^* and B^*. Spin coupling raises B^* to a higher excited state while A^* returns to the ground state. This process, observed mainly with rare-earth ions, can absorb infrared radiation and convert it to visible light.

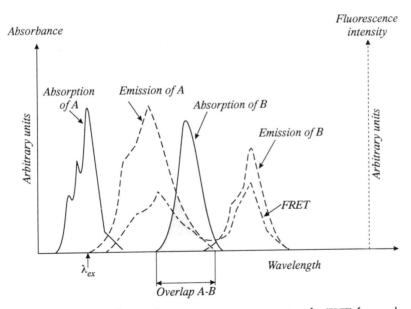

Figure 6.19 Optical spectra showing fluorescence resonant energy transfer (FRET) from molecule A to molecule B. This occurs by the overlap of the emission of A (donor) excited at λ_{exc} and the absorption of B (acceptor) with subsequent emission of fluorescence with a mixed spectrum of A and B. The quenching of the donor in a donor-acceptor system can enhance the acceptor fluorescence emission. The emission spectra of A and B are shown dotted.

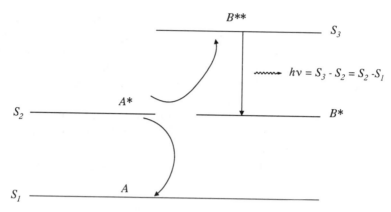

Figure 6.20 Energy transfer by spin coupling between molecules A and B.

5. Nonresonant process. Energy transfer can occur between two mismatched multi-dipoles via phonon exchange.

6.2.8 Nonlinear Effects

In linear optics, a linear relationship exists between the total polarization P (dipole moment per unit volume) and the electric field E of a propagating wave:

$$P = \chi \cdot E \tag{6.62}$$

where χ is the volume susceptibility or polarizability of the medium. This relationship assumes that the separation of positive and negative charges is proportional to the electric field. The refractive index (n) and the absorption coefficient of the medium are independent on the incident light intensity:

$$n = (1 + \chi)^{1/2} = (1 + P/E)^{1/2} \tag{6.63}$$

This linearity does not exist for susceptible optical materials that are subjected to very strong perturbations of an electric field, and the polarizability can be expanded in a power series of the electric field:

$$P(E) = \chi_1 E + \chi_2 E^2 + \ldots \chi_j E^j + \ldots \tag{6.64}$$

In practice, χ decreases in magnitude rapidly from 1 to j and only the first three terms of the electric field need be considered. The refractive index also becomes dependent on E and generates nonlinear optical effects:

$$n = (1 + \chi_1 + \chi_2 E + \ldots \chi_j E^{j-1} + \ldots)^{1/2} \tag{6.65}$$

The main nonlinear effect proportional to E^2 is the generation of harmonics at double or higher frequencies of the input optical wave, particularly when optimized conditions known as phase-matching occur (matching the input wave to the output wave). Optical systems often use nonlinear effects, such as the Kerr effect, frequency multiplexing, frequency generation (sum or difference), optical phase conjugation, and optical switching. These phenomena have not yet been exploited in chemical optical sensing. However, since the susceptibility of a compound depends on its composition and structure, the change of the χ coefficients by chemical interaction has been experimentally examined for a nonlinear optical pH sensor using the second-order susceptibility of Langmuir-Blodgett monomolecular films on solid substrates [29].

References

[1] Born, M., and E. Wolf, *Principles of Optics*, Oxford: Pergamon Press, 1975.

[2] Agrawal, G. P., *Non Linear Optics: Principles and Applications*, Boston: Academic Press, 1989.

[3] Synder, A. W., and J. D. Love, *Optical Waveguides Theory*, London: Chapman and Hall, 1983.

[4] Marcuse, D., *Theory of Dielectric Optical Waveguides*, San Diego: Academic Press, 1991.

[5] Marsh, J. H., and R. M. De la Rue, *Waveguide Optoelectronics, Vol. 226*, NATO ASI Series E Applied Sciences, Vol. 226, Dordrecht: Kluwer Academic Pub., 1990.

[6] Vassalo, C., *Optical Waveguide Concepts*, Amsterdam: Elsevier, 1991.

[7] Collin, R. E., *Field Theory of Guided Waves*, 2nd ed., New York: IEEE Press, 1992.

[8] Koshiba, M., *Optical Waveguide Analysis*, New York: McGraw Hill, 1992.

[9] Daly, J. C., ed., *Fiber Optics*, Boca Raton: CRC Press, 1984.

[10] Briley, B. E., *An Introduction to Fiber Optics System Design*, Amsterdam: North Holland, 1988.

[11] Miller, C. M., S. C. Mettler, and I. A. White, eds., *Optical Fiber Splices and Connectors: Theory and Methods*, New York: Marcel Dekker, 1986.

[12] Jeunhomme, L. B., *Single Mode Fiber Optics*, New York: Marcel Dekker, 1990.

[13] Dakin, J. C., and B. Culshaw, eds., *Optical Fiber Sensors, Vol. 1: Principles and Components*, 1988; *Vol. 2: Systems and Applications*, Norwood, MA: Artech House, 1989.

[14] Göpel, W., J. Hesse, and Z. N. Zemel, eds., *Sensors: A Comprehensive Survey, Vol. 6: Optical Sensors*, Weinheim: VCH, 1992.

[15] Lefevre, H., *The Fiber Optic Gyroscope*, Norwood, MA: Artech House, 1993.

[16] Halley, P., *Les Systemes à Fibres Optiques*, Paris: Eyrolles, 1985.

[17] Rogers, A. J., "Essential Optics," Chap. 3 in *Optical Fiber Sensors, Vol. 1: Principles and Components*, J. C. Dakin and B. Culshaw, eds., Norwood, MA: Artech House, 1988, pp. 25–106.

[18] Giles, I. P., D. Uttam, B. Culshaw, and D. E. N., Davies, "Coherent Optical Fiber Sensors With Modulated Laser Sources," *Electron. Lett.*, Vol. 19, 1983, pp. 14–15.

[19] Eckbreth, A. C.," Remote Detection of CARS Employing Fibre Optic Guides." *Appl. Opt.*, Vol. 18, 1979, pp. 3215–3216.

[20] Stewart, P., J. Norris, D. Clark, M. Tribble, I. Andonovic, and B. Culshaw, "Evanescent Field Absorption: The Sensitivity of Optical Waveguides," *Proc. SPIE–Int. Soc. Opt. Eng.*, Vol. 990, 1988, pp. 188–195.

[21] Gauglitz, G., A. Brecht, G. Kraus, and W. Nahm, "Chemical and Biochemical Sensors Based on Interferometry at Thin (Multi) Layers," *Sensors and Actuators*, Vol. B 11, 1993, pp. 21–27.

[22] Dodd, J. N., *Atoms and Light Interactions*, New York: Plenum Press, 1991.

[23] Wolfbeis, O. S., G. Boisdé, and G. Gauglitz, "Optochemical Sensors," Chap. 12 in *Sensors: A Comprehensive Survey, Vol. 2*, W. Göpel, J. Hesse, and J. N. Zemel, eds., 1991, pp. 573–645.

[24] Lakowicz, J. R., ed., *Topics in Fluorescence Spectroscopy, Vol. 2: Principles*, 1991; *Vol. 4: Design and Chemical Sensing*, New York: Plenum Press, 1994.

[25] Ropp, R. C., *Luminescence and the Solid State*, Amsterdam: Elsevier, 1991.

[26] Cheung, H. C., "Resonance Energy Transfer," Chap. 3 in *Topics in Fluorescence Spectroscopy, Vol. 2: Principles*, New York: Plenum Press, 1991, pp. 127–176.

[27] Sharma, A., and O. S. Wolfbeis, "Fibre Optic Oxygen Sensor Based on Fluorescence Quenching and Energy Transfer," *Appl. Spectrosc.*, Vol. 42, 1988, pp. 1009–1012.

[28] Christian, L. M., and W. R. Seitz, "An Optical Ionic-Strength Sensor Based on Polyelectrolyte Association and Fluorescence Energy Transfer," *Talanta*, Vol. 35, 1988, pp. 119–122.

[29] Draxler, S., I. Pflanzl, Z. Xiang, and M. E. Lippitsch, "Chemical Sensors Based on Non-linear Optics," *Sensors and Actuators*, Vol. B 11, 1993, pp. 129–131.

Chapter 7

Optical Fibers and Planar Waveguides

7.1 INTRODUCTION

Optical fibers and planar waveguides are the basic elements in optical chemical sensing, and an understanding of their characteristics is essential for optodes. This chapter summarizes the most important concepts of optical guided waves in fibers and waveguides [1,2]. More advanced aspects of the different modal effects (mode coupling, leaky modes, mode conversion) and their importance in chemical sensing have been reviewed by Parriaux [3].

7.2 RAY MODEL OF LIGHT PROPAGATION

An optical fiber can be thought of as a long rod of transparent material (a cylindrical dielectric structure) in which the rod or core has a higher refractive index (n_1) than that of the surrounding material or cladding (n_2), as shown in Figure 7.1. Total internal reflection of the light takes place at the walls of the rod, at the core-cladding interface, and the propagating light will be confined within the core. From Snell's laws, (6.24) and (6.25), the angle of reflection θ at the core-cladding interface is greater than the critical angle θ_c. Light, which is injected into the fiber at an angle θ_0 smaller than the limiting angle θ_{lim}, will be totally internally reflected at the core-cladding interface.

The numerical aperture (NA) is defined by this limiting value of the acceptance angle at the input to the fiber:

$$NA = n_0 \sin \theta_{lim} = (n_1{}^2 - n_2{}^2)^{1/2} = n_1(2\Delta)^{1/2} \qquad (7.1)$$

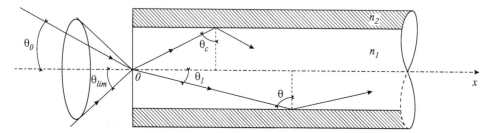

Figure 7.1 Basic structure of an optical fiber, with a cylindrical core of refractive index n_1 surrounded by a cladding of index n_2 $(n_2 < n_1)$. Light injected into the fiber at the limiting angle θ_{lim} arrives at the core-cladding interface at the critical angle θ_c. All rays entering the fiber at an angle smaller than the limiting angle will be reflected at the cladding and propagate in the core. The propagation angle θ_1 is related to θ_0 by Snell's law (6.24).

where Δ is the normalized difference of the refractive indexes between the core and cladding:

$$2\Delta = (n_1{}^2 - n_2{}^2)/n_1{}^2 \tag{7.2}$$

In silica optical fibers, the NA value is generally between 0.2 to 0.4. It may be as high as 0.8, or even higher in plastic and fluoride glass fibers. Increasing the refractive index difference between the core and the cladding of the fiber increases the acceptance angle and the numerical aperture. A large acceptance angle gives rise to a large number of rays with different reflection angles (i.e., different transmission modes), which propagate down the fiber.

7.2.1 Mode Propagation in Waveguides

The distance traveled by different rays, and hence their relative velocities along the fiber, depends on the angle of propagation (the reflection angle at the core-cladding interface) of each transmission mode. Another way of characterizing the progression of a light wave along the guide, taking into account the different transmission modes, makes use of the wave number (β):

$$\beta = n_1 k \sin \theta \tag{7.3}$$

See (6.12) for a definition of k. With the total internal reflection conditions given by the core and cladding ($\sin \theta \geq n_2/n_1$), the extreme values of the wave number are

$$n_1 k \geq \beta \geq n_2 k \tag{7.4}$$

A complete analysis of light transmission in waveguides normally requires solution of

Maxwell's equations [4]. However, a simplified treatment is given here with the main results. For a guide of diameter $2a$, a transverse whole wave number (q) is defined in terms of the group velocity (u) as

$$q = n_1 \cdot k \cdot \cos \theta = u/a \qquad (7.5)$$

Hence,

$$q^2 = k^2 n_1{}^2 - \beta^2 = u^2/a^2 \qquad (7.6)$$

Another parameter p, the transverse decay constant, is defined in terms of the longitudinal group velocity (v), with $p = v/a$ and

$$p^2 = \beta^2 - n_2{}^2 k^2 = v^2/a^2 \qquad (7.7)$$

A dimensionless parameter V, called the *normalized frequency*, which is characteristic of the light guide and inversely proportional to the wavelength of the light, can be derived from (7.6) and (7.7):

$$V^2 = u^2 + v^2 = a^2(q^2 + p^2) = a^2 k^2 (n_1{}^2 - n_2{}^2) \qquad (7.8)$$

which may be expressed as

$$V = (NA) \cdot (2\pi \cdot a/\lambda) \qquad (7.9)$$

The dispersion curves for a slab waveguide [5] are generated by the graphical solution of (7.8) using the u and v values as the rectangular axes. The intersections of these values with the V circular functions (where $u = v$) represent the solutions for the u and v parameters of the transverse electric (TE) modes, as shown in Figure 7.2.

The condition for only a single mode to exist in an optical fiber [4] is that V should be less than 2.405. Thus, the cutoff wavelength λ_c, which is defined as the wavelength above which the fiber only has a single-mode regime, is given by

$$\lambda_c = V \cdot \lambda/2.405 = (NA) \cdot (2\pi \cdot a/2.405) \qquad (7.10)$$

7.2.2 Polarization in Optical Fibers

The intensity distribution and polarization of any mode can be described by the LP (light pattern) notation with two indexes, one for the azimuthal integer (i_a), the second for the polarization state (s_p). This notation gives the modal field shape for any particular mode. For instance, LP_{01} is the fundamental mode of a single-mode fiber. At shorter wavelengths than the single-mode cutoff, with $2.4 < V < 3.8$, both the

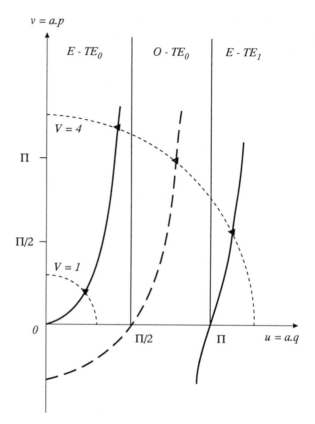

Figure 7.2 Graphical solution of the dispersion curves for TE modes in a slab waveguide. Triangles correspond to solutions for given V values.

LP_{01} ($i_a = 0$) mode and an antisymmetric mode called LP_{11} ($i_a = 1$) become guided. The representation of these modes and the intensity distribution of light in the fiber (the near-field pattern) are shown in Figure 7.3. At still shorter wavelengths with V greater than 3.832, other modes become guided, such as the LP_{02} and LP_{21}.

In special asymmetric fiber structures, it is possible to alter the polarization of the two orthogonal polarizations so that they are no longer degenerate. Such special fibers have either a noncircular core or a linear or twist strain in the core in order to deliberately introduce a birefringence in the fiber. These types of fibers are of great interest for optical sensors. However, they have been mainly applied for sensing physical parameters when combined with active effects such as electro-optic and magneto-optic effects.

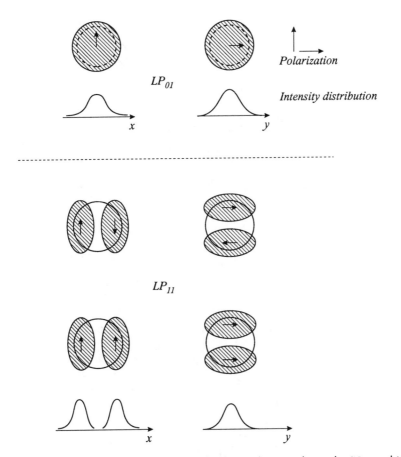

Figure 7.3 Light pattern and intensity distribution for the two lowest order modes (LP_{01} and LP_{11}) in a cylindrical waveguide. The polarization direction is shown by an arrow.

7.3 PLANAR WAVEGUIDES

7.3.1 Wave Propagation in Planar Guides

In chemical and biological sensors [6], planar waveguides are well adapted for evanescent wave sensing (see Chapter 8) using grating couplers and sensitive polymeric films (see Section 7.3.3). The simplest planar waveguide may be described as a symmetrical dielectric slab in which the center core of thickness $(2a)$ and refractive index n_1 are sandwiched between two cladding layers of refractive index n_2. Propagation of the light is along the z-axis, as shown in Figure 7.4(a).

Maxwell's equations applied to TE and transverse magnetic (TM) modes provide

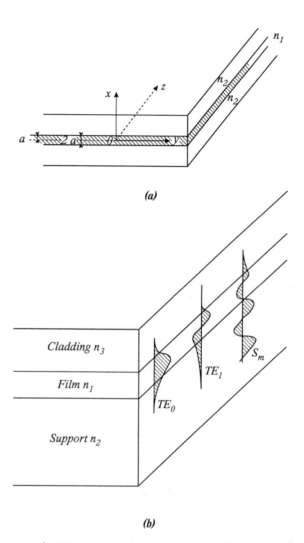

Figure 7.4 Planar waveguides: (a) a symmetrical structure with a film of refractive index n_1 and thickness $2a$ sandwiched in between two slabs ($n_1 > n_2$) and (b) an asymmetrical structure with three refractive indexes $n_1 > n_2 > n_3$. The patterns of two guided modes, TE_0 and TE_1, and a support mode S_m are shown.

Table 7.1
Eigenvalue Equations for a Planar Waveguide

	Even	Odd
TE mode	$u \cdot \tan u = v$	$u \cdot \cot u = -v$
TM mode	$u \cdot \tan u = (n_1/n_2)^2 \cdot v$	$u \cdot \cot u = -(n_1/n_2)^2 \cdot v$

solutions for the E and H fields. Within the core slab, for $x \geq \pm a$, the E_y and H_y components are [2]

$$E_y = A_e \cos(u/a)x + A_o \sin(u/a)x \tag{7.11}$$

$$H_y = B_e \cos(u/a)x + B_o \sin(u/a)x \tag{7.12}$$

and outside the core:

$$E_y = C_1 \cdot e^{-(v/a)x} \tag{7.13}$$

$$H_y = C_2 \cdot e^{-(v/a)x} \tag{7.14}$$

The same parameters u and v, which are used for an optical fiber, are also introduced in these equations. For the A and B constants, the subscripts e and o are used for even and odd values, respectively.

The eigenvalues (characteristic values) of a particular mode are obtained by matching the field components at the boundary. These eigenvalues for a planar guide are given in Table 7.1 for TE and TM modes.

As for an optical fiber, the number of modes increases as the normalized frequency V increases. Cutoff occurs at $v = 0$; hence, $u = V$ (see (7.8)), and for the TE mode $u = m(\pi/a)$, where m is the mode order, equal to 0, 1, or 2. The minimum thickness of the guide to ensure light guiding is

$$2a = m \cdot \lambda/2(n_1{}^2 - n_2{}^2)^{1/2} \tag{7.15}$$

Generalization of these concepts can be made for a nonsymmetrical planar waveguide (Figure 7.4(b)) in which the support has a refractive index n_2 different from the cladding refractive index n_3, with the usual condition $n_1 > n_2 > n_3$.

An asymmetric parameter (p_a) is then defined as

$$p_a = (n_2{}^2 - n_3{}^2) / (n_1{}^2 - n_2{}^2) \tag{7.16}$$

This parameter is equal to zero for a symmetrical waveguide ($n_2 = n_3$), and to $+\infty$ for $n_1 = n_2 > n_3$.

In evanescent sensors, the light is not confined within the planar waveguide itself, but penetrates into the surroundings and thus senses the chemical environment on the surface of the guide (see Chapter 8). The higher order modes have the greatest penetration outside the guide.

7.3.2 Nonlinearity in Planar Waveguides

If a film, support, and cladding exhibit a pronounced nonlinearity (subscript χ), the effective refractive index N_χ is dependent on the local optical intensity (I):

$$N_\chi(I) = n + n_\chi \cdot I \tag{7.17}$$

where n_χ is the nonlinear coefficient and n is related to the β and k terms (see (7.3)). The mode equation can take into account this effect of nonlinearity at the interface by introducing a correction coefficient, which contains a normalized parameter (Q) in terms of the field strength [7]. For instance, the effective refractive index of a nonlinear cladding of refractive index n_3 becomes

$$n_{\chi 3} = n_3 + 2Q/(a^2 \cdot k^2) \tag{7.18}$$

These nonlinear effects are also important for planar optical components (e.g., gratings, couplers).

7.3.3 Configurations of Planar Waveguides

Several configurations of planar devices are commonly found in chemical sensors, especially with fluorescent and evanescent wave spectroscopic techniques (see Chapter 8). The basic configuration consists of a sample layer of refractive index n_2 deposited onto a prism or a plane surface of refractive index n_1 (Figure 7.5(a)). This configuration has been extended to different planar waveguide geometries. In Figure 7.5(b) the light is totally reflected at the support interface (index n_2) and partially reflected at the sample layer (index n_3). In Figure 7.5(c), a thin polymeric film acting as waveguide adsorbs the analyte from the sample. Changes of refractive index and film thickness can be expressed as shifts in the waveguide propagation modes [8].

Another configuration has been tried for chemical sensors [9] with grating couplers fabricated directly on the planar optical waveguide (Figure 7.6). These sensors have a high sensitivity and respond to changes in submonolayers. They measure the refractive index of a sample covering the waveguide or the adsorption-desorption of molecules on the waveguide surface in the region of the grating coupler. In grating couplers, a reversal of the direction of wave propagation transforms reciprocally the incoupling into an outcoupling process. Thus, input grating couplers [10] and output grating couplers [11,12] can be employed in the same sensor.

For instance, a plane wave with an incidence angle θ can be coupled into the planar waveguide provided the input coupling condition is satisfied:

$$N = n_0 \sin \theta + \kappa \cdot (\lambda_0/\Lambda) = n_0 \sin \theta + G_v \cdot \lambda_0/2\pi \tag{7.19}$$

where N is the effective refractive index of the guided mode, n_0 the refractive index of air, λ_0 the vacuum wavelength, Λ the grating period, $\kappa = 0, \pm 1, \ldots$ the diffraction

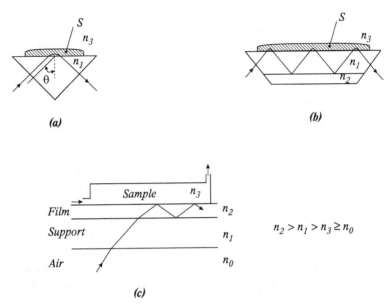

Figure 7.5 Configurations of practical waveguides for chemical sensors: (a) basic arrangement with prism; (b) simple planar waveguide; and (c) planar waveguide with thin film sensitive to the analyte. Grating couplers may be incorporated into the support. S is the sample.

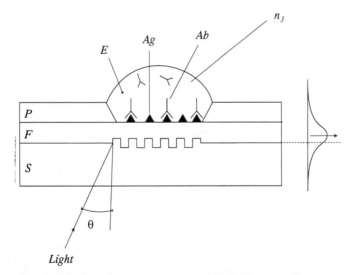

Figure 7.6 Planar waveguide with an integrated optic grating. The SiO_2-TiO_2 film F (refractive index n_1) is deposited onto a grating on a glass support (S) of refractive index n_2. The chemical sample E of refractive index n_3 is incorporated in a polymeric support P. Changes in the optical evanescent wave depend on the biological reaction (e.g., in a antigen-antibody reaction) that changes n_3.

order, and G_v is the grating vector of length $G_v = \kappa.(2\pi/\Lambda)$. This equation is also valid for the output coupling process.

The effective refractive index N depends on the waveguide parameters such as the refractive index of the guiding layer (n_1) and its thickness $(2a)$ and the refractive index of the support (n_2). For a sample of refractive index n_3, an effective index shift ΔN occurs for changes in n_3. Similarly, if the sample is in the form of a thin film, as a homogeneous adsorbed layer, on the guiding layer (as in Figure 7.5(b)), the effective index shift ΔN is due to changes in the film refractive index n_3 and its thickness $2a'$.

As an example, Figure 7.7 shows the sensitivity of the sensor for two different modes (TE_0 and TM_0) and two n_3 values. The sensitivity is high for very thin waveguides and a high value of $(n_2 - n_1)$. For a 0.1- to 0.2-μm-thick waveguide layer of SiO_2-TiO_2 on a glass substrate, the sensor can detect changes of $\Delta n_3 < 2.10^{-5}$ and $\Delta a' < 0.01$ nm in the thickness of the adsorbed layer [13].

An advantage of this configuration is that the grating coupler is integrated into the structure and forms an integral part of the sensor. If a prism coupler is used instead of a grating coupler, the input coupling condition is simplified with $N = n_p \sin \theta$, where $n_p > N$ is the refractive index of the prism. Examples of practical devices are given by a compact integrated sensor module based on a tapered-thickness

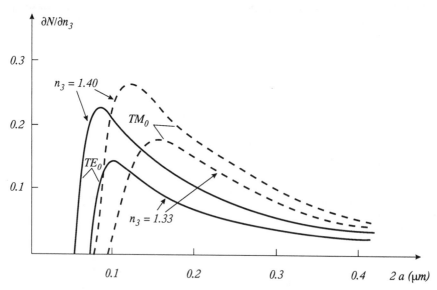

Figure 7.7 Theoretical sensitivity $\partial N/\partial n_3$ of a planar waveguide sensor for two different samples ($n_3 = 1.33$ and 1.40) and for two modes, TE_0 (solid line) and TM_0 (dotted line). The sample layer is put on a SiO_2-TiO_2 waveguide of variable thickness (2a) with a refractive index $n_1 = 1.75$ deposited on a glass substrate, $n_2 = 1.47$. (*After:* [9] with permission.)

waveguide with a spatially varying effective refractive index [14]. Similarly, a miniature sensor module has been reported with replicated chirped grating couplers on the waveguide [15].

The Kretschmann configuration [16] for SPR measurements (see Figure 8.14) is based on a prism geometry, covered by a very thin metallic film (about 50 nm) and overlaid with a dielectric polymer (index n_3) that is sensitive to the surroundings. The guided modes are analyzed by determining the minimum in the reflectivity at specific incident angles of the monochromatic light (see Section 8.8).

Many other configurations of planar waveguides have been adapted for chemical and biochemical sensors (see Chapter 10). An interesting concept is optoelectrochemical transduction, in which the planar waveguide is overlaid with an electrically modulated film. The absorption spectrum changes with the reducing-oxidizing (redox) state [17].

7.4 TYPES OF OPTICAL FIBERS

A large variety of optical fibers exists in terms of their structure, geometry, and materials. These parameters determine their performance characteristics (attenuation, dispersion) and their physical (e.g., refractive index), optical, and spectroscopic (e.g., range of wavelength) properties. A general classification of fibers and their refractive index profiles is given by Yeh [2]. The main types are multimode fibers, with either a step-index or graded-index profile, and single-mode fibers. Special fibers have been developed for sensing purposes. Up to now, most optodes (around 70%) for optical chemical sensing have been based on multimode fibers. Planar waveguides and single-mode fibers are used in only 25% and 5% of optodes, respectively. Multimode fibers have many advantages: good light transmission over short and medium distances and a wide range of optical components commercially available.

The most commonly used type of fiber for chemical sensors is the step-index fiber, which has a constant refractive index in the core region (index n_1) and in the cladding (index n_2) (see Figure 7.8(a)). The diameter of multimode fibers made of silica and fluoride glasses varies from 125 µm up to as large a diameter as 1.5 mm (with typical core sizes of 50, 100, 200, 400 µm, and so on). Plastic fibers are normally available with diameters of 0.5 and 1 mm.

A graded-refractive-index fiber has a parabolic refractive index profile with a maximum index at the center of the core, as shown in Figure 7.8(b). This refractive index profile reduces the modal dispersion by a factor of 100 times and increases the bandwidth of the fiber. A large number of modes are still present in the core, approximately half the number in a step-index fiber with a similar core size. In chemical sensing, this fiber can serve as a dispersive element in time domain optical spectrometry [18].

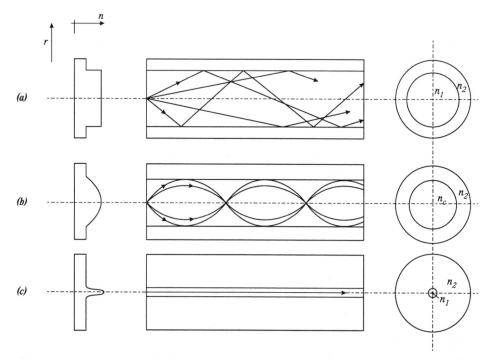

Figure 7.8 Main types of optical fiber: (a) step-index fiber, (b) graded-index fiber, and (c) monomode fiber. The figure shows the refractive index profile of the core and cladding (on left-hand side), the propagation of light rays in the fiber, and the cross-sectional structure (on right-hand side).

For a monomode fiber (Figure 7.8(c)), the numerical aperture is low, less than 0.2, and the fiber core size (typically 5 to 10 μm) is reduced to guarantee that only a single mode can propagate at the chosen wavelength, according to the criterion $V < 2.405$ (see (7.10)). Modal dispersion is eliminated with a single mode, although effects due to polarization, waveguide dispersion, and material dispersion must be taken into account.

7.5 SPECIAL FIBERS FOR SENSORS

7.5.1 Special Multimode Fibers

Special fiber geometries are often required. Noncoherent bundles of thin fibers are used with several types of extrinsic optode sensors. An extra layer (a coating that is sensitive to the analyte) and a membrane are added to the outside of the end of the fiber bundle.

The advantage of a bifurcated bundle (see Figure 2.11) is that a part of the bundle can be used for the input (exciting) light to the sensor and the other part for the light signal returning from the sensor. As a result, background fluorescence is dramatically reduced. The shape of the ends of the bundle can also be readily adapted to the shape of the source and detector.

Another type of fiber produced specifically for Raman or fluorescent spectrometry is a coaxial fiber with a double silica core [19] or multiple cores in multicomponent materials (e.g., silica and fluoride glasses, or silica and plastics), shown in Figure 7.9(a) [20]. These prototype fibers transmit the exciting and emitted light with different spectral characteristics over long distances. A special arrangement with a pierced concave mirror can separate the input and output light beams of the fiber (Figure 9.7).

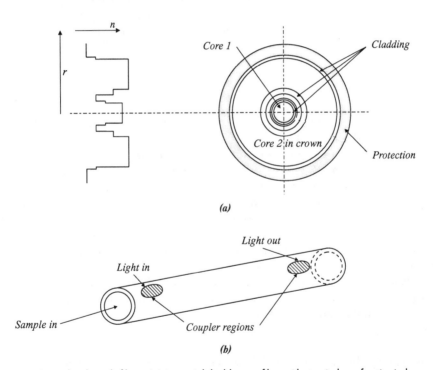

(a)

(b)

Figure 7.9 Special multimode fibers: (a) A coaxial double-core fiber with step-index refractive index profile for the central core (1) and ring (2). The figure shows the index profile (on left) and structure (on right). Several concentric cores can be made in the crown region. The material of the cores (1 and 2) can be different, and removal of the central core (e.g., in silica) leads to a hollow tube fiber (e.g., in fluoride glass) (*After:* [20]). (b) A capillary waveguide with light coupled in and out at the ends. The inner wall is coated with a sensitive polymer and changes in its color are monitored by an evanescent wave. (*After:* [21], with permission.)

Another concept is a capillary fiber with a tubular structure and a chemically sensitive coating on the inner surface. The sample cavity has a well-defined volume, as shown in Figure 7.9(b) [21].

7.5.2 Birefringent Fibers

In asymmetric fiber structures, it is possible to alter the polarization of the two orthogonal polarizations in a monomode fiber. These are then no longer degenerate. A number of special fibers have been developed with modified polarization properties for sensing applications, especially for physical measurements [22]. They have been used in chemical and biochemical sensors in particular cases, for example, in interferometric and fluorescent polarization sensors. In single-mode fibers, the fundamental (HE_{11}) mode is linearly polarized; however, fibers do not have a perfect geometry and the state of polarization varies along the length of the fiber. The degree of birefringence B is expressed as

$$B = |\beta_x - \beta_y| \cdot (2\pi/\lambda) \tag{7.20}$$

where β_x and β_y are the propagation constants of the x- and y- polarized modes, respectively, and λ is the wavelength.

Special fibers are needed when the state of polarization requires strict control, either for very low birefringence or for maintaining a permanent polarization.

Fibers can be fabricated with an almost negligible birefringence by carefully reducing the noncircularity and eliminating stresses in the core and cladding by using materials with equal thermal expansion coefficients. Alternatively, the birefringence of a fiber can be randomized along its length by *spinning* during fabrication by twisting the fiber rapidly during the drawing process to give a permanent twist to the fiber axis.

Polarization-maintaining fibers, which are useful in chemical sensing combined with a sensitive film, are characterized by their high birefringence and produce a stable state of linear polarization over short distances (<5m). Three ways are commonly used to introduce a strong linear birefringence in a fiber:

1. By making the core noncircular in shape. The core must be very small and the difference of the core and cladding refractive indexes must be large (Δn is typically 0.03 to 0.04), since the birefringence (typically 4.10^{-4}) is proportional to Δn^2.
2. By producing a linear birefringence with an elliptical cladding (Figure 7.10(a)) or an asymmetric cladding. This can be fabricated by machining the cladding before drawing the fiber. An example is the D-shaped fiber (Figure 7.10(b)).

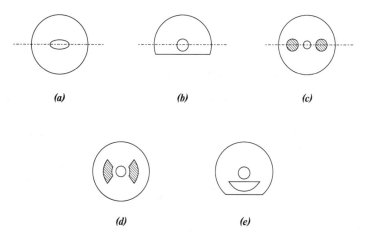

Figure 7.10 Special polarization-preserving single-mode fibers for sensors: (a) elliptical core; (b) D-shaped core; (c) PANDA type; (d) bow-tie type; and (e) hollow fiber. The fibers in a, c, d, and e are used for polarization control, and those in b and e in evanescent spectroscopy.

3. By asymmetric doping in the cladding, inducing a transverse strain across the core. Different geometrical shapes for the doping have been tried, such as a PANDA (polarization-maintaining and absorption-reducing) fiber, as in Figure 7.10(c), or a bow-tie form, as in Figure 7.10(d) [23].

For a polarization-maintaining fiber, the attenuation of the two polarization modes (along the major and minor axes of the fiber asymmetry) are equal, but the phase constants are very different. The limits of polarization holding are dictated by Rayleigh scattering and by the stability of the minor (orthogonally polarized) field component with random bends or curvature of the fiber. The resultant polarization crosstalk increases as a function of the fiber length. For elliptical birefringent fibers, in which the eigenmodes are elliptically polarized, a short beat length is a measure of the insensitivity to external perturbations.

In polarizing fibers, an added attenuation is introduced for one of the modes, resulting in linearly polarized light emerging from the fiber. Thus, the spectral variation of the attenuation (and hence the extinction coefficient) is different for x- and y- polarized modes.

A highly sensitive all-fiber technique for the detection of refractive index changes has been proposed using a differential phase modulation between the x- and y-polarized modes of a single-mode fiber [24]. A sensitive film is deposited on the side of a D-shaped fiber, or a fiber where the cladding has been removed on one side by polishing, to allow the evanescent field to penetrate the surrounding medium. This technique has been applied to a humidity optode.

7.5.3 Fibers for Evanescent Wave Spectroscopy

The evanescent field close to the core of the fiber can be used to sense the analyte surrounding the fiber. This means that the core must be exposed by reducing the thickness of the cladding, for example, with a D-shaped fiber. The thickness of the cladding left on the exposed side is adjusted during the production of the preform to control the strength of the evanescent effect for sensors, or alternatively to fabricate optical components such as all-fiber couplers and polarizers. A further development of the D-shaped fiber has led to the hollow section fiber (Figure 7.10(e)) with a single longitudinal hole close to the fiber core. The hole, which is machined in the preform, becomes an integral part of the structure of the fiber after drawing. For example, metal-glass fiber polarizers have been made from hollow section fibers [25] in which a metal is incorporated into the fiber close to the core.

The amount of evanescent power interacting with the analyte is a critical parameter and is related to the fraction (f_e) of light power in the cladding region (P_{clad}) compared to the total power (P_t) carried by the fiber:

$$f_e = P_{clad}/P_t = 1 - (P_{core}/P_t) \tag{7.21}$$

where P_{core} is the fraction of light power in the core. Substantial fractions (>50%) of power in the cladding can be achieved in single-mode fibers by choosing a low V parameter [26]. The f_e value is a maximum for modes near cutoff (low values of V). In weakly guiding $(n_1 \approx n_2)$ multimode fibers, the average value of f_e is $(4/3) \cdot (M)^{-1/2}$, since M is a value proportional to $2V^2$. In addition, large enhancements of the evanescent field can be achieved by coating the sensor with high-refractive-index coatings such as porous TiO_2 overlayers with $n = 2.3$ [27].

Evanescent wave chemical measurements have been made with discrete sensing cells or specially constructed fiber devices. As an example, a sensing cell with a polished section of a single-mode fiber has an attenuation that is a linear function of the refractive index of the surrounding solution [28]. Another example is a single-mode tapered fiber, which is particularly suitable for gas sensing. The taper serves as a mode filter and encourages light propagation in the sensitive cladding modes [29]. A tapered fiber has also been tried in biochemical analysis for calcium determination [30]. In another biosensing example, the penicillin was measured from an enzymatic reaction on a single-mode fiber [31].

In summary, single-mode D-fibers [32] and tapered fibers supporting a few modes are good candidates for evanescent wave spectroscopy. Multimode fibers with tapered, etched, or modified claddings are also valuable (see Section 8.7) [33]. A novel fiber design has also been proposed for increasing the fraction of total optical power carried by the evanescent field [34]. This fiber is drawn from a standard preform, on which two flats and a rounded vertex are produced by grinding and polishing. A very thin cladding is produced in the rounded vertex region.

7.5.4 Fibers for Interferometry

Single-mode fibers are the basic elements of interferometric sensors for chemical sensing. An early example is the interferometric sensor for hydrogen determination based on the strain induced in a hydrogen-absorbing metal film coated on a single-mode fiber [35]. Planar and integrated optic components are also interesting for interferometric chemical sensors. One problem for remotely read integrated optic sensors is the connection (or pigtailing) of fibers to the planar waveguide [36].

7.5.5 Doped Fibers and Lasing Fibers

Rare-earth doped fibers are common components in fiber lasers and amplifiers applied to optical communications. Incorporating rare-earth ions into the fiber core can also increase the Verdet constant, the Kerr effect, and nonlinear optical coefficients. These effects are employed in constructing sensors for physical measurements. These doped fibers are also of interest for chemical sensors in two ways—as fiber lasers and amplifiers for the optical instrumentation and for sensing based on absorption or fluorescence.

As a light source, fiber lasers offer an improvement over the usual laser diode or LED coupled to a fiber. Fiber lasers have very low thresholds and high gains, are compatible with single-mode and integrated optic components, and have the possibility of Q-switching and peak powers of up to 250W in short pulses (50 ns to 1 µs in duration). The most common rare-earth dopants in silica and fluoride glasses are shown in Figure 7.11 with their respective wavelength ranges [37].

In chemical sensing, doped fluoride glass fibers are valuable as sources (e.g., for gas measurements), and the laser transitions cover a wide range of wavelengths in the near-IR region. Dopants are easily introduced into fluoride glasses, and many transitions are observed in the visible region (e.g., with thulium, holmium, and praseodymium). In silica, the most common dopants are neodymium, erbium, thulium, and ytterbium.

As examples of the use of fiber lasers for chemical sensing, a silica fiber laser doped with Yb^{3+} is well suited for online measurement of neptunium ($\lambda = 980$ nm) in a nuclear environment [38]. Other dopants, such as lanthanides (U^{3+} with a laser transition of about 2.43 µm), are suitable for gas measurements in the near IR.

Plastic fibers doped with organic fluorescent materials have also been applied to sensing. A simple fiber humidity sensor consists of a polymer cladding doped with umbelliferon dye [39]. A special polycarbonate fiber with a mixed cladding of PMMA

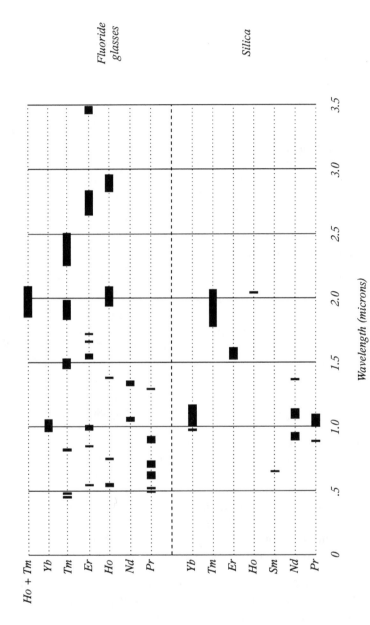

Figure 7.11 Laser transitions of different rare-earth ions in doped fluoride and silica fibers. (*After:* [37], with permission).

and polyvinylidene fluoride (PVDF) was developed as a light source for detecting gaseous NH_3 and HCl in concentrations less than 10 ppm [40].

7.6 FIBER MATERIALS

The wavelength range in chemical sensing is very large, extending from UV (about 0.2 μm) to IR (up to 20 μm). The choice of a suitable fiber material must take into account not only the optical transmission in the desired wavelength range, but also practical limitations such as chemical stability, mechanical flexibility, and even its behavior under nuclear radiation. SiO_2-based glasses, near-IR fluoride glasses, and plastics are the main materials for commercial optical fibers.

7.6.1 SiO$_2$-Based Fibers

Silica glass is the best material for single-mode fibers, mainly for medium- and long-distance applications. The most common single-mode fibers have a SiO_2 core doped with GeO_2 and/or P_2O_5, and a cladding doped with B_2O_3 and/or fluorine. These are the most common dopants that increase (e.g., with GeO_2, P_2O_5) or decrease (e.g., with B_2O_3 or fluorine) the refractive index of silica.

Apart from special fibers, two main types of multimode fibers are commonly employed in chemical sensing:

1. All-silica fibers are often preferred for extrinsic sensors. These fibers can resist high-temperatures, up to 350°C with polyimide jackets and up to 600°C with thermocoaxial jackets. They also survive well in nuclear environments with integrated doses up to 10^{10} rad, which induce losses of less than 50 dB · km^{-1} at 0.85 μm [41].
2. Plastic-clad silica (PCS) fibers have a pure silica core (sometimes doped) and a plastic cladding such as silicone rubber. This cladding can be easily removed, which is an advantage for refractometry and surface reaction measurements and for grafting chemical reactants directly onto the silica core.

The limiting fiber loss in the UV and visible spectrum regions is due to Rayleigh scattering, and in the IR region is due to phonon absorption by vibrations of the SiO_2 or GeO_2 molecules, as shown in Figure 7.12. An attenuation minimum occurs at about 1.55 μm. Other factors that affect losses are impurity metal ions and particularly the hydroxyl radical OH. Thus, the loss in dB · km^{-1} for only 1 ppm of OH in silica is approximately 6,000 to 8,000 at the fundamental absorption band at 2.73 μm, and for the overtone bands, 150 to 200 at 2.23 μm, 30 to 50 at 1.39 μm, 2 to 3 at 1.25 μm, 1 to 1.5 at 0.945 μm, and 0.05 at 0.725 μm. A small OH concentration up to 1,200 ppm improves the transmission in the UV region [42]. Nuclear radiation and defects due to fiber drawing and color centers restrict the shortest useful wavelength

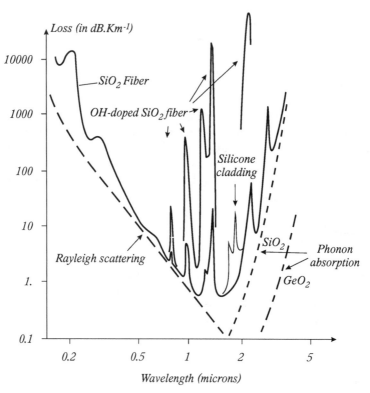

Figure 7.12   Typical losses of silica fibers in the 0.2- to 3-μm range. The lowest limiting loss in the visible region is determined by Rayleigh scattering (heavy dashed line), and in the near IR by absorption of vibrations of SiO_2 molecules or GeO_2 molecules (small dotted lines). The effect of the silicone cladding can be seen for a PCS type fiber. Fibers with a high OH content (1,000 ppm) have strong absorption bands due to OH vibrations.

for absorption and fluorescence measurements to about 350 nm for fibers up to 10m in length.

7.6.2 Infrared Fibers

At wavelengths longer than 1.7 μm, other glass materials must be used for Fourier transform infrared (FTIR) and near-IR applications, with an extended spectral range shown in Figure 7.13.

Fluoride glasses transmit between 1.7 and 4.5 μm, and up to 5 μm for some compounds [43,44]. Typical losses are 100 dB/km, but a theoretical minimum attenuation has been calculated of around 10^{-3} dB/km at 3.2 μm. In binary glass systems, the main compositions are MF_3-AlF_3 and MF_3-ZrF_4 where M is calcium,

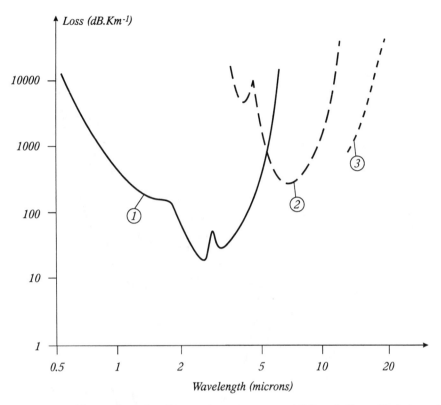

Figure 7.13 Typical losses of nonsilica fibers in the near-IR region: (1) fluoride fibers; (2) chalcogenide glasses; and (3) halogenide AgBr/Tl fibers.

strontium, barium, or lead. The ternary glass systems produce better materials of the RF_3-BaF_2-ZrF_4 type, where R is lanthanum, germanium, or gadolinium. Examples of more complex systems are ZBLA (with zirconium, barium, lanthanum, aluminum fluorides) and ZBLAN (ZBLA with NaF). Because of their low transition temperature (400 to 500°C), the increased attenuation of fluoride glasses is easy to anneal out after exposure to nuclear radiation [45]. The numerical aperture of fluoride fibers can usually be adapted to any application because of the wide range of possible core and cladding compositions. Fluoride glasses can also be doped with rare-earth metals to produce fluorescent or laser-amplifying fibers.

Chalcogenide glass fibers transmit between 5 and 11 μm (Figure 7.13). They contain at least two elements of the family arsenic, germanium, phosphorus, sulfur, selenium, and tellurium. The cladding material is a fluoride polymer, which results in high attenuation. Some interesting applications of chalcogenide fibers are evanescent spectroscopy [46] and chemical surface measurements [47]. Telluride fibers have been produced with $0.11 \text{ dB} \cdot \text{m}^{-1}$ at 6.6 μm [48].

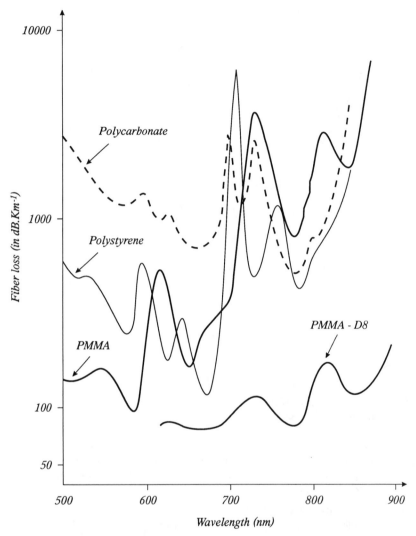

Figure 7.14   Typical losses of plastic optical fibers in the 500- to 900-nm range for different core materials: (a) polycarbonate, (b) polystyrene, (c) PMMA, and (d) deuterated-PMMA.

Halide glasses, such as AgBr, AgCl, KCl, CsBr, $ZnCl_2$, and iodide glasses have an extended wavelength range out to 15 to 20 µm. However, halide fibers have a high attenuation, are very sensitive to humidity, and generally have a high cost. Thallium bromoiodide (KSR-5) fibers are the best known in this category. Halide IR fibers are starting to be used for gas [49] and liquid [50] detection.

7.6.3 Plastic Fibers

Step-index multimode plastic fibers are an ideal material for many medical applications in which the effects of fiber absorption and dispersion are unimportant for short lengths. Their main advantages are a large-diameter core (450 to 900 µm for fiber diameters of 500 to 1,000 µm), a very thin cladding thickness (10 to 20 µm), which is easy to remove for sensing surface chemical reactions, a high acceptance angle ($NA = 0.5 - 0.6$ and $\theta_0 = 70°$) allowing efficient coupling with light sources, low weight, very low cost, and relatively good biocompatibility. In addition, their refractive index can be chosen over a wide range, typically 1.35 to 1.6.

The main drawbacks of plastic fibers are their restricted spectral range (Figure 7.14) with high attenuations in the red and near-IR spectrum due to strong CH absorption bands (several hundred $dB \cdot km^{-1}$). They have a limited temperature operating range, particularly for elevated temperatures. The operating range is typically -30 to +80°C. The two most common plastics for the fiber core are PMMA and polystyrene. Derivatives, such as fluoropolymers, are used for the lower refractive index cladding [51]. Special fiber materials include polycarbonate core fibers (refractive index 1.5 to 1.58) for high-temperature use up to 135°C. Deuterium-doped PMMA or polystyrene, in which the deuterium replaces the CH band with a CD band, shifts the spectral absorption to longer wavelengths and increases the spectral transparent range. However, the optical transmission of these fibers is less stable over time.

References

[1] Daly, J. C., ed., *Fiber Optics*, Boca Raton: CRC Press, 1984.

[2] Yeh, C., *Handbook of Fiber Optics. Theory and Applications*, New York: Academic Press, 1990.

[3] Parriaux, O., "Guided Wave Electromagnetism and Opto-Chemical Sensors," Chap. 4 in *Fiber Optic Chemical Sensors and Biosensors, Vol. 1*, O. S. Wolfbeis, ed., Boca Raton: CRC Press, 1991.

[4] Adams, M. J., *An Introduction to Optical Waveguides*, New York: Wiley, 1981.

[5] Jeunhomme, L. B., *Single Mode Fiber Optics*, New York: Marcel Dekker, 1990.

[6] Burgess, L. W., "Overview of Planar Waveguide Techniques," *Proc SPIE–Int. Soc. Opt. Eng.*, Vol. 1368, 1990, pp. 224–229.

[7] Bennion, I., and M. J. Goodwin, "Third-Order Nonlinear Guided-Wave Optical Devices," Chap. 8 in *Nonlinear Optics in Signal Processing*, R. W. Eason and A. Miller, eds., London: Chapman & Hall, 1993, pp. 286–321.

[8] Bowman, E. M., and L. W. Burgess, "Evaluation of Polymeric Thin Film Waveguides as Chemical Sensors," *Proc. SPIE–Int. Soc. Opt. Eng.*, Vol. 1368, 1990, pp. 239–250.

[9] Lukosz, W., and K. Tiefenthaler, "Sensitivity of Integrated Optical Grating and Prism Couplers as (Bio)Chemical Sensors," *Sensors and Actuators*, Vol. 15, 1988, pp. 273–284.

[10] Nellen, P. M., and W. Lukosz, "Integrated Optical Input Grating Couplers as Chemo- and Immunosensors," *Sensors and Actuators*, Vol. B 1, 1990, pp. 592–596.

[11] Lukosz, W., P. M. Nellen, C. Stamm, and P. Weiss, "Output Grating Couplers on Planar Waveguides as Integrated Optical Chemical Sensors," *Sensors and Actuators*, Vol. B 1, 1990, pp. 585–588.

[12] Clerc, D., and W. Lukosz, "Integrated Optical Output Grating Coupler as Biochemical Sensor," *Sensors and Actuators*, Vol. B 18/19, 1994, pp. 581–586.

[13] Tiefenthaler, K., and W. Lukosz, "Sensitivity of Grating Couplers as Integrated-Optical Chemical Sensors," *J. Opt. Soc. Am.*, Vol. B6, 1989, 2, pp. 209–220.

[14] Kunz, R. E., J. Edlinger, B. J. Curtis, M. T. Gale, L. U. Kempen, H. Rudigier, and H. Schütz, "Grating Couplers in Tapered Waveguides for Integrated Optical Sensing," *Proc. SPIE–Int. Soc. Opt. Eng.*, Vol. 2068, 1994, pp. 313–325.

[15] Kunz, R. E., J. Edlinger, P. Sixt, and M. T. Gale, "Replicated Chirped Waveguide Gratings for Optical Sensing Applications," *Sensors and Actuators*, Vol. A 46/47, 1995, pp. 482–486.

[16] Kretschmann, E., and H. Raether, "Radiative Decay of Non Radiative Surface Plasmons Excited by Light," *Naturforschung*, Vol. A 23, 1968, pp. 2135–2136.

[17] Piraud, C., E. K. Mwarania, J. Yao, K. O'Dwyer, D. J. Schiffrin, and J. S. Wilkinson, "Optochemical Transduction on Planar Optical Waveguides," *J. of Lightwave Technol.*, Vol. 10, 1992, pp. 693–699.

[18] Whitten, W. B., "Time Domain Optical Spectrometry With Fiber Optic Waveguides," *Appl. Spectrosc. Rev.*, Vol. 19, 1983, pp. 325–362.

[19] Fevrier, H., P. Saisse, P. Plaza, and N. Q. Dao, "Fibre optique et application de cette fibre à un dispositif optique pour effectuer à distance l'analyse chimique d'un corps," French Pat. Appl., 1985, EN 85.09725.

[20] Boisdé, G., G. Cardin, G. Mazé, and M. Poulain, "Conduit optique coaxial et procédé de fabrication de ce conduit," French Pat., 1991, B.F. 2,683,638.

[21] Weigl, B. H., and O. S. Wolfbeis, "Capillary Optical Sensors," *Anal. Chem.*, Vol. 66, 1994, pp. 3323–3327.

[22] Gambling, W. A., and S. B. Poole, "Optical Fibers for Sensors," Chap. 8 in *Optical Fiber Sensors*, J. C. Dakin, and B. Culshʳw, eds., Norwood, MA: Artech House, 1988, pp. 249–276.

[23] Birch, R. D., D. N. Payne, and M. P. Varnham, "Fabrication of Polarisation-Maintaining Fibres Using Gas-Phase Etching," *Electron. Lett.*, Vol. 19, 1982, pp. 866–867.

[24] Ecke, W., W. Haubenreisser, H. Lehmann, S. Schroeter, G. Schwotzer, and R. Willsch, "Phase Sensitive Fibre Optic Monoptodes for Chemical Sensing," *Sensors and Actuators*, Vol. B 11, 1993, pp. 475–479.

[25] Li, L., G. Wylangowski, D. N. Payne, and R. D. Birch, "Broadband Metal/Glass Single-Mode Fibre Polarizers," *Electron. Lett.*, Vol. 22, 1986, pp. 1020–1022.

[26] Gloge, D., "Weakly Guiding Fibers," *Appl. Opt.*, Vol. 10, 1971, pp. 2252–2258.

[27] MacCraith, B. D., "Enhanced Evanescent Wave Sensors Based on Sol-Gel-Derived Porous Glass Coatings," *Sensors and Actuators*, Vol. B 11, 1993, pp. 29–34.

[28] Lew, A., C. Depeursinge, F. Cochet, H. Berthou, and O. Parriaux, "Single-Mode Evanescent Wave Spectroscopy," *Proc. SPIE–Int. Soc. Opt. Eng.*, Vol. 514, 1986, pp. 71–74.

[29] Villaruel, C. A., D. D. Dominguez, and A. Dandridge, "Evanescent Wave Fiber Optic Chemical Sensor," *Proc. SPIE–Int. Soc. Opt. Eng.*, Vol. 798, 1987, pp. 225–229.

[30] Hale, Z. M., and F. P. Payne, "Demonstration of an Optimized Evanescent Field Optical Fiber Sensor," *Anal. Chim. Acta*, Vol. 293, 1994, pp. 49–54.

[31] Carlyon, E. F., C. R. Lowe, D. Reid, and I. Bennion, "A Single Mode Fiber Optic Evanescent Wave Biosensor," *Biosens. Bioelectron.*, Vol. 7, 1992, pp. 141–146.

[32] Stewart, G., J. Norris, D. F. Clark, and B. Culshaw, "Evanescent Wave Chemical Sensors: A Theoretical Evaluation," *Int. J.*, Optoelectron., Vol. 6, 1991, pp. 227–298.

[33] MacCraith, B. D., G. O'Keeffe, C. McDonagh, and A. K. McEvoy, "LED-Based Fibre Optic Oxygen Sensor Using Sol-Gel Coatings," *Electron. Lett.*, Vol. 30, 1994, pp. 888–889.

[34] Matejec, V., M. Chomat, M. Pospisilova, M. Hayer, and I. Kasik, "Optical Fiber With Novel Geometry for Evanescent Wave Sensing," *Sensors and Actuators*, Vol. B 29, 1995, pp. 416–422.

[35] Butler, M. A., "Optical Fiber Hydrogen Sensor," *Appl. Phys. Lett.*, Vol. 45, 1984, pp. 1007–1009.

[36] Gauglitz, G., and J. Ingenhoff, "Integrated Optical Sensors for Halogenated and Non-halogenated Hydrocarbons," *Sensors and Actuators*, Vol. B 11, 1993, pp. 207–212.

[37] Monerie, M., "Status of Fluoride Fiber Lasers," *Proc. SPIE–Int. Soc. Opt. Eng.*, Vol. 1581, 1991, pp. 2–13.

[38] Magne, S., "Etude d'un laser à fibre dopée Ytterbium. Spectroscopie laser de fibres dopées," Thesis, University St-Etienne (France), 1993.

[39] Muto, S., A. Fukasawa, M. Kamimura, F. Shinmura, and H. Ito, "Fiber Humidity Sensor Using Fluorescent Dye-Doped Plastics," *Jpn. J. Appl. Phys.*, Part 2, Vol. 28, 1989, pp. 1065–1066; *Chem. Abstr.*, Vol. 111, 105343c.

[40] Sawada, H., A. Tanaka, and N. Wakatsuki, "Plastic Optical Fiber Doped With Organic Fluorescent Materials," *Fujitsu Sci. Tech. J.*, Vol. 25, 1989, pp. 163–169; *Chem. Abstr.*, Vol. 111, 204992v.

[41] Boisdé, G., F. Blanc, P. Mauchien, and J. J. Perez, "Fiber Optic Chemical Sensors in Nuclear Plants," Chap. 14 in *Fiber Optic Chemical Sensors and Biosensors, Vol. 2*, O. S. Wolfbeis, ed., Boca Raton: CRC Press, 1991, pp. 135–149.

[42] Wolfbeis, O. S., G. Boisdé, and G. Gauglitz, "Optochemical Sensors," Chap. 12 in *Sensors: A Comprehensive Survey, Vol. 2*, W. Göpel, J. Hesse, and J. N. Zemel, eds., Weinheim: VCH, 1991, pp. 573–645.

[43] Poulain, M., M. Poulain, and J. Lucas, *Material Research Bulletin, Vol. 10*, 1975, p. 243.

[44] Aggarwal, I. D., and G. Lu, *Fluoride Glasses and Fiber Optics*, San Diego: Academic Press, 1991.

[45] Abgrall, A., M. Poulain, G. Boisdé, V. Cardin, and G. Mazé, "Infrared Study of Gamma Irradiated Fluoride Optical Fibers," *Proc. SPIE–Int. Soc. Opt. Eng.*, Vol. 618, 1986, pp. 63–69.

[46] Heo, J., M. Rodrigues, S. J. Saggesse, and G. H. Sigel, "Remote Fiber Optic Chemical Sensing Using Evanescent Wave Interactions in Chalcogenide Glass Fibers," *Appl. Opt.*, Vol. 30, 1991, pp. 3944–3951.

[47] Kellner, R., and K. Taga, "New Developments in the Field of Chemical Infrared Fibre Sensors," *Proc. SPIE–Int. Soc. Opt. Eng.*, Vol. 1510, 1991, pp. 232–241.

[48] Sanghera, J. S., and I. D. Aggarwal, "Development of Low Loss IR Optical Fibers for Optical Sensors," *Proc. SPIE–Int. Soc. Opt. Eng.*, Vol. 2367, 1995, pp. 99–108.

[49] Messica, A., A. Greenstein, A. Katzir, U. Schiessl, and M. Tacke, "Fiber Optic Evanescent Wave Sensor for Gas Detection," *Opt. Lett.*, Vol. 19, 1994, pp. 1167–1169.

[50] Bunimovich, D., E. Belotserkovsky, and A. Katzir, "Fiber Optic Evanescent Wave Infrared Spectroscopy of Gases and Liquids," *Rev. Sci. Instrum.*, Vol. 66, 1995, pp. 2818–2820.

[51] Marcou, J., *Les Fibres Plastiques*, Paris: Masson, 1994.

Chapter 8

Optical Measurement Techniques

8.1 INTRODUCTION

This chapter briefly describes the principles of the measurement techniques and associated instrumentation for optical sensing. A large number of books and articles cover general optical techniques and instrumentation [1–3] and more specialized works on spectroscopy [4,5] and chemical fiber-optic sensors [6].

The general elements of a chemical sensing system are shown schematically in Figure 8.1 for a fiber-optic configuration. However, optical fibers are not indispensable in extrinsic optochemical sensing (see dotted lines in Figure 1.4). Light from a bulk optics source (e.g., halogen lamp, gas laser; see Chapter 9) or preferably a solid-state source [7], such as laser diode or LED, is injected into a fiber with suitable transforming optics that collimate and focus the light beam. The input optics may provide many other functions, such as wavelength selection (by filters or gratings), modulation of the light (by an electro-optic modulator), and polarization and phase control for interferometers. Common components for these functions are reviewed in Chapter 9. The light is transmitted by a fiber to the optode (the sensing head; see Chapter 10). Additional coupling elements may be added that divide the light into several channels for multiplexing a number of optodes and to provide a reference channel. Techniques of wavelength and time multiplexing are described in Section 9.3, and their application to networks is demonstrated in Section 12.4.

Light from the return fiber is transformed by the receiving optics, which may contain functions similar to those of the injection optics. The photodetector, usually a single-element *pin* photodiode (PD) or a charge-coupled device (CCD) array (Section 9.2), can be synchronized to a pulsed light source, and phase sensitive detection is often used with a lock-in amplifier. This reduces noise and the effect of ambient illumination.

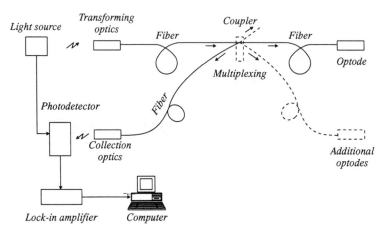

Figure 8.1   Schematic diagram of a general optical instrumentation fiber system for chemical sensing, showing the main optical elements. Multiplexing allows multipoint measurement with several optodes.

8.2 OPTICAL MODULATION SCHEMES

Optical instrumentation systems for chemical sensing can be conveniently classified into five main categories similar to those of physical sensors [8], according to the way the light is modulated: intensity, wavelength, polarization, phase, and time modulation (see Table 8.1). The modulation scheme also dictates the type of detection technique and frequently the requirements of the source as well.

8.2.1 Intensity (Amplitude) Modulation

Intensity or amplitude modulation is the simplest method. Typical examples are absorption, fluorescence (Figure 8.2(a)), or reflectance measurements with a probe at a fixed wavelength. The signal is directly related to the number of transmitted photons, and the sensitivity is determined by the noise characteristics of the photodetector (see Section 9.2). A typical performance of an optical instrumentation system allows long-term stable measurement of changes in relative intensity ($\Delta I/I$) of 10^{-2} to 10^{-3}. Much higher sensitivities are possible, but slow drifts (due to aging of the source, temperature effects, and other factors) generally limit the performance. In order to compensate for intensity variations in the output of the source, for unknown attenuation losses in the fiber transmission line and connectors, stray fluorescence, and photobleaching, it is desirable to introduce a reference channel for calibration. This may be done by adding a separate reference fiber and a second photodetector, or alternatively by using wavelength modulation.

Table 8.1
Optical Modulation Schemes and Detection Techniques

Type of Information	Physical Mechanism	Detection Circuitry	Main Limitations
Intensity (I)	Modulation of number of photons transmitted after absorption or emitted by luminescence, scattering, or refractive index changes	Analog detection	Normalization for source intensity variations and variable line and connector losses
Wavelength (λ)	Spectral-dependent variations of absorption and fluorescence	Amplitude comparison at two (or several) fixed wavelengths, or continuous wavelength scan	Wavelength-dependent line losses; suitable scanned-wavelength sources or spectrometers
Phase (ϕ)	Interference between separate paths in Mach-Zehnder, Michelson, Fabry-Perot interferometers	Fringe counting, or fractional phase-shift detection	Stability and measurement of small phase shifts; coherent sources and detection methods
Polarization (p)	Changes in the rotation of polarization	Polarization analyzer and amplitude comparison	Random polarization changes in line (induced birefringence in fibers and components)
Time resolved (t time domain, ω frequency domain)	Transient behavior (lifetime) of luminescence or absorption	Analysis of time-decay amplitude signal	Modal time dispersion in fibers; low intensity signal

8.2.2 Wavelength Modulation

Wavelength modulation provides additional spectral information over simple intensity measurements. A continuous wavelength scan is often done with a grating spectrometer and CCD array in a compact fiber-optic spectrometer (Section 8.4.2). Alternatively, a number of fixed wavelengths can be selected by interference filters or by sources (LEDs or laser diodes) that are modulated at different electrical frequencies to distinguish their signals at a single photodetector (Figure 8.2(b)). Wavelength-tunable solid-state sources are also becoming available [9]. Simple referencing is done at a wavelength unchanged by the chemical signal. Errors occur from wavelength-dependent line losses and crosstalk between the different wavelength bands.

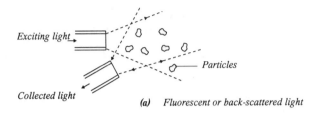

Exciting light

Particles

Collected light

(a)   Fluorescent or back-scattered light

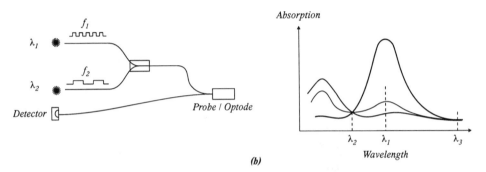

(b)

Figure 8.2 Typical examples of light modulation with a two-fiber probe: (a) intensity modulation for measurement of fluorescent or scattering compounds; and (b) dual-wavelength arrangement (for indicator-based sensors). λ_1 measures the signal absorption. The reference is obtained from λ_2 at the isobestic point or λ_3 for the background absorption.

The most common reference methods in spectrometry are photon detection at two wavelengths in absorption, scattering, or luminescence with intensity-ratio normalization, and the two-wavelength excitation technique in fluorescence (see Figure 4.5).

8.2.3 Phase Modulation

Phase modulation is used less often in chemical sensing, but is of increasing interest, particularly with integrated-optic systems; it is employed in other types of fiber-optic sensors for physical measurements with high performance and sensitivity. Two coherent light beams in different fibers or different paths interfere with one another, and one beam is modulated by the chemical environment. Phase techniques offer high accuracy and sensitivity, and phase shifts of 10^{-6} to 10^{-7} radians can be measured over a very wide dynamic range. Errors arise from random variations of phase in the fiber (for instance, by bending effects) and optical components and noise errors in the source and detector. Sophisticated detection methods are employed. Three main types of interferometer are found in chemical fiber-optic sensors (see Section 8.10).

8.2.4 Polarization Modulation

Polarization modulation is used in special sensors and also occurs with optically active chemical compounds (e.g., dextro and laevo forms). In a typical system, a polarizer produces a controlled state of polarization (linear, circular, elliptical) before the light enters the optode. A change in polarization occurs depending on the chemical interaction and is measured by an analyzer, which can take many forms; for example, in Figure 8.3, two orthogonal analyzers with separate detectors measure the rotation of the polarization produced by the optode. Errors arise from random polarization changes due to intrinsic polarization variations in the fiber or stress-induced birefringence. Instead of using conventional fibers, high birefringence monomode fibers can be used to maintain a constant state of polarization (see Chapter 7). However, such fibers are expensive and are generally limited to distances of 5m.

8.2.5 Time Modulation

Time modulation is essentially a subclass of intensity or wavelength modulation, but since it is important in fluorescence measurements, it is included here as a separate class. In time domain (TD) fluorometry, a pulsed light source (laser, flash lamp, laser diode) generates a short pulse (typical pulse duration of 1 ns to 1 μs) that excites fluorescence in the optode (Figure 8.4(a)). The fluorescent decay signal is measured as a function of time, and the decay curve determines the lifetime of the chemical species. The method is autoreferencing, since no absolute intensities need to be measured, and it is therefore insensitive to absorption and scattering losses. However, the low-intensity signal requires sophisticated averaging to improve the signal-to-noise ratio. Often, high-intensity UV excitation is used with special large-core fibers, and additional errors may arise from modal dispersion effects at long distances. In frequency domain (FD) fluorometry (Figure 8.4(b)), the phase shift ($\Delta\varphi$) is measured when the sample is excited with amplitude modulated light (modulation on the order

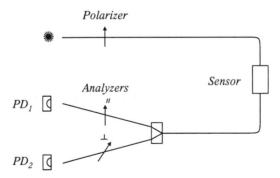

Figure 8.3 Typical example of light modulation with an optode that causes the polarization to rotate.

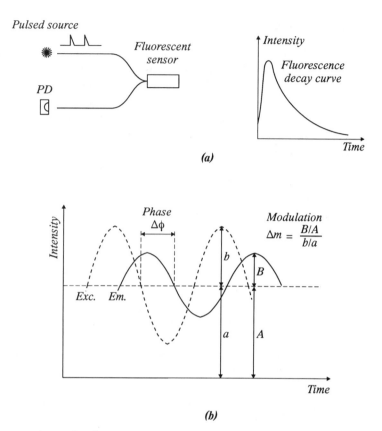

Figure 8.4   Typical examples of time modulation: (a) time-resolved system to measure the fluorescence decay time and (b) FD fluorometry principles (*After:* [10], with permission) showing the phase angle $\Delta\phi$ and the modulation Δm parameters. Both these parameters are a function $f(\omega,\tau)$ of the modulation frequency ω and the lifetime τ.

of 50 ps) [10]. Continuous wave laser diodes are well suited for these lifetime measurements.

8.3 CHEMICAL MEASUREMENT TECHNIQUES AND SIGNAL CHARACTERISTICS

Table 8.2 presents a summary of the main chemical measurement techniques, the types of light modulation, and the main modulation parameter. The most common techniques are absorption, scattering, and luminescence. All the techniques modulate the intensity of the light received by the detector, but in some cases the total intensity collected (by several detectors) remains constant (interferometry, polarimetry).

Table 8.2
Principal Chemical Measurement Techniques and Types of Light Modulation

Chemical Measurement Technique	Light Modulation	Parameter Modulation
Absorption, scattering, reflectance	I	
Colorimetry, spectrometry (absorption, Raman, SERS), fluorimetry, luminescence	I, λ	
Refractometry	I	n
Interferometry	I, φ	n, L
Spectral interferometry	I, φ, λ	n, L
Polarimetry	I, **p**	
Ellipsometry	I, **p**	L
Photokinetics	**I**, t	
Reflectometry	**I**, t	L
Combined refractometry and reflectometry	**I**, t	n
Evanescent spectroscopy, SPR	**I**, λ	n
Lifetime-based fluorimetry	I, λ, **t**	
Flight-time fluorimetry	I, λ, **t**	n
Polarized lifetime-based fluorescence	**I, λ, p, t**	
Photoacoustic	**I**, ν	P

Notes: Abbreviations are as follows: SERS = surface-enhanced Raman spectroscopy. Light modulation: I = intensity, λ = wavelength, φ = phase, p = polarization, ν = optical frequency, and t = time. (The main type of modulation is indicated in bold type.) Physical parameter modulation: L = path length, n = refractive index, P = pressure.

Refractive index (n), path length (L), and pressure (P) are the most common physical parameters that create modulation of the light. Recently the swelling property of polymers (size change) has also been used as a modulation parameter for measuring water [11] and pH in organic solvents [12] or determining the fluorescence of gels [13] or the spectral reflectance of thin silver islands [14,15].

The signal characteristics for these different measurement techniques are summarized as a function of the main experimental variable (e.g., wavelength or time) in Figure 8.5 and of the analyte or quencher concentration in Figure 8.6. Techniques such as absorption, intensity, and fluorescence measurements have well-defined spectral bands (Figure 8.5(a,b)). The relationship between the refractive index (n) and the wavelength λ (Figure 8.5(e)) is usable in time-of-flight techniques, reflectometry, and spectral interferometry. For fluorescence, the definitions of intensity decay and lifetime have been given in Section 6.2.6.1.

For direct absorption (Figure 8.6(a)), fluorescence (Figure 8.6(b,c)) and refractive index measurements (Figure 8.6(d)) without an indicator, a linear relationship is observed for moderate analyte concentrations, with nonlinear behavior

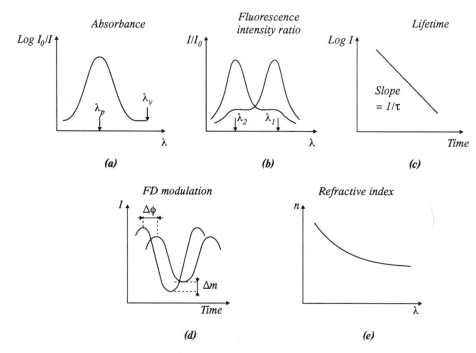

Figure 8.5 Signal characteristics for various measurement techniques as a function of the wavelength (λ) or time: (a) absorption at peak (λ_p) and valley (λ_v) wavelengths; (b) ratio of two fluorescent intensities at λ_1 and λ_2; (c) fluorescent decay time (lifetime) measurement; (d) FD modulation for fluorescence (φ is the phase and m the amplitude modulation); and (e) refractive index.

occurring for high analyte concentrations due to complexation effects or the inner filter effect (IFE) in fluorescence (Figure 8.6(b)). I_0/I and τ_0/τ vary linearly with the concentration of a quencher (Figure 8.6(e)) according to the Stern-Volmer equation (see Section 8.6). In the absence of any IFE, quenching and complexation effects, the lifetime (τ) (Figure 8.6(f)) is independent of the analyte concentration [16]. For reflectance measurements (Figure 8.6(g)), the signal (R) decreases with analyte concentration, in the form of a sigmoid curve with the logarithm of analyte concentration, and the function f(R) of (8.8) increases linearly with analyte concentration.

For indicator-based systems, the mass action law normally applies. A plot, against the logarithm of analyte concentration, for absorbance, fluorescence, fluorescence ratio intensity, resonant energy transfer, and reflectance signals (but not the decay time) show a sigmoid curve form in all cases in which the dye interacts with the analyte in the ground state, independently of the spectroscopic technique. For instance, in the FD technique, the phase angle and modulation can be expressed as a function of the logarithm of analyte concentration (Figure 8.6(h)) as discussed by

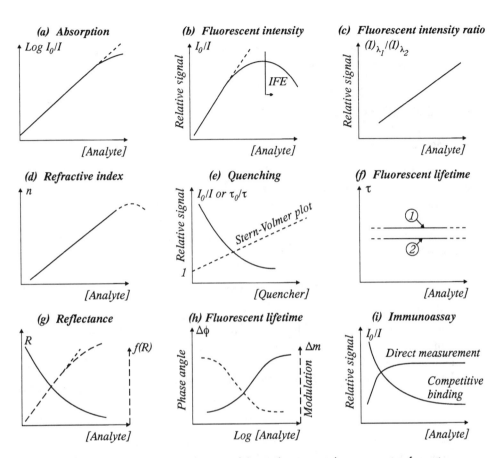

Figure 8.6 Signal characteristics as a function of the analyte or sample concentration for various measurement techniques without indicator (a to g): (a) absorption or intensity; (b) intensity of fluorescence, including IFEs, over a large concentration range of the analyte; (c) ratio of the fluorescent intensities at two wavelengths; (d) refractive index; (e) fluorescence quenching; (f) fluorescent lifetime in the absence of quenching and IFEs (Compounds 1 and 2 can be distinguished by their different lifetimes.); (g) reflectance (R and f(R) shown dotted; see (8.6) and (8.8)); (h) fluorescent lifetime for an indicator-based system in the frequency domain with phase modulation (ϕ is the phase and m the amplitude.); and (i) fluorescent intensity in immunoassays with labeling.

Szmacinski and Lakowicz [10]. This is valid for systems with two decay times. Figure 8.6(i) shows the relative signals for immunoassay with a fluorescent label in direct and competitive systems.

8.4 ABSORBANCE MEASUREMENTS

8.4.1 Measurements

The physical basis for absorption (and fluorescence) has already been explained on the basis of electronic transitions on a Jablonski energy level diagram (Figure 6.14). The change in light intensity due to absorption is determined by the number of absorbing species in the optical path and is related to the concentration C of the absorbing species via the Beer-Lambert relationship expressed logarithmically (as in (6.53)):

$$A = \log T = \log I_0/I = \varepsilon \cdot d \cdot C \tag{8.1}$$

where A is the absorbance, T is the transmittance defined as I_0/I, I_0 is the incident light and I the transmitted light, ε is the molar absorptivity, and d is the path length (the length of the measurement cell) traversed by the light. ε has typical values of 1 to 200 for metal ions and 10^4 to 10^5 $M^{-1} \cdot cm^{-1}$ for dyes. The absorbance is the sum of all the absorptions of different components in the analyte:

$$A = d \cdot (\varepsilon_1 \cdot C_1 + \varepsilon_2 \cdot C_2 + \ldots \varepsilon_n \cdot C_n) \tag{8.2}$$

However, the law may not hold if chemical association or dissociation takes place, particularly for high concentrations, and accurate measurements are better made in dilute solutions. An example of the nonlinear behavior at high concentrations due to the formation of complexes is shown in Section 12.2.1.

The main limitations of absorption techniques are the small dynamic range of the measuring instrument (normally 30 dB) and interference from background absorptions. The latter problem can be eliminated by measurement of two or more wavelengths, where one of the wavelengths acts as an internal reference. When employing absorption techniques with optical-fiber sensors, the medium used to support the selective chemistry, and indeed the selective chemistry itself, should be optically transparent.

8.4.2 Fiber-Optic Absorption Spectrometers

Spectral measurements are normally made with prisms, gratings, or wavelength-selective filters. Special colorimeters or spectrometers have been developed for optical-fiber sensors, with appropriate optics to match the fiber numerical aperture (normally 0.4 or greater) and the small size of the fiber core (0.1 to 1 mm). Commercial grating spectrophotometers and CCD arrays allow measurements over a wide spectral range (typically 250 to 1,000 nm) and in real time [17]. Suitable light sources are quartz halogen lamps and xenon flash lamps; The latter has a high output in the UV.

Spectrophotometers can also be integrated with the processing electronics and, where required, allow multiplexing of several optodes using a single centralized instrument by means of star couplers or mechanical switches to sequentially pole an array of fibers connected to different probes. For the multisensing approach, multifiber arrays have been demonstrated with 144 fibers (12 fibers in 12 rows) coupled directly to a CCD array [18]. The crosstalk between neighboring fibers is less than 3% of the peak signal. Commercial miniaturized spectrometers are available for direct mounting on a slot-in PC circuit card [19].

Portable photometers generally use LEDs as light sources and solid-state detectors and have a modular construction [7]. Simple instruments monitoring several wavelengths, as in Figure 8.7, can cover multiple applications (e.g., for both a photometer and pH meter [20]; see also Part IV).

8.4.3 Fourier Transform Infrared and IR Spectroscopy

FTIR spectroscopy is a powerful technique for determining IR spectra by measuring the whole spectrum and deconvoluting using Fourier transform (FT) techniques [21]. FTIR spectroscopy can identify many chemical bonds, such as C-H, C=O, N-H, and O-H, which have strong characteristic absorptions occurring in the IR (2.5 to 20 μm) and weaker overtones in the near IR (NIR) (from 0.8 to 2.5 μm). FTIR instruments coupled to fiber optics are now widely available [22–27]; silica glass fibers transmit light in the NIR, and special fibers (silver halides, fluoride glasses, sapphire) in the IR.

Remote Raman measurements for liquid and solid samples can be made with a modified FTIR spectrometer [28]. Attenuated total internal reflectance (ATR) measurements in the IR have been made with a special cell using a silver bromide/silver chloride polycrystalline fiber, 0.9 mm in diameter and 100 mm long, immersed in the sample solution [29]. A glucose sensor has been demonstrated by the formation of gluconic acid from immobilized glucose oxidase on an ATR crystal [30].

A number of important applications of fiber probes with IR spectroscopy have emerged: gas sensing particularly explosive hydrocarbons such as CH_4 with overtone spectroscopy in the NIR (see Chapter 12), octane number determination of gasoline with multivariate calibration [31], measurement of water in low concentrations (typically 1,000 to 100 ppm or lower) present in solvents through its absorption lines at $5,181 \text{ cm}^{-1}$ and $6,802 \text{ cm}^{-1}$, determination of oil in water by measuring the C-H absorption at $2,950 \text{ cm}^{-1}$.

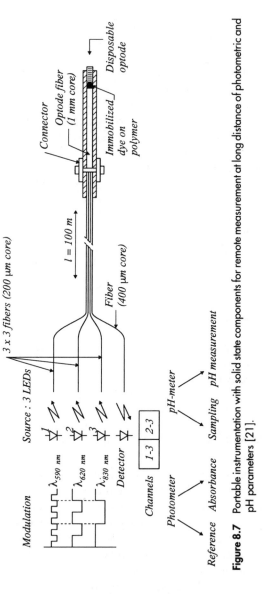

Figure 8.7   Portable instrumentation with solid state components for remote measurement at long distance of photometric and pH parameters [21].

8.5 SCATTERING

8.5.1 Types of Scattering

Unlike the processes of absorption and luminescence, scattering of light need not involve a transition between quantized energy levels in atoms or molecules. Instead, a randomization in the direction of the light radiation occurs. In Mie scattering, the intensity of the scattered radiation can be related to the concentration of the scattering particles. The size of the particle, with a radius R, in relation to the wavelength λ of the light determines the intensity I_s of the scattered signal (see Figure 8.8):

$$I_s = K \cdot R^6/\lambda^4 \quad \text{when } R < \lambda \tag{8.3}$$

$$I_s = K \cdot R^4/\lambda^2 \quad \text{when } R \approx \lambda \tag{8.4}$$

$$I_s = K \cdot R^2 \quad \text{when } R > \lambda \tag{8.5}$$

In both Rayleigh and Mie scattering, the polarization of the particle remains constant in contrast to Raman scattering, which can be observed if intense light sources such as lasers are used (see Section 6.2.1).

8.5.2 Reflectance Measurements

When the medium or the selective chemistry used in an optical sensor is opaque or when it transmits light only weakly, then measurement of the intensity of the reflected light may be used. Reflection takes place when light infringes on a boundary surface, and two distinct types of reflection are possible. The first is a *mirror* type or specular reflection, which occurs at the interface of a medium with no transmission through it (see Figure 6.6). The second type is diffuse reflection, where the light penetrates the medium and subsequently reappears at the surface after partial absorption and multiple scattering within the medium. Specular reflection can be minimized or eliminated through proper sample preparation or optical engineering.

The optical characteristics of diffuse reflectance have been recognized to be dependent on the composition of the system. Several models for diffuse reflectance have been proposed based on the radiative transfer theory, and all these models consider that the incident light is scattered by particles within the medium. The most widely used is the Kubelka-Munk theory, in which it is assumed that the scattering layer is infinitely thick. The reflectance R is related to the absorption coefficient K and the scattering coefficient S by the equation

$$f(R) = (1 - R)^2/2R = K/S \tag{8.6}$$

where $f(R)$ is the Kubelka-Munk function.

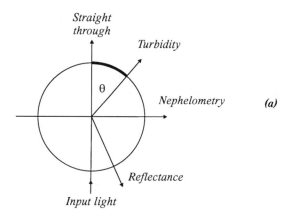

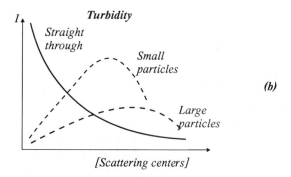

Figure 8.8 Scattering techniques: (a) turbidity is measured at varying angles of θ, nephelometry at 90°, and reflectance is obtained when the medium or interface is opaque and (b) the turbidity signal shows the relative intensity of scattered light as a function of the number of scattering centers for small and large particles.

The absorption coefficient K can be expressed in terms of the molar absorptivity (ε) and the concentration C of the absorbing species as

$$K = \varepsilon \cdot C \tag{8.7}$$

Equation (8.6) then becomes

$$f(R) = \varepsilon C / S = kC \tag{8.8}$$

where S is assumed to independent of concentration. Equation (8.8) is analogous to the Beer-Lambert equation (8.1) and holds true within a range of concentrations for solid solutions in which the absorber is incorporated with the scattering particles and for systems in which the absorber is absorbed on the surface of a scattering particle. Absolute reflectance values are evaluated using standard reflectance samples such as $BaSO_4$ [32].

Alternative models have been compared [33] such as the Pitts-Giovanelli theory, valid for scattering particles suspended in an absorbing aqueous medium [34], and the Rozenberg model, applicable to mixtures of absorbing and nonabsorbing powders [35].

In reflectance, the reagent support is generally a film attached to the fiber end [36] or a resin (e.g., Amberlite), as shown in Figure 8.7. The instrumentation uses solid-state components for the source and detectors [37]. A dry reagent strip has been demonstrated for potassium measurement in serum [38] and chlorine gas sensing [39]. Reflectance is a technique well adapted to tissue diagnosis [40] and noninvasive in vivo determinations [41] (see also Section 11.3.7). Special cells have been developed for the mid-IR region [42].

8.5.3 Raman Measurements

The basis of the Raman effect (see Section 6.2.1) is the emission of a photon with an energy close to that of the exciting or laser pump light due to the creation, or annihilation, of phonons (quantized vibrations). Transitions occur at lower energy for Stokes radiation and at higher energy for anti-Stokes radiation. The intensity of the scattered Raman signal is very weak compared to the pump source (10^{-4} to 10^{-6} smaller than fluorescence) and is separated typically by 1,000 to 5,000 cm^{-1}. This requires sophisticated instrumentation and a high-resolution spectrometer (with holographic grating) to separate the Raman lines from the exciting source [43]. Excitation is normally with an argon (488, 514.5 nm) or krypton (647.1 nm) ion laser, or a Nd-YAG (yttrium-aluminium garnet doped with Nd_2O_3) laser (1.06 µm) in the NIR. Diode-bar-pumped Nd/YAG [44] and diode lasers [45,46] have also been used. CCD detectors or a camera are well adapted for multichannel determination.

Many types of optodes have been described [47,48] (see Chapter 10) with long path-length cells in order to increase the sensitivity [49] or eliminate the background noise [50,51]. Long fibers, up to 100m long, with a graded-index lens have been employed for the detection of weak scatterers such as Fe_2O_3 [52].

Aqueous media are very suitable for observing Raman scattering, since water, despite being a strong absorber in the IR, has a low Raman effect. Raman spectroscopy is valuable in measuring highly polarizable bonds, such as C=C, C≡C, C≡N, S-S, which have weak IR transitions but strong Raman lines [6]. Nearly all organic molecules exhibit Raman spectra.

The main drawbacks of fiber-optic Raman spectrometry are the low signal strength, background noise from scattering and luminescence in the fiber [53], and the need for sophisticated and expensive instrumentation. Fluorescence of interfering species can be suppressed by phase-resolved Raman spectroscopy, a technique that matches the momenta of photons of two beams along a set direction under phase-matching angle conditions [54]. FT NIR spectrometry can compensate for low-level Raman signals [55].

Commercial FT NIR spectrometers have been modified for fiber-optic Raman measurements [28]. Portable Raman systems are being developed [56], and fiber-optic Raman spectroscopy is well adapted for in situ monitoring in different fields [57].

8.5.4 Surface-Enhanced Raman Spectroscopy

SERS is a technique for producing a large increase in the Raman signal by up to 10^7 times, and the theoretical aspects have been well discussed [58]. Typically, a thin layer of silver (50 to 100 nm) is deposited on a roughened surface and the analyte is evaporated from solution onto the silver layer. Many other supports have been tried: metal-coated microspheres, silver electrodes, silver and gold sol solutions, and silver and indium island films. The optical instrumentation is the same as a conventional Raman system, and the type of fiber must be carefully chosen to eliminate stray background noise. The sample and fiber geometry is important; a geometry with the exciting and collection fibers on opposite sides of the SERS support is superior to one with the fibers on the same side [59]. Detection limits of 1 ng or lower are possible.

FT and NIR Raman techniques have been extended to SERS using Nd/YAG or diode lasers, such as GaAlAs emitting at 785 nm [60]. Limitations are imposed by background Raman noise, which is dependent on fiber length. Miniaturized optodes and associated optics are valuable. Applications for drug detection [61] and for in situ remote environmental monitoring are in progress [62].

Surface-enhanced resonance Raman scattering (SERRS) combines SERS with resonance excitation with an increased signal comparable to fluorescence spectroscopy. Raman measurements using SERRS have been made of indicators (e.g., eriochrome black T) anchored on a silver support of an abrasively roughened fiber [63]. Fiber-optic pH sensors have also been reported [64].

8.6 LUMINESCENCE

8.6.1 Characteristics of Fluorescence

Measurement of fluorescence is an extremely sensitive technique capable of determining very low analyte concentrations (e.g., trace impurities). It is well suited to optical sensing, and a single fiber can carry both the exciting and fluorescent light, since the wavelength of the emitted luminescence λ_{em} is different from that of the exciting radiation λ_{ex}. The main variables are the quantum efficiency (η), the intensity of the fluorescence and its time dependence characterized by the fluorescent lifetime (τ), and the anisotropy and anisotropic decay.

In the absence of a quencher (see below) and with weakly absorbing species ($A_\lambda < 0.05$), the response of an optical fluorescence sensor is linear with analyte concentration C:

$$F_0 = 2.3k \cdot P_0 \cdot \eta_0 \cdot \varepsilon_\lambda \cdot d \cdot C \tag{8.9}$$

where F_0 is the intensity of measured fluorescence, P_0 is the intensity of the incident light (at λ_{ex}), η_0 is the quantum yield of fluorescence, ε_λ is the molar absorptivity (at λ_{ex}), d is the optical path length in the sample, and k is a proportionality factor for the particular sensor configuration and instrumental factors (e.g., the spectral band-width of the fluorimeter). The variation of η as a function of time is related to the fluorescence F (see Section 6.2.6):

$$F_0/F = \eta_0/\eta = \tau_0/\tau \tag{8.10}$$

Quenching is defined as a collisional process in which a quencher with a concentration $[Q]$ decreases the fluorescent quantum efficiency of the fluorophore. The collisional deactivation constant k_c (see Chapter 6) is proportional to $[Q]$:

$$k_c = K_{sv} \cdot [Q] = k_q \cdot \tau_0 \cdot [Q] \tag{8.11}$$

where K_{sv} is the Stern-Volmer constant, k_q the quenching rate constant, and τ_0 is the fluorescence lifetime (decay rate constant) in the absence of the quencher. Hence,

$$F_0/F = \eta_0/\eta = 1 + K_{sr} \cdot [Q] \tag{8.12}$$

The typical variation of the intensity of fluorescence with time is shown in Figure 8.9 with or without a quencher present. For static quenching (where the interaction occurs in the ground state), only the emitted fluorescent intensity is changed, and not τ_0. In contrast, for dynamic quenching (interaction in the excited state), τ is also reduced. This behavior can be written in terms of the quenching:

$$\tau_0/\tau = 1 + K_{sr} \cdot [Q] \tag{8.13}$$

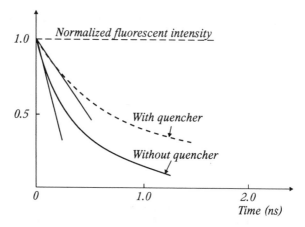

Figure 8.9 The effect of quenching on fluorescence, shown by the typical variation of the fluorescent decay curve with and without a quencher.

Many indicators are quenched by more than one analyte, and the overall quenching process must include all of their respective contributions (subscript i) in the Stern-Volmer equation [6]:

$$F_0/F = 1 + \Sigma K_i \cdot [Q_i] \tag{8.14}$$

If several compounds are present and fluoresce simultaneously, their fluorescence emission spectra can be mutually attenuated by absorption of the exciting radiation. This is known as the inner filter effect (IFE) and is different from fluorescent quenching, due to the presence of a quencher. This effect is widely known in conventional fluorescence and a correction factor is applied [65,66]. The IFE is particularly influenced by the geometry of the optode (e.g., by the difference between the path lengths for excitation and emission).

In fluorescence resonance energy transfer (FRET) elucidated first by Förster [67], the energy transfer (E_T) of the electronic energy between a donor A and an acceptor B (see Chapter 6) can be determined from the relative fluorescent yield and lifetime in the absence (F and τ) and in the presence of the acceptor (F_a and τ_a):

$$E_T = 1 - (F_a/F) = 1 - (\tau_a/\tau) \tag{8.15}$$

and expressed in terms of the critical separation distance R_0 and distance r between A and B:

$$E_T = R_0^6/(R_0^6 - r^6) \tag{8.16}$$

The corresponding rate constant K_{ET} is

$$K_{ET} = R_0^6/(r^6 \cdot \tau_d) \tag{8.17}$$

where τ_d is the lifetime of the donor.

8.6.2 Practical Measurements

The main types of fluorescent sensors are based on measurement of intensity, lifetime (TD and FD), energy transfer, IFE, and polarization. The simplest devices are fluorometers working at two or more discrete wavelengths. A typical arrangement for fluorospectrometry is shown in Figure 9.7 (Chapter 9) with a holographic grating and diode array as detector, or similar configurations [68]. The devices are significantly simpler in solid-state instruments, and lock-in amplification provides an excellent discrimination against ambient light noise [69]. The optode remains the critical component for enhancing the fluorescent signal (see Chapter 10) [70].

Measuring the intensity ratio at two wavelengths (see Figures 4.5 and 8.5(b)) is commonly used. The ratio of the intensity in the absence and in the presence of an analyte can also give a linear response function of the Stern-Volmer type [71].

A sensor based on lifetime measurements is self-referencing because the decay time is independent of the fluorophore concentration and source/detector variations have almost no influence. Multiexponential analysis and time-correlated single-photon counting is often employed for lifetime analysis (TD type) [72]. However, temporal broadening can occur from modal or material dispersion in the optical fiber [73]. Simple linear curve fitting has been reported for monoexponential decays [74]. A reference fluorophore solution to determine the instrument response function has been proposed for multiexponential decays [75]. A schematic diagram of a time-resolved multichannel spectrofluorimeter is shown in Figure 12.7.

At present, FD fluorescence is relatively straightforward [76]. The indicator exhibits different lifetimes in the free and bound forms. If these forms are described by monoexponential emission, the phase shift (ϕ_f) is

$$\phi_f = \arctan(s_f/c_f) = \arctan(2\pi f \tau) \tag{8.18}$$

where s_f and c_f are the sine and cosine transform, respectively, of the time-dependent emission and f the modulation frequency. The demodulation factor (m_f) is

$$m_f = (s_f^2 + c_f^2)^{1/2} \tag{8.19}$$

This technique has made substantial progress for subnanomolar detection with a dynamic range greater than 1,000 [77]. An example of FD instrumentation is given in Figure 11.18 [78].

Energy transfer between a donor and an acceptor is a new technique for sensing chemical and biochemical compounds, and some important examples are given in Table 8.3 for analytes, donor/acceptor couples, and other quantities.

Table 8.3
Examples of Fluorescent Resonance Energy Transfer Sensors With Fiber Optics

Analyte	Donor/Acceptor	Measurement Type	Comments	References
pH	Eosine/phenol red	Intensity		[79]
pH	HCA/CDF or Fl	Intensity	Increase of $\Delta I/\Delta pH$	[80]
CO_2	Sulforhodamine/thymol blue	Lifetime (phase change)	Stokes/anti-Stokes detection	[81]
Oxygen	Pyrene/perylene	Intensity	Stern-Volmer plots	[82]
SO_2	Pyrene/perylene	Intensity	Stern-Volmer plots	[83]
Potassium	MEDPIN/carbocyanine	Intensity		[84]
Zinc	Coumarin, Fl and Rh derivatives/azosulfamide	Lifetime (phase change)	Labeled enzyme	[85]
Ionic strength	Fluorescein/Texas red	Two-wavelength ratio	Labeled dextran	[86]
Mannosides	Coumarin/TRITC	Fluorescence ratio	Langmuir-Blodgett films	[87]
Glucose	FITC/rhodamine	Intensity	Labeled dextran/concanavalin	[88,89]
Amylase activity	Fl/procion red	Intensity	Labeling	[90]
Phenytoin (drug)	B-Phycoerythrin/Texas red	Fluorescence ratio	Immunosensor	[91,92]

Note: Abbreviations are as follows: FITC = fluorescein isothiocyanate; TRITC = tetramethyl rhodamine isothiocyanate, Fl = fluorescein; Rh = rhodamine; HCA = hydroxymethylcoumarin; CDF = carboxy dimethyl fluorescein

The IFE has been used for the measurement of potassium [93] and ammonia [94]. An interesting enhancement of the sensitivity of a pH-fluorescent indicator ($\Delta I/\Delta pH$) has been demonstrated by double IFE using two pH-sensitive absorbers mixed with a fluorophore possessing complementary $pK_a's$ and spectral overlaps [95]. In this way, the entire fluorescence intensity compressed into a narrow pH range has been applied to small pH changes (<0.01 pH units).

Fluorescence polarization methods have been demonstrated for immunoassays [96–99] but have not been extended much into other areas [100].

Fluorimetry in the red and NIR spectral regions is a promising technique for the future [101]. It has the advantage of various factors such as the high scattering that occurs in tissues, serum, or sea water. In addition, new fluorophores have a high quantum yield, and laser diodes and low-loss IR fibers are available.

8.6.3 Other Luminescent Sensors

Phosphorescent indicators have a longer lifetime than fluorophores; however, the instrumentation for phosphorescence remains similar to that for fluorimetry.

Phosphorescent indicators such as ruthenium (II) complexes with $\tau_0 = 0.4$ to 6 μs [102,103] or metalloporphyrins with $\tau_0 = 50 - 1,000$ μs [104] are good examples for oxygen determination by quenching. A model has been developed for describing the nonexponential luminescent decay in a luminophore immobilized into a polymer [105]. These indicators can be employed for others sensors, such as a CO_2 sensor based on luminescence quenching by proton transfer to an excited ruthenium-polypyridyl complex [106] and a sensor for relative humidity in air [107].

Chemiluminescence and bioluminescence are methods for detecting the chemical and biochemical reaction directly [108,109]. The luminescent signal is integrated over a known time and may be measured by a simple luminometer equipped with optical fibers [110].

8.7 EVANESCENT WAVE SPECTROSCOPY

Evanescent wave spectroscopy and SPR (see Section 8.8) are techniques for the measurement of a chemical reaction on the surface of a waveguide, which can be in the form of a slab guide, a planar integrated-optic guide, or an optical fiber. The waveguide may be coated with a reagent, such as an indicator or an antibody, depending on the type of sensor.

As already explained in Section 7.3, light traveling in a guide is confined to the core, but the light energy associated with bound guided modes also penetrates into the surrounding lower refractive index medium (see Figures 8.10 and 7.4). The electric field amplitude of the light, E, decays exponentially with distance x into the rarer medium:

$$E = E_0 \cdot \exp(-x/d_p) \tag{8.20}$$

where E_0 is the electric field at the surface of the guide. The depth of penetration, d_p, is defined as the distance for the electric field amplitude to fall to $1/e$ of this value at the surface [111]:

$$d_p = \lambda / \{2\pi n_1 \left[\sin^2\theta - (n_1/n_2)^2\right]^{1/2}\} \tag{8.21}$$

where n_1 and n_2 are the indexes of the guide and the surrounding medium, and θ is the angle of propagation in the guide. Higher order modes (with higher θ) have a larger penetration depth and the intensity of the electric field E_0 at the surface is higher. Penetration depths are typically 50 to 1,000 nm for visible light ($d_p < \lambda$ [112]), which implies that the light probes many monolayers at the surface of the guide.

Evanescent wave spectroscopy is a highly promising technique for optical sensing. Measurement of a chemical reaction by light traveling inside the guide may be due to scattering by molecules on the surface by absorption or fluorescence changes, and

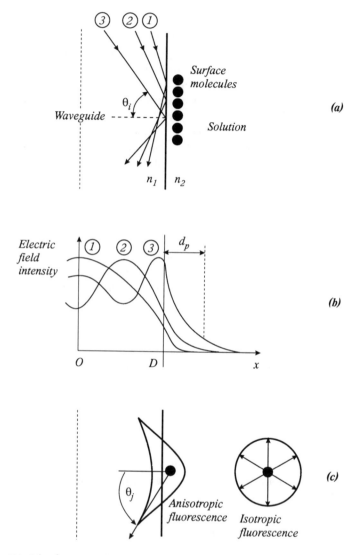

Figure 8.10   Principle of evanescent wave spectroscopy: (a) Light guided inside the waveguide measures surface molecules by the evanescent wave. Three rays of light are shown with different incident angles. (b) The electric field intensity of the three rays falls off exponentially outside the waveguide. The highest order mode, 3, with the highest incidence angle, θ_i, has the largest penetration depth d_p into the solution. D is the dimension of the waveguide or radius of the fiber. (c) The angular distribution of emission from fluorescent molecules is isotropic in solution, but preferentially couples into the waveguide by *tunneling* from surface molecules.

is done by means similar to those for measurement of a solution in a conventional spectroscopy. A fluorescent indicator (e.g., attached as a marker in antibody/antigen reactions) is often used [113,114]. Normally, fluorescence created in solution radiates isotropically, but fluorescence from molecules close to the surface is preferentially coupled into the guide by tunneling (Figure 8.10(c)) [115]. This physical effect increases the intensity of fluorescence collected by the guide by a factor of 5 to 10 and is a valuable method of distinguishing fluorescence of the surface reaction from background fluorescence in the solution.

Two typical instrumentation systems are shown as examples. In the first (Figure 8.11), a low-cost xenon flash lamp is used as a powerful UV source to excite fluorescent molecules. Light in the waveguide makes multiple reflections, increasing the sensitivity of the measurement. Back-coupled fluorescent light is separated from the exciting beam by a dichroic beam splitter and measured by a photodetector. The signal is normalized for the excitation flash by a reference detector. The design of the waveguide is critical, as well as the method of injecting the light, which is often by means of a prism coupler. Compact disposable cell designs have been developed for biomedical analysis with molded plastic slab waveguides, including input and output coupling prisms and sample cuvette (see Sections 10.3 and 10.4 for specific probe designs). The guide is often precoated with analyte and indicator (e.g., an antibody with fluorescent marker). A dynamic measurement is made; that is, the antibody-antigen binding reaction is recorded as a function of time after the introduction of the sample. This allows errors due to stray light, background noise, and fluorescence from

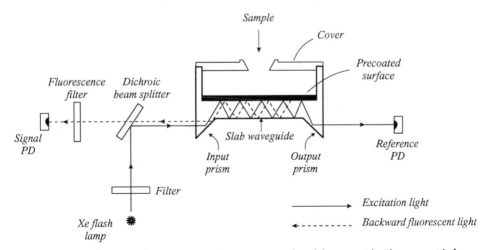

Figure 8.11   Instrumentation for an evanescent wave sensor with a slab waveguide. The waveguide forms part of an integrated cuvette design, including input and output prisms. The reagent is precoated onto the waveguide top surface. Fluorescence excited by light from a xenon flash lamp is collected in the backward direction to reduce stray light.

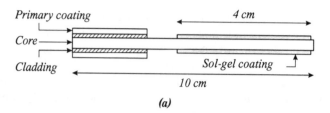

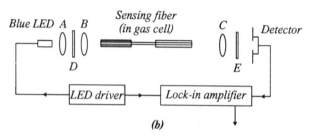

Figure 8.12 Evanescent wave sensor with an optical fiber for oxygen determination: (a) sensing fiber, and (b) instrumentation. Light from a blue LED is injected by lenses (A, B) and a short-wavelength pass filter (D) into the fiber. Fluorescent light from the fiber is collected by a lens (C) and passed through a long wavelength pass filter (E) to the detector. (*After:* [118], with permission.)

solution to be accounted for. Planar waveguides are suitable for these measurements [116]. Plastic materials such as PMMA permit easy fabrication of molded conical sensors [117].

The second example of typical instrumentation is for an optical fiber probe design (Figure 8.12) [118]. It is based on intensity measurement for oxygen determination, with a sol-gel coating on the optical fiber. Fluorescent decay analysis can be also used [119]. Evanescent wave sensors are being developed for the NIR domain [120] with FT-NIR [121], and for the IR with multimode chalcogenide [122] and silver halide fibers coupled to an ATR cell [123]. Single-mode fibers [124] and tapered fibers [125,126] have been tried in biosensing, and D-shaped fibers in gas sensing [127,128].

8.8 SURFACE PLASMON RESONANCE

Surface plasmon techniques are an extension of evanescent wave spectroscopy. The difference is that the waveguide is covered with a thin metal film. SPR can occur at a dielectric-metal layer interface when light that is totally reflected within an underlying dielectric induces a collective oscillation in the free-electron plasma at the metal film boundary (Figure 8.13). The conditions for this to occur are that the momentum of the photons (K_x) in the plane of the film should match that of the surface plasmons (K_{sp}) on the opposite surface of the metal/film [129]. This occurs at a defined critical incident angle of the light, θ_{sp}. The resulting effect is to produce a large change in the reflection coefficient at this resonance angle (Figure 8.14). The momentum K_{sp} of the surface plasmons is a function of the dielectric constants ε_m and ε_2 of the metal and sample layers, respectively:

$$K_{sp} = (\overline{\omega}/c) \cdot (1/\varepsilon_2 + 1/\varepsilon_m)^{-1/2} = (\overline{\omega}/c) \cdot (\varepsilon_1 \cdot \sin \theta_i) \qquad (8.22)$$

where ε_1 is the dielectric constant of the support (prism), θ_i the incidence angle of the light, $\overline{\omega}$ the optical frequency, and c the velocity of light in free space.

A typical sensor employs a silver film of 40- to 50-nm thickness evaporated onto a glass plate or prism. The sample is placed on the silver layer. The intensity of the electric field of the evanescent wave within the sample layer is enhanced by two orders of magnitude compared to the value if it were placed directly on the dielectric interface without the metal film present. Changing the sample properties shifts the resonance angle, providing a highly sensitive means of monitoring small changes due to the surface reaction.

The most typical measurement arrangement is the Kretschmann [130] configuration (Figure 8.14(b)). A laser is focused onto the sensor (a thin plate) mounted on a prism, and the reflectance is determined as a function of the incident angle of the light. A sharp minimum occurs in the reflectance at the critical resonance angle θ_{sp} needed to excite the surface plasmon. Measurement of the shift of the angle is directly related to concentration changes in the surface layer. Buildup of successive monolayers of the sample alters the resonance angle (Figure 8.14(c)). The angle can be measured by mechanically scanning the input or output beam, or by employing a cylindrical prism with a CCD array to register the output beam (Figure 8.14(d)) [131]. The typical angular resolution in determining the angle θ is better than 10^{-2} degrees.

Two types of sample layers are found in SPR sensors. The first type is

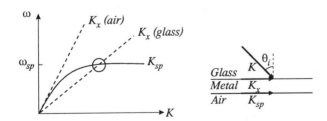

Figure 8.13   Schematic diagram of SPR: Dispersion curves for light K_x in air and glass, and surface plasmons K_{sp} in the metal film. The resonant condition of matching K_x and K_{sp} is circled.

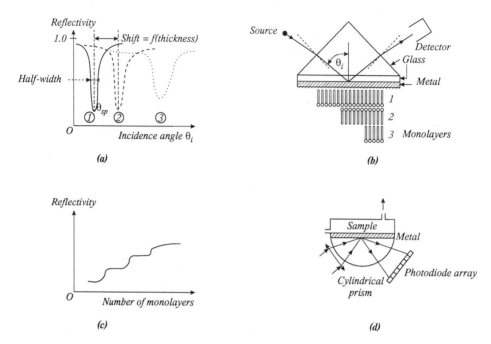

Figure 8.14   SPR: (a) ATR curves showing the resonant dip for films with 1, 2, and 3 monolayers (the shift in resonance angle is a function of film thickness); (b) instrumentation for the Kretschmann arrangement, producing the resonant condition by varying the input angle (θ_i) of the light; (c) curve of the reflectivity with increasing number of monolayers at a fixed angle; and (d) a measurement cell for liquid samples with a cylindrical prism and PD array for direct measurement of the resonant angle.

characterized by dielectric (ε) changes in the sample that are independent of wavelength (e.g., antibody-antigen reactions). In the second type, the measurand induces changes $\varepsilon(\lambda)$ at specific wavelengths (e.g., with various immobilized chromophores) [132].

Added layers [133] and guiding multilayer structures [134,135] have been assessed for increasing the sensitivity (see also Chapter 10). Computer simulation programs are valuable for optimizing measurements [136]. Novel arrangements have been tried, such as a single plasmon structure containing an array of reference guides excited at a single wavelength [137]. The array consists of parallel ridge guides of unequal height that support different modes, giving an extended measurement range.

SPR has been used for biochemical analysis [138], such as glucose and urea [139], and for immunoassay reactions and gas sensing [131], such as ammonia in air [136]. The technique has been extended with fiber optics [140] using sol-gel coatings [141,142].

8.9 REFRACTOMETRY

The determination of the refractive index of liquids (and gases) is usually based on Snell's law (see (6.24)) and the measurement of some parameter related to the angle of refraction at the waveguide-to-liquid interface. The optical probe is designed to determine either the critical angle at total internal reflection, the change in the angle at noncritical reflection, or a related condition such as resonant coupling of a grating coupler (see Chapter 7). The simplest measurement is made by determining the light intensity transmitted (or reflected) by the optical probe, since light that exceeds the critical angle is lost into the liquid. The transmitted light I is therefore a function of the ratio of the refractive indexes (r) of the liquid (n_l) and probe (n_p):

$$I = f(n_l/n_p) = f(r) \tag{8.23}$$

The sensitivity S to small changes in the liquid is

$$S = \partial I / \partial n_l = (1/n_p) \cdot (df/dr) \tag{8.24}$$

The refractive index is generally a strong function of temperature T (typically 10^{-3} to $10^{-4}/°C$), and the material must be chosen to minimize the temperature variation:

$$dI/dT = n_l/n_p \cdot [n_l'/n_l - n_p'/n_p] \cdot df/dr \tag{8.25}$$

where n_l' and n_p' are the derivatives with respect to temperature. This shows that the variation (drift) with temperature is related to the sensitivity of the probe, and can be reduced by matching the temperature variation of the probe material with the liquid being measured. Sensitivities of 10^{-5} to 10^{-6} are possible, but temperature correction is critical.

Many fiber and probe geometries are possible (see Section 10.3), and different materials can be chosen for the probe (e.g., glasses and plastics) with suitable refractive indexes.

Integrated-optic (IO) refractive index sensors have been built with planar waveguides and grating couplers. The waveguide can be part of a flow-through cell; liquid and gas molecules are absorbed as a surface layer or they may penetrate into a porous layer covering the waveguide [143,144]. Figure 8.15 shows a grating coupler covered with a tapered high-index film used as a miniature IO refractometer [145]. The expanded laser beam will only excite a waveguide mode for a certain waveguide thickness, and the position depends on the tapered film. The position, measured with a CCD, changes with the refractive index of a liquid on the waveguide. The cell has a sensitivity of 2×10^{-4} units. Ion-selective membranes have been tried for IO refractometry [146], extending the possibilities and selectivity of these techniques.

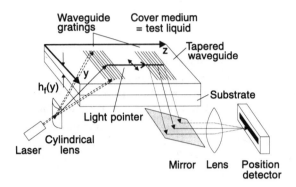

Figure 8.15   Integrated-optic sensor for refractive index measurement. Changing the refractive index of the liquid sample produces an angular variation of the light beam due to the tapered waveguide. (*Source:* [145]. Reprinted with permission.)

8.10 INTERFEROMETRY

8.10.1 Interferometers

In a fiber interferometer, a physical or chemical parameter introduces a phase difference between two coherent beams of light which interfere. The phase difference is measured by fringe counting (one fringe = 360-degree phase) or incremental phase (with a sensitivity up to 10^{-5} of a degree). In chemical sensors, there are three main types of interferometers.

1. Two-beam interferometer illustrated by Mach-Zehnder and Michelson configurations (Figure 8.16). The two beams travel different paths in a signal and a reference fiber.
2. Single-path interferometer demonstrated by Sagnac and Fabry-Perot configurations (Figure 8.16(c)). The two beams travel in the same fiber. A miniature version of a Fabry-Perot cavity is shown in Figure 8.16(d) [147].
3. Multimode interferometer in which different modes in a multimode fiber can interfere with one another. This method has problems with stability and is little used.

In a Mach-Zehnder interferometer, the phase (ϕ) of the light traveling in one arm of the interferometer is given by

$$\phi = k \cdot n \cdot L \qquad (8.26)$$

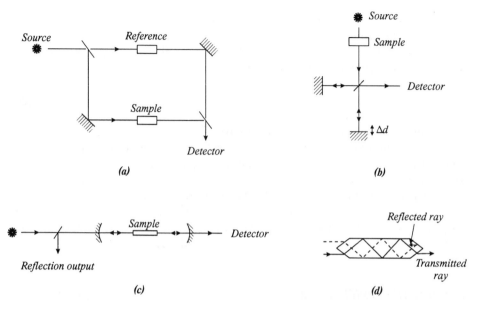

Figure 8.16 Principal types of interferometer for chemical sensing: (a) Mach-Zehnder arrangement; (b) Michelson arrangement; (c) Fabry-Perot interferometer; (d) fiber version of a Fabry-Perot cavity (*After:* [147]).

where n is the refractive index along the path length L, and k is a constant. The phase is modulated by variation of some physical parameter, such as pressure *(P):*

$$d\phi/dP = k \cdot n \cdot L \left[(1/L) \cdot \delta L/\delta P + (1/n) \cdot \delta n/\delta P\right] \qquad (8.27)$$

The first term is due to the change in length. The second term is the refractive index change, which is normally the sum of two terms due to longitudinal stretching and radial compression of the fiber.

8.10.2 Examples of Interferometric Sensors

The Mach-Zehnder arrangement has been used for a hydrogen gas sensor with an optical fiber coated with a palladium metal film [148,149]. Absorption of the gas introduces a strain in the film that is transferred to the fiber. Similarly, a nonspecific sensor for flammable gases has been demonstrated in which a platinum metal-coated fiber catalyzes the oxidation of the gas, and the temperature increase is measured with the interferometer [150]. Changes of the refractive index of a sensitive film by adsorption of gases such as CO_2, SO_2, and NO_2 have been demonstrated with Mach-Zehnder and Fabry-Perot configurations [151].

Fiber-optic interferometry is a valuable method for measurement of gases and transparent films (thickness d and refractive index n) by evaluating the interference spectra [152]. The optical thickness $(n \cdot d)$ can be correlated with the order m of the interference spectrum at a given wavelength:

$$2n \cdot d = m \cdot \lambda \qquad (8.28)$$

This method has been extended to the analysis of multilayer films [153,154].

Integrated-optic interferometric sensors have been built with planar waveguides using Köster prisms as beam splitter and recombiner in a Mach-Zehnder arrangement [155]. A differential interferometer for refractometry is based on SiO_2-TiO_2 waveguiding films [156,157]. A molded IO Mach-Zehnder interferometer has been proposed for low-cost mass production [158]. IO Mach-Zehnder interferometers have also been exploited for a glucose sensor [159], an immunosensor with spectral interference measurement [160], and for pesticide determination [161].

8.11 PHOTOACOUSTIC MEASUREMENTS

The technique of photoacoustic spectroscopy is based on passing a modulated beam of light through a sample (e.g., a gas contained in a cell) and measuring the acoustic emission (pressure wave) generated [162]. This is due to excited molecules that return to their ground (unexcited) state through collisions, and release the absorbed energy to heat, creating pressure variations in the gas. The instrumentation consists of a powerful light source (laser), which is tuned to one of the absorption wavelengths of the sample. The photoacoustic cell, which holds the sample, is equipped with a sensitive pressure sensor, usually a microphone. This can also be a fiber-optic hydrophone, which may be wrapped around the sample cell, or a miniaturized integrated-optic microphone [163].

A fiber Michelson interferometer with photoacoustic techniques has been applied to noncontact measurement of subnanometer surfaces in methanol and water with a pressure sensitivity of 0.1 $Pa/Hz^{1/2}$ [164]. Pulsed photoacoustic sensors with laser diodes have been exploited for the detection of pentane and methanol [165] and crude oil contamination (100 to 1,000 $mg.L^{-1}$ range) in water [166].

Correlation spectroscopy is a general absorption method to detect a gas in a measurement cell by comparison to a reference cell used as a matched optical filter [167]. Light is scanned sequentially across the measurement cell (e.g., White type), and the reference gas cell is periodically modulated by an acoustic resonator. Application of a high-voltage electric field to molecules, which possess a permanent electric dipole, broadens or splits the fine spectral lines; this is known as the Stark effect. Modulation methods that act directly on the gas include Stark and pressure modulation (Figure 8.17(a)). In an indirect approach, an angle modulation of the light

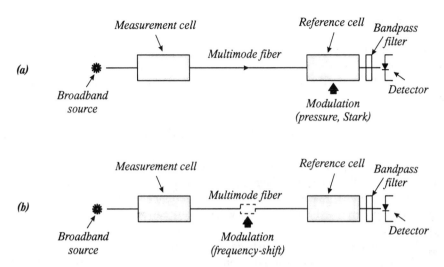

Figure 8.17   Schematic representation of correlation spectroscopy for gas measurement: (a) with pressure or electric field (Stark) modulation; and (b) angle modulation of light using frequency-shift modulation. (*After:* [167], with permission.)

by either a frequency or a phase shift gives a redistribution of the optical spectrum as it passes between the measurement and reference gas cells (Figure 8.17(b)).

Recent experimental techniques, which have been demonstrated for remote gas measurement with optical fibers, are Stark modulation (see Section 12.3.1), a combination of a broadband optical source with a Michelson interferometer, and the use of chopped-light sources [168].

8.12 OTHER OPTOCHEMICAL SENSING TECHNIQUES

8.12.1 Grating Light Reflection Spectroscopy

The technique of grating light reflection spectroscopy has been applied recently to chemical sensing [169]. It relies on the modulation of the light propagation that is incident on a diffraction grating. This grating is fabricated on a planar wall from a highly reflective chrome mask layer deposited on a glass slide. Features in the reflection spectrum can be directly related to the complex dielectric constant of the sample solution. For this technique, which is different from SPR, gratings can be realized with any metal.

8.12.2 Ellipsometry

Ellipsometry [170] determines the change in polarization state of light at oblique incidence caused by reflection from a film. The measured quantity is the complex reflectance ratio (ρ) of the reflection coefficients R_p and R_s for p- and s-polarized light, respectively:

$$\rho = R_p/R_s = \tan(\psi) \exp(i\Delta\varphi) \qquad (8.29)$$

where ψ is the amplitude ratio of the signals and $\Delta\varphi$ the phase shift (the so-called ellipsometric angle) of the orthogonal components of the polarized light. These measurements allow the determination of the thickness and refractive index of the film independently [171]. Cylindrical waveguides can be used in evanescent wave sensing [172].

Ellipsometry has been applied to gas (NO_x) sensing using copper-phthalocyanine films [173]. However, this sensor is irreversible. In a new approach, a differential phase modulation between two orthogonal polarization modes interacts in a single-mode fiber via an evanescent field [174]. Polarization modulation is obtained by changing the refractive index of a thin film exposed to acetone vapor.

8.12.3 Optical Time Domain Reflectometry

Optical time domain reflectometry (OTDR) is an *optical radar* system in which a light pulse is transmitted down a fiber and the backscattered light (from scattering centers along the fiber) produces a return echo signal [175]. The signal propagation time depends on the speed of the light, and hence on the fiber length. Thus, the variation in attenuation and its spatial distribution along the fiber can be measured simultaneously. This is a very powerful technique and the position of specific points of increased attenuation can also be located along the fiber. This leads to the concept of a distributed sensor in which the whole fiber length is sensitive to an outside parameter; for example, the attenuation may change as a function of temperature, allowing a spatial temperature profile to be measured. This technique and derived methods such as polarization OTDR (POTDR) and frequency OTDR (FOTDR), are mainly used for physical parameters (e.g., temperature, strain) and are also suitable for sensor networks (see Chapter 12).

Few chemical sensors have been demonstrated with ODTR techniques. Examples are refractive index determination [176], measurement of dyes immobilized onto fibers [177], and a quasidistributed luminescent sensor with a fiber Bragg grating [178]. Figure 8.18 shows a microbend fiber-optic sensor with a water-sensitive polymer (hydrogel) deposited onto a central support [179]. In the presence of water, the hydrogel swells and exerts a microbending force on the graded-index fiber. The attenuation increase is proportional to the force. The swelling is totally reversible

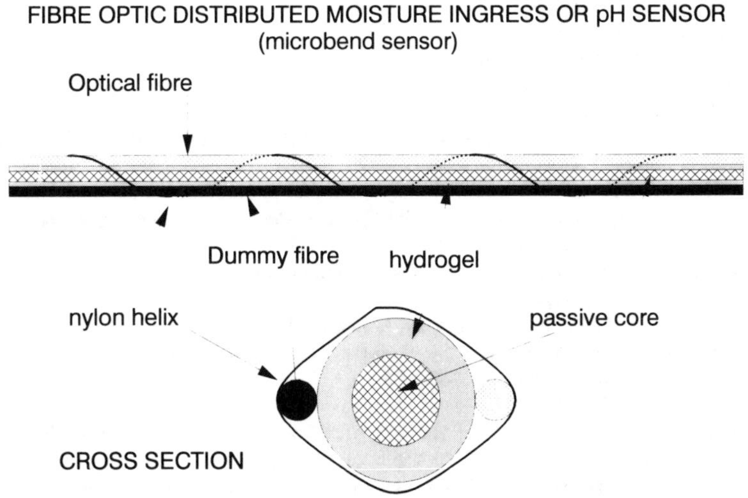

FIBRE OPTIC DISTRIBUTED MOISTURE INGRESS OR pH SENSOR
(microbend sensor)

Optical fibre

Dummy fibre   hydrogel

nylon helix   passive core

CROSS SECTION

Figure 8.18 Side view and cross section of a microbend fiber-optic sensor for water detection with a water-sensitive hydrogel layer. The passive core in the middle is on the order of 2 to 3 mm in diameter and the hydrogel layer on the order of 50 μm in thickness. The attenuation can be measured by OTDR. (*After:* [179], with permission.)

when the water is removed. This sensor can also be adapted for distributed pH measurements.

References

[1] Chester A. N., S. Martelucci, and A. M. Scheggi, eds., *Optical Fibers Sensors*, Dordrecht: Martinus Nijhoff, 1987.

[2] Dakin, J. P., and B. Culshaw, eds., *Optical Fibers Sensors, Vol. 1, Principles and Components; Vol. 2, Systems and Applications*, Norwood, MA: Artech House, 1988 and 1989.

[3] Göpel, W., J. Hesse, and J. N. Zemel, eds., *Sensors: A Comprehensive Survey, Vols. 2* and *3: Chemical and Biochemical Sensors; Vol. 6: Optical Sensors*, Weinheim: VCH, 1991.

[4] Schulman, S. G., ed., *Molecular Luminescence Spectroscopy, Vol. 1*, 1985; *Vol. 2*, 1988; *Vol. 3*, 1991, New York: Wiley.

[5] Lakowicz, J. R., ed., *Topics in Fluorescence Spectroscopy, Vol. 1: Techniques*, 1990; *Vol. 2: Principles; Vol. 3: Biochemical Applications; Vol. 4: Probe Design and Chemical Sensing*, 1994, New York: Plenum Press.

[6] Wolfbeis, O. S., "Spectroscopic Techniques," Chap. 2 in *Fiber Optic Chemical Sensors and Biosensors, Vol. 1*, O. S. Wolfbeis, ed., Boca Raton: CRC Press, 1991, pp. 25–60.

[7] Taib, M. N., and R. Narayanaswamy, "Solid State Instruments for Optical Fibre Chemical Sensors: A Review," *Analyst*, Vol. 120, 1995, pp. 1617–1625.

[8] Harmer, A., "Optical Fibre Sensors," in *Encyclopedia of Systems and Control*, M. Singh, ed., London: Pergamon Press, 1986, pp. 3455–3471.

[9] Mantz, A. W., "A Review of the Applications of Tunable Diode Laser Spectroscopy at High Sensitivity," *Microchem. J.*, Vol. 50, 1994, pp. 351–364.

[10] Szmacinski, H., and J. R. Lakowicz, "Lifetime Based Sensing," Chap. 10 in *Topics in Fluorescence Spectroscopy, Vol. 4: Probe Design and Chemical Sensing*, J. R. Lakowicz, ed., New York: Plenum Press, 1994, pp. 295–333.

[11] Bai, M., and W. R. Seitz, "A Fiber Optic Sensor for Water in Organic Solvents Based on Polymer Swelling," *Talanta*, Vol. 41, 1994, pp. 993–999.

[12] Seitz, W. R., K. Hassen, V. Conway, Z. Shakhsher, and S. Pan, "Structure Effects on Mechanical and Swelling Properties of Amine Modified Polystyrene Beads for pH Sensing," *Proc. Electrochem. Soc.*, Vol. 93–7, 1993, pp. 74–80.

[13] MacCurley, M. F., "An Optical Biosensor Using a Fluorescent Swelling Sensing Element," *Biosens. Bioelectron.*, Vol. 9, 1994, pp. 585–592.

[14] Schalkhammer, T., C. Lobmaier, F. Pittner, A. Leitner, H. Brunner, and F. R. Aussenegg, "The Use of Metal-Island-Coated pH Sensitive Swelling Polymers for Biosensor Applications," *Sensors and Actuators*, Vol. B 24/25, 1995, pp. 166–172.

[15] Aussenegg, F. R., H. Brunner, A. Leitner, C. Lobmaier, T. Schalkhammer, and F. Pittner, "The Metal Island Coated Swelling Polymer Over Mirror System (MICSPOMS): A New Principle for Measuring pH," *Sensors and Actuators*, Vol. B 29, 1995, pp. 204–209.

[16] Gruber, W., P. O'Leary, and O. S. Wolfbeis, "Detection of Fluorescence Lifetime Based and Solid State Technology and Its Applications to Optical Oxygen Sensing," *Proc. SPIE–Int. Soc. Opt. Eng.*, Vol. 2388, 1995, pp. 148–158.

[17] Velluet, M. T., F. Blanc, and P. Vernet, "Fiber Optic Spectrophotometer With Photodiode Linear Array," *Proc. SPIE–Int. Soc. Opt. Eng.*, Vol. 990, 1988, pp. 78–83.

[18] Piccard, R., and T. Vo-Dinh, "A Multi-Optical Fiber Array With Charge Coupled Device Image Detection for Parallel Processing of Light Signals and Spectra," *Rev. Scient. Instrum.*, Vol. 62, 1991, pp. 584–594.

[19] Korth, H. E., "A Computer Integrated Spectrophotometer for Film Thickness Monitoring," *Journal de Physique*, Vol. 44, C10, 1983, pp. 101–104.

[20] Boisdé, G., F. Blanc, and X. Machuron-Mandard, "pH Measurements With Dyes Co-immobilization on Optodes Principles and Associated Instrumentation," *Intern. J. Optoelectron.*, Vol. 6, 1991, pp. 407–423.

[21] Strauch, B., "Fourier Transform Infrared Spectroscopy," Chap. 11 in *Instrumentation in Analytical Chemistry, Vol. II*, J. Zyka, ed., New York: Ellis Horwood, 1994, pp. 209–231.

[22] Archibald, D. D., C. E. Miller, L. T. Lin, and D. E. Honigs, "Remote Near-IR Reflectance Measurements With the Use of a Pair of Optical Fibers and a Fourier Transform Spectrometer," *Appl. Spectrosc.*, Vol. 42, 1988, pp. 1549–1558.

[23] Saggese, S. J., M. R. Shahriari, and G. H. Sigel, "Evaluation of an FTIR Fluoride Optical Fiber System for Remote Sensing of Combustion Products," *Proc. SPIE–Int. Soc. Opt. Eng.*, Vol. 1172, 1989, pp. 2–12.

[24] Matson B. S., and J. W. Griffin, "Infrared Fiber Optic Sensors for the Remote Detection of Hydrocarbons Operating in the 3.3 to 3.36 Micron Region," *Proc. SPIE–Int. Soc. Opt. Eng.*, Vol. 1172, 1989, pp. 13–26.

[25] Druy, M. A., P. J. Glatkowski, and W. A. Stevenson, "Fiber Optic Remote Fourier Transform Infrared (FTIR) Spectroscopy," *Proc. SPIE–Int. Soc. Opt. Eng.*, Vol. 1584, 1989, pp. 48–52.

[26] Stuart, A. D., "Some Applications of Infrared Optical Sensing," *Sensors and Actuators*, Vol. B 11, 1993, pp. 185–193.

[27] Delgue, M., "Real Time Non Destructive Quality Control Using FTIR and Optical Fibers," *Quim. Ind., (Madrid)*, Vol. 40, 1993, pp. 14–18.

[28] Archibald, D. D., L. T. Lin, and D. E. Honigs, "Raman Spectroscopy Over Optical Fibers With the Use of a Near-IR FT Spectrometer," *Appl. Spectrosc.*, Vol. 42, 1988, pp. 1558–1563.

[29] Simhony, S., A. Katzir, and E. M. Kosower, "Fourier Transform Infrared Spectra of Organic Compounds in Solution and as Thin Layers Obtained by Using an Attenuated Total Internal Reflectance Fiber Optic Cell," *Anal. Chem.*, Vol. 60, 1988, pp. 1908–1910.

[30] Weigl, R., and R. Kellner, "Development and Performance of a Novel IR-ATR-Based Glucose Sensor System," *Proc. SPIE–Int. Soc. Opt. Eng.*, Vol. 1145, 1989, pp. 134–138.

[31] Parisi, A.F., L. Nogueivas, H. Prieto, "On-line Determination of Fuel Quality Parameters Using Near-Infrared Spectrometry With Fiber Optics and Multivariate Calibration," *Anal. Chim. Acta*, Vol. 238, 1990, pp. 95–100.

[32] Wendlandt, W. W., and H. G. Hecht, *Reflectance Spectroscopy*, New York: Wiley, 1966.

[33] Hecht, H. G., "A Comparison of the Kubelka-Munk, Rozenberg, and Pitts-Giovanelli Methods of Analysis of Diffuse Reflectance for Several Model Systems," *Appl. Spectrosc.*, Vol. 37, 1983, pp. 348–354.

[34] Hecht, H. G., "Comparison of Continuum Models in Quantitative Diffuse Reflectance Spectrometry," *Anal. Chem.*, Vol. 48, 1976, pp. 1775–1779.

[35] Hecht, H. G., "Quantitative Analysis of Powder Mixtures by Diffuse Reflectance," *Appl. Spectrosc.*, Vol. 34, 1980, pp. 161–164.

[36] Andres, R. T., and F. Sevilla, "Fiber Optic Reflectometric Study on Acid-Base Equilibria of Immobilized Indicators: Effects on the Nature of Immobilizing Agents," *Anal. Chim. Acta*, Vol. 251, 1991, pp. 165–168.

[37] Guthrie, A. J., R. Narayanaswamy, and N. A. Welti, "Solid State Instrumentation for Use With Optical Fibre Chemical Sensors," *Talanta*, Vol. 35, 1988, pp. 157–159.

[38] Ng, R. H., K. M. Sparks, and B. E. Stetland, "Colorimetric Determination of Potassium in Plasma and Serum by Reflectance Photometry With a Dry Chemistry Reagent," *Clin. Chem.*, Vol. 38, 1992, pp. 1371–1372.

[39] Momin, S. A., and R. Narayanaswamy, "Optosensing of Chlorine Gas Using a Dry Reagent Strip and Diffuse Reflectance Spectrophotometry," *Anal. Chim. Acta*, Vol. 244, 1991, pp. 71–79.

[40] Swetland, H. J., "Carotene Reflectance and the Yellowness of Bovine Adipose Tissues Measured With a Portable Fiber Optic Spectrophotometer," *J. Sci. Food Agric.*, Vol. 46, 1989, pp. 195–200.

[41] Ono, K., M. Kanda, J. Hiramoto, K. Yotsuya, and N. Sato, "Fiber Optic Reflectance Spectrophotometry System for in Vivo Tissue Diagnosis," *Appl. Opt.*, Vol. 30, 1991, pp. 98–105.

[42] Moser, W. R., S. R. Berard, R. J. Melling, and R. J. Burger, "A New Spectroscopic Technique for in Situ Chemical Reaction Monitoring Using Mid-Range Infrared Optical Fibers," *Appl. Spectrosc.*, Vol. 46, 1992, pp. 1105–1112.

[43] Strauch, B., "Laser Raman Spectrometry," Chap. 12 in *Instrumentation in Analytical Chemistry, Vol. II*, J. Zyka, ed., New York: Ellis Horwood, 1994, pp. 232–261.

[44] Zimba C. G., and J. F. Rabolt, "Raman Spectroscopy With Fiber Optic Probes and a Diode-Bar-Pumped Nd:YAG Laser," *Appl. Spectrosc.*, Vol. 45, 1991, pp. 162–165.

[45] Allred, C., and R. McCreery, "Near Infrared Raman Spectroscopy of Liquids and Solids With a Fiber Optic Sampler, Diode Maser and CCD Detector," *Appl. Spectrosc.*, Vol. 44, 1990, pp. 1229–1231.

[46] Angel, S. M., T. J. Kulp, and T. M. Vess, "Remote Raman Spectroscopy at Intermediate Ranges Using Low Power CW Lasers," *Appl. Spectrosc.*, Vol. 46, 1992, pp. 1085–1091.

[47] McCreery, R., M. Fleischman, and P. Hendra, "Fiber Optic Probe for Remote Raman Spectrometry," *Anal. Chem.*, Vol. 55, 1983, pp. 146–148.

[48] Plaza, P., N. Dao, M. Jouan, H. Fevrier, and H. Saisse, "Simulation et optimisation des capteurs à fibres optiques adjacentes," *Appl. Opt.*, Vol. 25, 1986, pp. 3448–3454.

[49] Schwab, S., and R. McCreery, "Remote, Long Pathlength Cell for High Sensitivity Raman Spectroscopy," *Appl. Spectrosc.*, Vol. 41, 1987, pp. 126–130.

[50] Myrick, M., and S. Angel, "Elimination of Background in Fiber Optic Raman Measurements," *Appl. Spectrosc.*, Vol. 41, 1990, pp. 126–130.

[51] Ma. J., and Y. S. Li. "Optical Fiber Raman Probe With Low Background Interference by Spatial Optimization." *Appl. Spectrosc.*, Vol. 48, 1994, pp. 1529–1531.

[52] Schoen. C. L., T. F. Cooney. S. K. Sharma, and D. M. Carey. "Long Fiber Optic Remote Raman Probe for Detection and Identification of Weak Scatterers," *Appl. Opt.*, Vol. 31, 1992, pp. 7707–7715.

[53] Boisdé. G.. W. Carvalho. P. Dumas, and V. Neuman. "Fluorescence Emission by Molecular Impurities in Irradiated and Non irradiated Silica." *Proc. SPIE–Int. Soc. Opt. Eng.*, Vol. 506, 1984, pp. 196–201.

[54] Bright. F. V., "Multicomponent Suppression of Fluorescent Interferents Using Phase Resolved Raman Spectroscopy." *Anal. Chem.*, Vol. 60, 1988. pp. 1622–1623.

[55] Schrader. B.. A. Hoffmann, A. Simon, and J. Sawatzki. "Can a Raman Renaissance Be Expected via the Near-Infrared Fourier Transform Technique?" *Vibrational Spectrosc.*, Vol. 1, 1991, pp. 239–249.

[56] Vess. T. M., and S. M. Angel. "Near Visible Raman Instrumentation for Remote Multi-point Process Monitoring Using Optical Fibers and Optical Multiplexing." *Proc. SPIE–Int. Soc. Opt. Eng.*, Vol. 1637, 1992, pp. 118–125.

[57] Dao. N. Q., and M. Jouan, "The Raman Laser Fiber Optics (RLFO) Method and Its Applications." *Sensors and Actuators.* Vol. B 11, 1993, pp. 147–160.

[58] Chang. R. K. and T.E. Furtak, eds., *Surface Enhanced Raman Scattering.* New York: Plenum Press, 1982.

[59] Bello. J. M., and T. Vo-Dinh, "Surface-Enhanced Raman Scattering Fiber-Optic Sensor." *Appl. Spectroscopy.* Vol. 44, 1990, pp. 63–69.

[60] Angel, S. M., M. N. Ridley, K. Langry, T. J. Kulp, and M. L. Myrick, "New Developements and Applications of Fiber Optic Sensors." Chap. 23 in *Chemical Sensors and Instrumentation*, R. W. Murray, R. E. Dessy, W. R. Heineman, J. Janata, and W. R. Seitz. eds., Washington, D.C., ACS Symp. Series, Vol. 403, 1989, pp. 435–363.

[61] Angel, S.M., J.N. Roe, B.D. Andresen, M.L. Myrick, and F.P. Milanovich. "Development of a Drug Assay Using Surface-Enhanced Raman Spectroscopy," *Proc. SPIE-Int. Soc. Opt. Eng.*, Vol. 1201, 1990, pp. 489–493.

[62] Angel. S. M., and M. L. Myrick, "Near Infrared Surface-Enhanced Raman Spectroscopy: New Developments and Applications." *Pract. Spectrosc.*, Vol. 13, 1993, pp. 225–245.

[63] Mullen. K. I., and K. T. Carron, "Surface-Enhanced Raman Spectroscopy With Abrasively Modified Fiber Optic Probes." *Anal. Chem.*, Vol. 63, 1991, pp. 2196–2199.

[64] Mullen. K. I., D. X. Wang, W. L. Crane, and K. T. Carron, "Determination of pH With Surface-Enhanced Raman Fiber Optic Probes." *Anal. Chem.*, Vol. 64, 1992, pp. 930–936.

[65] Leese, R. A., and E. L. Wehry, "Corrections for Inner Filter Effects in Fluorescence Quenching Measurements via Right Angle and Front Surface Illumination." *Anal. Chem.*, Vol. 50, 1978, pp. 1193–1197.

[66] Ratzlaff, E. H., R. G. Haufmann, and S. R. Crouch, "Absorption Corrected Fiber Optic Fluorometer." *Anal. Chem.*, Vol. 56, 1984, pp. 342–347.

[67] Cheung, H. C., "Resonance Energy Transfer." Chap. 3 in *Topics in Fluorescence Spectroscopy, Vol. 2: Principles*, J. R. Lakowicz, ed., New York: Plenum Press, 1991, pp. 127–176.

[68] Thompson, R. B., "Fluorescence Based Fiber Optic Sensors." Chap. 7 in *Topics in Fluorescence Spectroscopy, Vol. 2: Principles*, J. R. Lakowicz, ed., New York: Plenum Press, 1991, pp. 345–365.

[69] Hauser, P. C., and S. S. S. Tan, "All Solid State Instrument for Fluorescence Based Fibre Optic Chemical Sensors." *Analyst*, Vol. 118, 1993, pp. 991–995.

[70] Komives, C., and J. S. Schultz, "Fiber Optic Fluorometer Signal Enhancement and Application to Biosensor Design," *Talanta*, Vol. 39, 1992, pp. 429–441.

[71] He, X., and G. A. Rechnitz, "Linear Response Function for Fluorescence Based Fiber Optic CO_2 Sensors." *Anal. Chem.*, Vol. 67, 1995, pp. 2264–2268.

[72] Bright, F. V., G. H. Vickers, and G. M. Hieftje, "Use of Time Resolution to Eliminate Bilirubin Interference in the Determination of Fluorescein," *Anal. Chem.*, Vol. 58, 1986, pp. 1225–1227.

[73] Vickers, G. H., R. M. Miller, and M. Hieftje, "Time Resolved Fluorescence With an Optical Fiber Probe," *Anal. Chim. Acta*, Vol. 192, 1987, pp. 145–153.

[74] Papkovsky, D. B., J. Olah, and I. N. Kurochin, "Fibre Optic Lifetime Enzyme Biosensor," *Sensors and Actuators*, Vol. B 11, 1993, pp. 525–530.

[75] Brown, R. S., J. D. Brennan, and U. J. Krull, "An Optical Fiber Based Spectrometer for Measurement of Fluorescent Lifetimes," *Microchem. J.*, Vol. 50, 1994, pp. 337–350.

[76] Szmacinski, H., and J. R. Lakowicz, "Optical Measurements of pH Using Fluorescence Lifetimes and Phase Modulation Fluorometry," *Anal. Chem.*, Vol. 65, 1993, pp. 1668–1674.

[77] Thompson, R. B., and M. W. Patchan, "Fluorescence Lifetime Based Biosensing of Zinc: Origin of the Broad Dynamic Range," *J. Fluoresc.*, Vol. 5, 1995, pp. 123–130.

[78] Vo-Dinh,T., T. Nolan, Y. F. Cheng, M. J. Sepaniak, and J. P. Alarie, "Phase Resolved Fiber Optics Fluoroimmunosensor," *Appl. Spectrosc.*, Vol. 44, 1990, pp. 128–132.

[79] Jordan, D. M., D. R. Walt, and F. P. Milanovich, "Physiological pH Fiber Optic Chemical Sensor Based on Energy Transfer," *Anal. Chem.*, Vol. 59, 1987, pp. 437–439.

[80] Gabor, G., S. Chadha, and D. R. Walt, "Sensitivity Enhancement of Fluorescent pH Indicators Using pH-Dependent Energy Transfer," *Anal. Chim. Acta*, Vol. 313, 1995, pp. 131–137.

[81] Sipior, J., S. Bambot, M. Romauld, G. M. Carter, J. R. Lakowicz, and G. Rao, "A Lifetime Based Optical CO_2 Gas Sensor With Blue or Red Excitation and Stokes or Anti-Stokes Detection," *Anal. Biochem.*, Vol. 227, 1995, pp. 309–318.

[82] Sharma, A., and O. S. Wolfbeis, "Fiberoptic Oxygen Sensor Based on Fluorescence Quenching and Energy Transfer," *Appl. Spectrosc.*, Vol. 42, 1988, pp. 1009–1011.

[83] Sharma, A., and O. S. Wolfbeis, "Fiber Optic Fluorosensor for Sulfur Dioxide Based on Energy Transfer and Exciplex Quenching," *Proc. SPIE–Int. Soc. Opt. Eng.*, Vol. 990, 1988, pp. 116–120.

[84] Roe, J. N., F. C. Szoka, and A. S. Verkman, "Fibre Optic Sensor for the Detection of Potassium Using Fluorescence Energy Transfer," *Analyst*, Vol. 115, 1990, pp. 353–358.

[85] Thompson, R. B., and M. W. Patchan, "Lifetime Based Fluorescence Energy Transfer Biosensing of Zinc," *Anal. Biochem.*, Vol. 227, 1995, pp. 123–128.

[86] Christian, L. M., and W. R. Seitz, "An Optical Ionic Strength Sensor Based on Polyelectrolyte Association and Fluorescence Energy Transfer," *Talanta*, Vol. 35, 1988, pp. 119–122.

[87] Siegmund, H. U., and A. Becker, "A New Way of Biosensing Using Fluorescence Energy Transfer and Langmuir-Blodgett Films," *Sensors and Actuators*, Vol. B 11, 1993, pp. 103–108.

[88] Meadows, D., and J. S. Schultz, "Fiber Optic Biosensors Based on Fluorescence Energy Transfer," *Talanta*, Vol. 35, 1988, pp. 145–150.

[89] Meadows, D., and J. S. Schultz, "Design, Manufacture and Characterization of an Optical Fiber Glucose Affinity Sensor Based on an Homogeneous Fluorescence Energy Transfer Assay System," *Anal. Chim. Acta*, Vol. 280, 1993, pp. 21–30.

[90] Zhujun, Z., W. R. Seitz, and K. O'Connel, "Amylase Substrate Based on Fluorescence Energy Transfer," *Anal.Chim. Acta*, Vol. 236, 1990, pp. 151–156.

[91] Miller, W. G., and F. P. Anderson, "Antibody Properties for Chemically Reversible Biosensor Applications," *Anal. Chim. Acta*, Vol. 227, 1989, pp. 135–143.

[92] Astles, J. R., and W. G. Miller, "Measurement of Free Phenytoin in Blood With a Sel-Contained Fiber Optic Immunosensor," *Anal. Chem.*, Vol. 66, 1994, pp. 1675–1682.

[93] He, H., H. Li, G. Mohr, B. Kovacs, T. Werner, and O. S. Wolfbeis, "Novel Type of Ion Selective Fluorosensor Based on the Inner Filter Effect: An Optrode for Potassium," *Anal. Chem.*, Vol. 65, 1993, pp. 123–127.

[94] Werner, T., I. Klimant, and O. S. Wolfbeis, "Optical Sensor for Ammonia Based on the Inner Filter Effect of Fluorescence," *J. Fluorescence*, Vol. 4, 1994, pp. 41–44.

[95] Gabor, G., and W. R. Walt, "Sensitivity Enhancement of Fluorescent pH Indicators by Inner Filter Effects," *Anal. Chem.*, Vol. 63, 1991, pp. 793–796.

[96] Noeller, H., "Polarization Fluoroimmunoassay Apparatus," U.S. Pat., 4,451,149, 1984.

[97] Klein, C., H. G. Batz, B. Draeger, H. J. Guder, R. Herrmann, H. P. Josel, U. Nägele, R. Schenk, and B. Vogt, "Fluorescence Polarization Immunoassays," Chap. 17 in *Fluorescence Spectroscopy*, O. S. Wolfbeis, ed., Berlin: Springer-Verlag, 1992, pp. 245–258.

[98] Hemmilä, I., "Progress in Delayed Fluorescence Immunoassay," Chap. 18 in *Fluorescence Spectroscopy*, O. S. Wolfbeis, ed., Berlin: Springer-Verlag, 1992, pp. 259–266.

[99] Ozinskas, A. J., "Principles of Fluorescence Immunoassays," Chap. 14 in *Topics in Fluorescence Spectroscopy, Vol. 4: Probe Design and Chemical Sensing*, ed. by J. R. Lakowicz, New York: Plenum Press, 1994, pp. 449–496.

[100] Dowling, S. D., and W. R. Seitz, "Effect of Metal-Ligand Ratio on Polarization of Fluorescence From Metal-8-Quinoliol Complexes," *Spectrochim. Acta*, Vol. 40 A, 1984, pp. 991–993.

[101] Thompson, R. B., "Red and Near-Infrared Fluorometry," Chap. 6 in *Topics in Fluorescence Spectroscopy, Vol. 4: Probe Design and Chemical Sensing*, New York: Plenum Press, 1994, pp. 151–181.

[102] Moreno-Bondi, M. C., O. S. Wolfbeis, P. Leiner, and B. P. H. Schaffar, "Oxygen Optrode for Use in a Fiber Optic Glucose Biosensor," *Anal. Chem.*, Vol. 62, 1990, pp. 2377–2380.

[103] Klimant, I., P. Belser, and O. S. Wolfbeis, "Novel Metal Organic Ruthenium (II) Diimin Complexes for Use as Longwave Excitable Luminescent Oxygen Probes," *Talanta*, Vol. 41, 1994, pp. 885–891.

[104] Papkovsky, D. B., "Luminescent Porphyrins as Probes for Optical (Bio)Sensors," *Sensors and Actuators*, Vol. B 11, 1993, pp. 293–300.

[105] Draxler, S., M. E. Lippitsch, I. Klimant, H. Kraus, and O. S. Wolfbeis, "Effects of Polymer Matrices on the Time-Resolved Luminescence of a Ruthenium Complex Quenched by Oxygen," *J. Phys. Chem.*, Vol. 99, 1995, pp. 3162–3167.

[106] Orellana, G., M. C. Moreno-Bondi, E. Segovia, and M. D. Marzuela, "Fiber Optic Sensing of a Carbon Dioxide Based on Excited State Proton Transfer to a Luminescent Ruthenium (II) Complex," *Anal. Chem.*, Vol. 64, 1992, pp. 2210–2215.

[107] Papkovsky, D. B., G. V. Ponomarev, S. F. Chernov, A. N. Ovchinnikov, and I. N. Kurochkin, "Luminescent Lifetime Based Sensor for Relative Air Humidity," *Sensors and Actuators*, Vol. B 22, 1994, pp. 57–61.

[108] Burr, J. G., ed., *Chemiluminescence and Bioluminescence*, New York: Marcel Dekker, 1985.

[109] Van Dyke, K., ed., *Bioluminescence and Chemiluminescence: Instruments and Applications*, Vols. 1 and 2, Boca Raton: CRC Press, 1985.

[110] Blum, L. J., and S. M. Gautier, "Bioluminescence and Chemiluminescence Based Fiberoptic Biosensors," Chap. 10 in *Biosensor: Principles and Applications*, L. J. Blum, and P. R. Coulet, eds., New York: Marcel Dekker, 1991.

[111] Sutherland, R. M., C. Dähne, J. F. Place, and A. R. Ringrose, "Immunoassays at a Quartz-Liquid Interface: Theory, Instrumentation and Preliminary Application to the Fluorescent Immunoassay of Human Immunoglobulin G," *J. Immunol. Methods*, Vol. 74, 1984, pp. 253–265.

[112] Love, W. F., L. J. Button, and R. E. Slovacek, "Optical Characteristics of Fiber Optics Evanescent Wave Sensors," in *Biosensors With Fiberoptics*, D. L. Wise, and L. B. Wingard, eds., Clifton, NJ: Humana Press, 1991, pp. 139–180.

[113] Sutherland, R. M., C. Dähne, J. F. Place, and A. R. Ringrose, "Optical Detection of Antigen-Antibody Reactions at a Glass-Liquid Interface," *Clin. Chem.*, Vol. 30, 1984, pp. 1533–1538.

[114] Place, J. F., R. M. Sutherland, and C. Dähne, "Opto-Electronic Immunosensors: A Review of Optical Immunoassays at Continuous Surfaces," *Biosensors*, Vol. 1, 1986, pp. 321–334.

[115] Lee, E. H., R. E. Bremmer, J. B. Fenn, and R. K. Chang, "Angular Distribution of Fluorescence From Liquids and Monodispersed Spheres by Evanescent Excitation," *Appl. Opt.*, Vol. 63, 1979, pp. 862–868.

[116] Sloper, A. N., J. K. Deacon, and M .T. Flanagan, "A Planar Indium Phosphate Monomode Waveguide Evanescent Field Immunosensor," *Sensors and Actuators*, Vol. B 1, 1990, pp. 589–591.

[117] Slovacek, R. E., W. F. Love, and S. C. Furlong, "Application of a Plastic Evanescent Wave Sensor to Immunological Measurements of CKMB," *Sensors and Actuators*, Vol. B 29, 1995, pp. 67–71.

[118] MacCraith, B. D., G. O'Keeffe, C. McDonagh, and A. K. McEvoy, "LED-Based Fibre Optic Oxygen Sensor Using Sol-Gel Coating," *Electron. Lett.*, Vol. 30, 1994, pp. 888–889.

[119] MacCraith, B. D., C. M. McDonagh, G. O'Keeffe, E. T. Keyes, J. G. Vos, B. O'Kelly, and J. F. McGilp, "Light Emitting Diode Based Oxygen Sensing Using Evanescent Wave Excitation of a Dye Doped Sol-Gel Coating," *Opt. Eng.*, Vol. 33, 1994, pp. 3861–3866.

[120] Golden, J. P., L. C. Shriver-Lake, N. Narayanan, G. Patonay, and F. S. Ligler, "A Near IR Biosensor for Evanescent Wave Immunoassays," *Proc. SPIE–Int. Soc. Opt. Eng.*, Vol. 2138, 1994, pp. 241–245.

[121] De Grandpré, M. D., and L. W. Burgess, "A Fiber Optic FT-NIR Evanescent Field Absorbance Sensor," *Appl. Spectrosc.*, Vol. 44, 1990, pp. 273–279.

[122] Heo, J., M. Rodrigues, S. J. Saggesse, and G. H. Sigel, "Remote Fiber Optic Chemical Sensing Using Evanescent Wave Interactions in Chalcogenide Glass Fibers," *Appl. Opt.*, Vol. 30, 1991, pp. 3944–3951.

[123] Margalit, E., H. Dodiuk, E. M. Kosower, and A. Katzir, "Silver Halide Fiber Optic Evanescent Wave Spectroscopy for in Situ Monitoring of the Chemical Processes in Adhesive Curing," *Surf. Interface Anal.*, Vol. 15, 1990, pp. 473–478.

[124] Carlyon, E. E., C. R. Lowe, D. Reid., and I. Bennion, "A Single Mode Fibre Optic Evanescent Wave Biosensor," *Biosens. Bioelectron.*, Vol. 7, 1992, pp. 141–146.

[125] Golden, J. P., L. C. Shriver-Lake, G. P. Anderson, R. B. Thompson, and F. S. Ligler, "Fluorometer and Tapered Fiber Optic Probes for Sensing in the Evanescent Wave," *Opt. Eng.*, Vol. 31, 1992, pp. 1458–1462.

[126] Hale, Z. M., and F. P. Payne, "Demonstration of an Optimized Evanescent Field Optical Fiber Sensor," *Anal. Chim. Acta*, Vol. 293, 1994, pp. 49–54.

[127] Muhammad, F. A., and G. Stewart, "D-Shaped Optical Fibre Design for Methane Gas Sensing," *Electron. Lett.*, Vol. 28, 1992, pp. 1205–1206.

[128] Jin, W., G. Stewart, B. Culshaw, S. Murray, and D. Pinchbeck, "Parameter Optimization in a Methane Detection System Using a Broadband Source and Interferometric Signal Processing," *Proc. SPIE–Int. Soc. Opt. Eng.*, Vol. 2068, 1994, pp. 109–119.

[129] Liedberg, B., C. Nylander, and I. Lundström, "Surface Plasmon Resonance for Gas Detection and Biosensing," *Sensors and Actuators*,Vol. A 4, 1983, pp. 299–304.

[130] Kretschmann, E., "The Determination of the Optical Constants of Metals by Excitation of Surface Plasmons," *Z. Physik*, Vol. 241, 1971, pp. 313–324.

[131] Liedberg, B., I. Lundström, and E. Stenberg, "Principles of Biosensing With an Extended Coupling Matrix and Surface Plasmon Resonance," *Sensors and Actuators*, Vol. B 11, 1993, pp. 63–72.

[132] Van Gent, J., "Surface Plasmon Resonance Based Chemo-Optical Sensors," Thesis, University of Twente, Netherlands, 1990.

[133] Homola, J., "Model of a Chemo-Optical Sensor Based on Plasmon Excitation in Thin Silver Films," *Sensors and Actuators*, Vol. B 11, 1993, pp. 481–485.

[134] Matsubara, K., S. Kawata, and S. Minami, "Multilayer System for High Precision Surface Plasmon Resonance Sensor," *Opt. Lett.*, Vol. 15, 1990, pp. 75–77.

[135] Van Gent, J., P. V. Lambeck, R. J. Bakker, T. J. A. Popma, E. J. R. Sudhölter, and D. N. Reinhoudt, "Design and Realization of a Surface Plasmon Resonance Based Chemo-Optical Sensor," *Sensors and Actuators*, Vol. A 25/27, 1991, pp. 449–452.

[136] Van Gent, J., P. V. Lambeck, H. J. M. Kreuwel, G. J. Gerritsma, E. J. R. Sudhölter, D. N. Reinhoudt, and T. J. A. Popma, "Optimization of a Chemo-Optical Surface Plasmon Resonance Based Sensor," *Appl. Opt.*, Vol. 29, 1990, pp. 2843–2849.

[137] Kreuwel, H., "Planar Waveguide Sensors for the Chemical Domain," Thesis, University of Twente, Netherlands, 1988.

[138] Lofas, S., M. Malmqvist, I. Ronnberg, E. Stenberg, B. Liedberg, and I. Lundström, "Bioanalysis With Surface Plasmon Resonance," *Sensors and Actuators,* Vol. B 5, 1991, pp. 79–84.

[139] Namira, E., T. Hayashida, and T. Arakawa, "Surface Plasmon Resonance Study for the Detection of Some Biomolecules," *Sensors and Actuators,* Vol. B 24/25, 1995, pp. 142–144.

[140] Garces, I., C. Aldea, and J. Mateo, "Four Layer Chemical Fibre Optic Plasmon Based Sensor," *Sensors and Actuators,* Vol. B 7, 1992, pp. 771–774.

[141] Johnston, K. S., R. C. Jorgenson, S. S. Yee, and A. Russel, "Characterization of Porous Sol-Gel Films on Fiber Optic Surface Plasmon Resonance Sensors," *Proc. SPIE–Int. Soc. Opt. Eng.,* Vol. 2068, 1994, pp. 87–93.

[142] Collino, R., J. Therasse, F. Chaput, J. P. Boilot, and Y. Levy, "Sol-Gel Glasses as New Waveguides for Optical Immunosensors," *Mater. Res. Soc. Symp. Proc.,* Vol. 346, 1994, pp. 1033–1038.

[143] Clerc, D., and W. Lukosz, "Integrated Optical Output Grating Coupler as Refractometer and (Bio)-Chemical Sensor," *Sensors and Actuators,* Vol. B 11, 1993, pp. 461–465.

[144] Lukosz, W., "Integrated Optical Chemical and Direct Biochemical Sensors," *Sensors and Actuators,* Vol. B 29, 1995, pp. 37–50.

[145] Kunz, R. E., and I. U. Kempen, "Miniature Integrated Optical Refractometer Chip," *Proc. SPIE–Int. Soc. Opt. Eng.,* Vol. 2208, 1994, pp. 124–129.

[146] Freiner, D., R. E. Kunz, D. Citterio, U. E. Spichiger, and M. T. Gale, "Integrated Optical Sensors Based on Refractometry of Ion Selective Membranes," *Sensors and Actuators,* Vol. B 29, 1995, pp. 277–285.

[147] Welker, D. J., and M. G. Kuzyk, "Optical and Mechanical Multistability in a Dye-Doped Polymer Fiber Fabry-Perot Waveguide," *Appl. Phys. Lett.,* Vol. 66, 1995, pp. 2792–2794.

[148] Butler, M. A., "Optical Fiber Hydrogen Sensor," *Appl. Phys. Lett.,* Vol. 45, 1984, pp. 1007–1009.

[149] Butler, M. A., and D. S. Ginley, "Hydrogen Sensing With Palladium Coated Optical Fibers," *J. Appl. Phys.,* Vol. 64, 1988, pp. 3706–3712.

[150] Farahi, F., P. Akharan-Leilabady, J. D. C. Jones, and D. A. Jackson, "Optical Fiber Flammable Gas Sensor," *J. Phys. E.,* Vol. 20, 1987, pp. 432–434.

[151] Brandenburg, A., R. Edelhäuser, and F. Hutter, "Integrated Optical Gas Sensors Using Organically Modified Silicates at Sensitive Films," *Sensors and Actuators,* Vol. B 11, 1993, pp. 361–374.

[152] Kraus, G., and G. Gauglitz, "Application and Comparison of Algorithms for the Evaluation of Interferograms," *Fresenius J. Anal. Chem.,* Vol. 344, 1992, pp. 153–157.

[153] Brecht, A., G. Gauglitz, and W. Nahm, "Interferometric Measurements Used in Chemical and Biochemical Sensors," *Analusis,* Vol. 20, 1992, pp. 135–140.

[154] Gauglitz, G., A. Brecht, G. Kraus, and W. Nahm, "Chemical and Biochemical Sensors Based on Interferometry at Thin (Multi)-Layers," *Sensors and Actuators,* Vol. B 11, 1993, pp. 21–27.

[155] Lukosz, W., "Principles and Sensitivities of Integrated Optical and Surface Plasmon Sensors for Direct Affinity Sensing and Immunosensing," *Biosens. Bioelectron.,* Vol. 6, 1991, pp. 215–225.

[156] Stamm, C., and W. Lukosz, "Integrated Optical Difference Interferometer as Refractometer and Chemical Sensor," *Sensors and Actuators,* Vol. B 11, 1993, pp. 171–181.

[157] Stamm, C., and W. Lukosz, "Integrated Optical Difference Interferometer as Biochemical Sensor," *Sensors and Actuators,* Vol. B 18/19, 1994, pp. 183–187.

[158] Kunz, R. E., G. Duveneck, and M. Ehrat, "Sensing Pads for Hybrid and Monolithic Integrated Optical Immunosensors," *Proc. SPIE–Int. Soc. Opt. Eng.,* Vol. 2331, 1994, pp. 2–17.

[159] Liu, Y., P. Hering, and M. O. Scully, "An Integrated Optical Sensor for Measuring Glucose Concentration," *App. Phys. Bi. Photophys. Laser Chem.,* Vol. B 54, 1992, pp. 18–23.

[160] Brecht, A., J. Ingenhoff, G. Gauglitz, "Direct Monitoring of Antigen-Antibody Interactions by Spectral Interferometry," *Sensors and Actuators,* Vol. B 6, 1992, pp. 96–100.

[161] Schipper, E. F., R. P. H. Kooyman, R. G. Heideman, and J. Greve, "Feasibility of Optical Waveguide

Immunosensors for Pesticide Detection: Physical Aspects," *Sensors and Actuators*, Vol. B 24/25, 1995, pp. 90–93.

[162] Hess, P., and J. Pelzl, eds., *Photoacoustic and Photothermal Phenomena*, Berlin: Springer Series Opt. Sci., Vol. 58, 1987.

[163] Pliska, P., and W. Lukosz, "Integrated Optical Acoustical Sensors," *Sensors and Actuators*, Vol. A 41/42, 1994, pp. 93–97.

[164] Hand, D. P., S. Freeborn, P. Hodgson, T. A. Carolan, K. M. Quan, H. A. MacKenzie, and J. D. C. Jones, "Optical Fiber Interferometry for Photoacoustic Spectroscopy in Liquids," *Opt. Lett.*, Vol. 20, 1995, pp. 213–215.

[165] MacKenzie, H. A., G. B. Christison, P. Hodgson, and D. Blanc, "A Laser Photoacoustic Sensor for Analyte Detection in Aqueous Systems," *Sensors and Actuators*, Vol. B 11, 1993, pp. 213–220.

[166] Hodgson, P., K. M. Quan, H. A. MacKenzie, S. S. Freeborn, J. Hannigan, E. M. Johnston, F. Greig, and T. D. Binnie, "Applications of Pulsed Photoacoustic Sensors," *Sensors and Actuators*, Vol. B 29, 1995, pp. 339–344.

[167] Edwards, H. O., and J. P. Dakin, "Gas Sensors Using Correlation Spectroscopy Compatible With Fibre Optic Operation," *Sensors and Actuators*, Vol. B 11, 1993, pp. 9–19.

[168] Dakin, J. P., H. O. Edwards, and B. H. Weigl, "Progress With Optical Gas Sensors Using Correlation Spectroscopy," *Sensors and Actuators*, Vol. B 29, 1995, pp. 87–93.

[169] Anderson, B., A. Brodsky, and L. Burgess, "Application of Grating Light Reflection in Analytical Sensors," *Proc. SPIE–Int. Soc. Opt. Eng.*, Vol. 2293, 1994, pp. 80–96.

[170] Azzam, R. M. A., and N. M. Bashara, *Ellipsometry and Polarized Light*, Amsterdam: North Holland, 1977.

[171] Striebel, C., and G. Gauglitz, "Spektrale Ellipsometrie als Optische Messtechnik Kombiniert mit Rechnerischer Simulation," *GIT Fachz. Lab.*, Vol. 2, 1992, pp. 144–148.

[172] Wang, J., "Cylindrical Waveguide Evanescent Field Ellipsometry," *Opt. Eng.*, Vol. 31, 1992, pp. 1432–1435.

[173] Märtensson, J., H. Arwin, and I. Lundström, "Thin Films of Phthalocyanines Studied With Spectroscopic Ellipsometry: An Optical Gas Sensor?" *Sensors and Actuators*, Vol. B 1, 1990, pp. 134–137.

[174] Lehmann, H., M. E. Lippitsch, W. Ecke, W. Haubenreisser, R. Willsch, and D. Raabe, "Phase Sensitive Polarimetric Sensing in the Evanescent Field of Single Mode Fibres," *Sensors and Actuators*, Vol. B 29, 1995, pp. 410–415.

[175] Dakin, J. P., "Distributed Optical Fiber Sensor Systems," Chap. 15 in *Optical Fiber Sensors, Vol. 2*, B. Culshaw, and J. P. Dakin, eds., Norwood, MA: Artech House, 1991, pp. 575–598.

[176] Takeo, T., and H. Hattori, "U-Shaped Fiber Optic Refractive Index Sensor and Its Application," *Proc. SPIE–Int. Soc. Opt. Eng.*, Vol. 1544, 1991, pp. 282–286.

[177] Kvasnik, F., and A. D. McGrath, "Distributed Chemical Sensing Utilising Evanescent Wave Interactions," *Proc. SPIE–Int. Soc. Opt. Eng.*, Vol. 1172, 1989, pp. 75–82.

[178] Meltz, G., W. W. Morey, and J. R. Dunphy, "Fiber Bragg Grating Chemical Sensor," *Proc. SPIE–Int. Soc. Opt. Eng.*, Vol. 1587, 1992, pp. 350–361.

[179] Michie, W. C., B. Culshaw, I. MacKenzie, M. Konstantakis, N. B. Graham, C. Moran, F. Santos, E. Bergqvist, and B. Carlstrom, "Distributed Sensor for Water and pH Measurements Using Fiber Optics and Swellable Polymeric Systems," *Opt. Lett.*, Vol. 20, 1995, pp. 103–105.

Chapter 9

Light Sources, Photodetectors, and Optical Components

9.1 LIGHT SOURCES

The range of light sources available for sensors is very wide. They can be continuous, modulated in intensity or frequency (from dc to megahertz frequencies), and pulsed, either with a single narrow pulse or train of them (from 10^{-12} to 1 sec). Three main categories can be distinguished for chemical sensing applications: filament and gas discharge lamps, LEDs, and lasers [1–3]. A source can be characterized by its directional spectral emission and the ratio of its luminance compared to that of black body radiation at the same temperature.

9.1.1 Lamp Sources

Filament (incandescent) lamps and gas discharge lamps have a very broad spectral range with incoherent emission. They are called *white light* sources.

The most common is the tungsten lamp. The filament is contained in a quartz or glass bulb full of inert gas. The highest filament temperatures are obtained with the tungsten-halogen lamp; iodine is added to the bulb atmosphere and undergoes a recycling tungsten-iodine compound, which decreases the rate of filament evaporation and avoids darkening the bulb. The spectral emission is typically 320 to 2,600 nm (2.6 μm) with maximum luminance occurring at around 1 μm. Smaller size filaments improve the light coupling efficiency into an optical fiber. The lifetime is normally 10^3 to 10^5 hours. Low-voltage lamps, driven by batteries, are useful for portable instruments.

Gas discharge tubes consist of two electrodes inside a bulb filled with a neutral gas. The gas determines the spectral emission range. Continuous-wave arc lamps are generally preferred in the UV region. Deuterium has a continuum UV emission decreasing strongly with increasing wavelength and is only of limited use with optical fibers. Xenon gas arc lamps have a continuous spectral emission in the visible region. The short arc mercury vapor lamp is the highest radiance source, but it has discrete lines in the UV and visible regions. These sources require a transparent quartz window in the UV. A comparison of the relative irradiance of these lamps coupled into a UV-transmitting optical fiber is shown in Figure 9.1. The usable wavelength range in the UV region decreases with increasing length of the fiber. The spectral sensitivity of the detector also introduces limitations in the spectral response curve above 700 nm, resulting in a general bell-shaped response curve.

Flash lamps are filled with inert gases (xenon, krypton) in a bulb containing electrodes for producing intense electron discharges. Xenon lamps have a high radiance between 300 and 1,200 nm and are well suited for fluorescence applications. However, the discharge occurs over an extended length and moves position

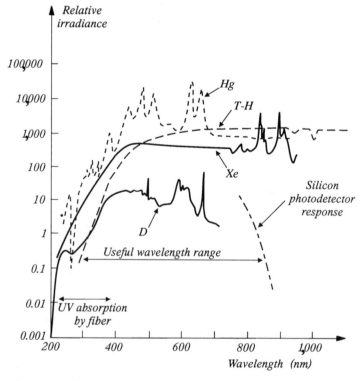

Figure 9.1   Relative spectral irradiance of various sources (lamps) coupled into a 10m length of silica fiber. (D: deuterium source; T-H: tungsten-halogen; Hg: mercury vapor; Xe: xenon arc lamp.) Spectral absorption by the fiber reduces the output, more noticeably in the UV region. The sensitivity of a silicon photodetector decreases the system response in the NIR (dotted curve on the right).

with each flash. Thus, coupling to an optical fiber is less efficient and requires special optical components. The short arc compact lamp and low-cost photographic flash bulb with appropriate optics are good sources for the near-UV region.

9.1.2 Light-Emitting Diodes

LEDs are spontaneous emission light sources usable for incoherent sensors. They emit light into a wide Lambertian cone with a very short coherence length (>30 μm) and generally with poor spatial coherence [4]. Characteristic optical parameters include the shape and peak wavelength of the emission spectrum, the size of the emitting source, the angle of emission (numerical aperture), and the luminous distribution over this emission angle. The wavelength of the emitted light depends on the basic material and is inversely proportional to the band gap energy: SiC (480 nm), GaP (540 nm), GaAsP (640 nm), GaAlAs (650 to 850 nm), GaAs (950 nm), and InGaAs (1.2 to 1.7 μm).

The two basic structures are the surface-emitting and edge-emitting types. They are commercially available with microlenses and coupled directly to a multimode optical fiber. The high-radiance surface-emitting LED produces light by recombination of electron-hole pairs inside a *pn* junction (Figure 9.2(a)). The edge-emitting LED has a heterojunction structure (Figure 9.2(b)) in which a charged zone (about 100 nm) is surrounded by confinement potential wells (50 nm). This structure is similar to a laser diode but without the lasing effect. A superluminescent diode (SLD) produces a strong population inversion and provides a significant gain by stimulated emission amplification.

LED sources have narrow spectral widths (20 to 40 nm) and are available from 460 nm to the NIR region. Their output power (typically 1 to 10 mW) decreases for shorter wavelength LEDs. The advantages are low cost, long lifetime, small dimensions, and ease of amplitude or frequency modulation. Thus, they are used in simple sensors (e.g., pH sensors). However, they exhibit drifts with temperature and aging, which are large drawbacks for chemical sensing specifications.

9.1.3 Laser Sources

Many types of lasers are commercially available [5] and widely used in spectroscopy [6] and fiber-optic instrumentation. The light from a laser is generated in a cavity between partially reflecting mirrors acting as an optical resonator that is designed for the lowest order transverse electromagnetic mode (TEM_{00}). The main characteristics of the laser, such as the wavelength, the divergence half angle of the beam, the waist (radius of a Gaussian beam in the propagation direction), and mode, are specified by the manufacturer.

9.1.4 Gas Lasers

The He-Ne laser emits continuous light, produced by a plasma discharge creating atomic collisions, at 632.8 nm and, in the green, at 543.5 nm, which makes it suitable

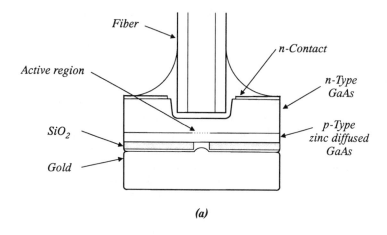

(a)

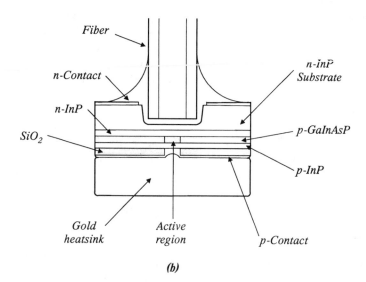

(b)

Figure 9.2 Cross section of LEDs coupled to an optical fiber: (a) surface-emitting diode fabricated in GaAs and (b) heterostructure edge-emitting diode.

for measuring rhodamine compounds. He-Ne lasers are small in size and are low-cost but are limited in power. They require an external modulator for time decay measurements [7]. Their main advantage is that they are much cheaper than other gas lasers. However, they are unusable in Raman spectroscopy because their light power is limited.

Argon and argon-krypton ion lasers, excited through electron collisions, generate large optical powers, but require water cooling (or in some cases air cooling) and are expensive. They are suitable for fiber-optic sensors because of a low beam divergence (<1 mrad). However, damage can occur with very high injected power at the entrance face of the fiber, and the beam is normally slightly defocused. A number of wavelengths in the visible (457.9, 488, and 514.5 nm) and near UV (351.1 nm) are available for fluorescence excitation of organic molecules or for SERS. In particular, the 488-nm band is frequently used with fluorescein labels.

Other lasers include the N_2 laser (337 nm) as the exciting source for uranium solutions or as a pump for dye lasers. The CO_2 laser (or mixture of $CO_2 + N_2 + He$) emits at 10.6 μm for IR applications.

Major drawbacks of gas lasers (others than He-Ne lasers) are their cost and their complex operating conditions. Also, they induce strong unwanted luminescence and Raman bands in optical fibers, a well-known background noise in chemical sensing [8–10].

9.1.5 Dye Lasers

Dye lasers can be pulsed or continuous. Light is generated in the optical cavity by an organic dye dissolved in a suitable solvent. Dye lasers pumped by Nd-YAG lasers generate radiation over a wide wavelength range, from 0.3 to 4 μm. The lasers are costly and relatively unstable; thus, in sensing they are used only for research and not commercial sensing applications.

9.1.6 Solid-State Lasers

The ruby laser (Al_2O_3 doped with Cr_2O_3), emitting at 694.3 nm, and Nd-YAG laser, emitting at 1.06 μm, have optical cavities formed by the crystal itself. Nd-YAG lasers are used in NIR spectroscopy, particularly for SERS and Fourier transform Raman spectroscopy (FTRS). By pumping an optical fiber with a Q-switched Nd-YAG laser, a set of Raman peaks extending into a continuous region is obtained and has been employed for laser photoacoustic sensing [11].

Laser diodes are heterostructure semiconductors, similar to LEDs. They have a resonant cavity resulting in multiple passes through the gain region that overcome the losses at each reflection by a large enough population inversion. The simplest form is a Fabry-Perot cavity, which has mirrors at each end of the lasing cavity. Most diode lasers have been developed for optical communications as continuous or pulsed sources, with continuous output power of 1 to 10 mW into a monomode fiber and kilowatt peak pulse power. As for LEDs, the emission wavelength depends on the material: around 670 nm for GaInP, 750 to 905 nm for GaAlAs, and 1,100 to 1,600 nm for InGaAsP. Diode lasers are valuable sources for interferometric sensors due to their ease of modulation and coupling with optical fibers, their stability, and their long lifetime. In phase modulation fluorometry (see Chapter 8), laser diode sources can be modulated by the driving current and allow a very simple instrumentation. However, their use for fluorescence time decay sensing requires red or NIR fluorophores [12]. Further developments for lasers in the visible region are needed for chemical sensing.

9.2 PHOTODETECTORS

9.2.1 Principles

A photodetector [13,14] is an element that detects radiant energy in a wavelength band to produce an output electrical signal proportional to the amount of light energy absorbed. It is an essential element in the optical-electronic transduction process. The responsivity (R) is given by

$$R = V/P_0 \tag{9.1}$$

where V is the output voltage (in volts) and P_0 the incident light power (in watts). The noise equivalent power (NEP) is defined from the detector noise level (V_n) compared to the minimum detectable output signal (V_{min}) as

$$NEP = V_n/V_{min} \tag{9.2}$$

The detectivity (D) is the reciprocal of the NEP (in W^{-1}) and the specific detectivity (D^*) measured in $cm \cdot Hz^{1/2} \cdot W^{-1}$ is

$$D^* = D \cdot (A_s \cdot \Delta f)^{1/2} = (A_s \cdot \Delta f)^{1/2}/NEP \tag{9.3}$$

where A_s is the sensitive area of the detector and Δf the electrical frequency bandwidth. The sensitivity (S_λ) at a given wavelength λ is expressed in $A \cdot W^{-1}$:

$$S_\lambda = I_\lambda/\Phi_0 \tag{9.4}$$

where I_λ is the photocurrent of the detector and Φ_0 the luminous flux.

Different types of photodetectors for chemical applications, their characteristics, various noise mechanisms, and signal-to-noise models have been analyzed [1]. Photodetectors can be divided into three classes: thermal, photoemissive, and solid-state photodetectors. Thermal detectors have a relatively broad flat spectral range and are generally used as power meters. They are not discussed here. The whole range of electrical frequencies in optical sensors can be covered in three categories [15]: dc to 100 kHz for low-cost discrete sensors, 100 kHz to 40 MHz for heterodyne sensor systems, and 40 MHz to several gigahertz for OTDR techniques and especially for distributed sensors.

9.2.2 Photoemissive Detectors

Photoemissive detectors are based on the emission of electrons from a photocathode in a semiconductor. Photomultiplier tubes (PMT) are the best known photoemissive detectors, consisting of a gas-filled phototube with built-in electron multiplier. Photoelectrons are accelerated by an electric field across multiple plates (dynodes), which have a high secondary emission coefficient (ξ). The amplification G of a photomultiplier having n dynodes of elementary gain (g) and a collection efficiency (p) between the photocathode and the first dynode is

$$G = p \cdot (\xi \cdot g)^n \qquad (9.5)$$

This gain can attain 10^6 to 10^7 or 10^8 for pulsed gains of a tube with 10 to 12 dynodes. Photomultipliers are best used for photon counting. Their sensitivity and wavelength range depend on the photocathode material (Figure 9.3). They are often used in standard spectrometers in the near-UV and visible regions up to 1 μm. In practice, photocathodes are identified by an S-number given by the manufacturer (e.g., S1 for an AgOCs photocathode).

9.2.3 Solid-State Photodetectors

In solid-state photodetectors, the absorption of a photon in a semiconducting element generates a mobile hole-electron pair. Hence, light can induce either a change in resistivity of a photoconductive element (photoconductor) by electrons generated in the conduction band and holes in the valence band, or a voltage variation across a *pn* junction in a photovoltaic detector (a reverse biased *pin* diode, avalanche diode).

Photoconductors are divided into three classes: intrinsic (e.g., PbS, CdTe, undoped silicon or germanium), extrinsic (e.g., doped germanium or silicon), and free carrier devices (e.g., InSb) according to the photon energy level and the possible types of transition within the material (Figures 9.4 and 9.5). The energy gap (E_g in electron volts) of an intrinsic photoconductor determines the long-wavelength cutoff limit (λ_g):

$$\lambda_g = 1.24/E_g \qquad (9.6)$$

In extrinsic photoconductors, a dopant can act as an electron donor or as an acceptor. A flow of current is established from an external source by the change in resistivity of the semiconductor. However, these extrinsic photoconductors must be operated at

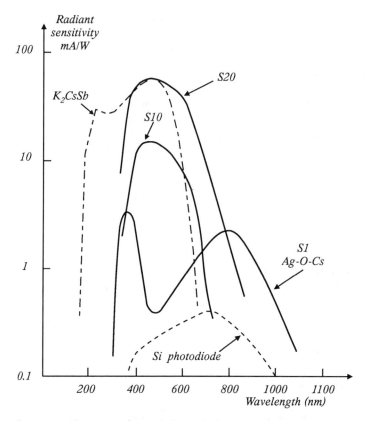

Figure 9.3   Relative spectral response of several photomultipliers: (S1) Ag-O-Cs type; (S10) BiAgOCs type with lime glass window; (S20) trialkali KNa₂SbCs type with lime glass window and K₂CsSb type with fused quartz window. Also shown for comparison is the response curve for a solid-state silicon photodiode.

cryogenic temperatures to avoid undesirable effects due to thermally induced carrier generation. This is a major disadvantage for sensing applications.

The best known junction photodetectors are reverse-biased *pin* diodes. An inner photoelectric effect occurs in a semiconductor when the incident photons are absorbed in the material creating an electron-hole pair in a biased *pn* junction to produce an open-circuit voltage or a short-circuit current. The junction can be a *pin* type by sandwiching an intrinsic semiconductor between *p*- and *n*-layers. A depletion zone, of a few microns thick, is formed in which electrons and holes diffuse (Figure 9.5). The diffusion of charges is due to the concentration difference of electrons and holes across the junction. Consequently, the current voltage characteristic is nonlinear. In the absence of incident light, a reverse bias current (dark current) is also observed.

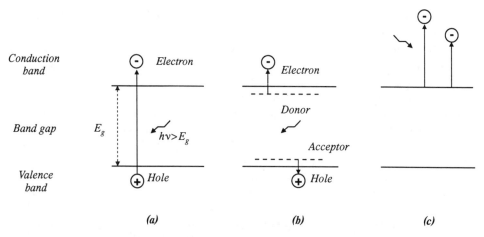

Figure 9.4 Energy level diagram showing possible transitions in a solid-state photoconductor. A photon with an energy greater than the energy gap E_g of the material will create an electron-hole pair: (a) an intrinsic photoconductor, with transitions occurring from the valence band to the conduction band; (b) an extrinsic photoconductor with transitions either from deep donor states to the conduction band or from the valence band to deep acceptor states; and (c) intraband transitions within the conduction band (free carrier photoconduction).

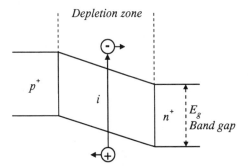

Figure 9.5 Energy level diagram of a *pin* junction photodiode showing the depletion zone and the band gap. Electrons and holes, created by absorption of photons, diffuse out of the intrinsic region under the influence of an electric field.

Hence, the total measured diode current (I_d) is the sum of the dark current (I_0) and current due to photons (I_λ) defined in (9.4):

$$I_d = I_0 + I_\lambda = I_0 + S_\lambda \cdot \Phi_0 \tag{9.7}$$

The most common photodiode materials are silicon, germanium, GaAs, InAs, and InSb. The minimum detectable power for a silicon photodiode is about 10^{-13}W to

10^{-15}W . A comparison of the spectral sensitivity of a PMT and a silicon photodiode is given in Figure 9.3.

The normal type of junction is a *pn* type in homojunction detectors; however, a number of detectors have a heterojunction structure such as GaAs-Ga/Al/As, PbTe-Pb/Sn/Te, or Ga/In/As/P systems. In addition, the *pn* junction can be separated by a weak electron donor, providing additional electrons that collide with atoms in the lattice, creating an internal gain or multiplying effect. These detectors (e.g., InGaAs type) are called *avalanche photodiodes* (APD). However, the characteristics change with temperature and they require compensation for drift. A metal-semiconductor interface can also be added in a configuration called a Schottky barrier.

Table 9.1 summarizes the performance of the most common detectors used in sensing at ambient temperature. More information can be obtained in specialized books [14,16,17].

Linear diode arrays (a number of diodes in a single line) and solid-state imaging arrays (diodes in a two-dimensional matrix), called CCDs, offer interesting possibilities for multiwavelength detection. In these devices, the free charges generated by the absorbed photons are stored in the depletion region of each diode. A charge transfer is produced by synchronizing a voltage pulse (clock) for a chosen integration time, and the charge is shifted sequentially from diode to diode. This results in a readout at the

Table 9.1

Performance at Ambient Temperatures of Some Solid-State Photodetectors

Type	Cutoff (μm)	D^* (cm · Hz$^{1/2}$ · W^{-1})	Response time (μs)	Range (μm)
Photoconductors				
PbS	2.3	1.10^{11}	300	1–3 (cooled)
PbSe	5	1.10^{10}	2	1–8 (cooled)
TL$_2$S	1.1	2.10^{12}	500	
Hg/Cd/Te	5.5	4.10^{9}	0.3	2–20
InSb	7.3	4.10^{8}	0.2	
InAs	3.8	1.10^{8}	0.2	
Junction Detectors				
GaAs	0.9	4.10^{11}	1,000	0.6–0.7
GaAs/AlGaAs				0.8–1.1
Si	1.1	2.10^{12}	1,000	0.2–1.0
Ge	1.8	4.10^{11}		0.5–1.8
InAs	3.4	7.10^{9}	2	1–3
InGaAs/InP	2.6			1.1–2.1

Note: D^* = Specific detectivity. See (9.3).

end of the diode array, giving a multiplexed signal for each pixel (individual diode) as a video line readout (shown in Figure 9.6). The number of pixels is usually a multiple of 128 (256, 512, 1,024, or 2,048) with pixel (diode) sizes of 13 or 27 μm.

Multiwavelength spectrometry is possible using a monochromator, which disperses the light onto a diode array (Figure 9.7). Each pixel detects the light in a small wavelength range, allowing simultaneous measurement of a broad wavelength spectrum [18]. Diode arrays used for spectroscopy have very tall and narrow pixels (typically 25 μm × 2.5 mm for a linear array) to be well adapted to the slit and grating geometry of the monochromator. Further developments are in progress for

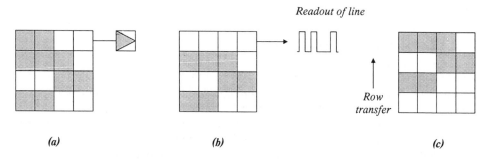

(a)   *(b)*   *(c)*

Figure 9.6   Schematic diagram of charge transfer in a matrix diode array: (a) the integration period, in which each diode accumulates charge depending on the amount of light received; (b) sequential readout of line 1; and (c) transfer of all rows to the next higher row ready for readout of line 2.

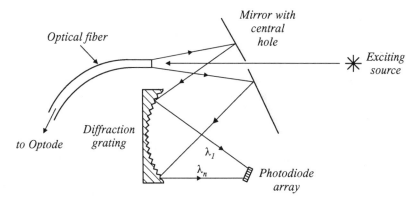

Figure 9.7   Use of a photodiode array for multiwavelength fluorescent spectrometry. Light from the exciting source is coupled through a hole in the mirror into the fiber. Fluorescent light returning from the optode is reflected from the mirror onto a diffraction grating, which disperses the light onto the photodiode array.

x-ray-sensitive silicon CCD chips, NIR arrays, and cooled detectors, which offer new opportunities for laboratory and sensing systems [19].

9.3 PASSIVE COMPONENTS

This section presents a brief summary of passive optical components that are important in chemical sensor technology. More detailed descriptions can be found in books on conventional and modern optics [20,21], optical fibers [22–24], optical processing [25,26], and fiber-optic sensors [27]. Passive optical components have fixed optical properties, as opposed to active components (such as sources, detectors, and modulators), that can be controlled by external means.

9.3.1 Conventional Components

The simplest components are lenses and mirrors that can adapt the light beam at the input and output of an optical fiber, for absorption, reflectance, and fluorescence measurements. A typical arrangement for fluorescence sensing, shown in Figure 9.7, has a mirror, either flat, elliptical, or spherical, containing a hole for injecting the exciting light into the input fiber. Emitted light from the sensing fiber is focused onto a photodiode array by a mirror and a spherical or parabolic holographic grating.

Dichroic mirrors are interference filters that act as couplers or beam splitters and separate the incident and emitted light. Interferometric filters, such as Mach-Zehnder, Michelson, and Fabry-Perot types, are frequently adapted for fiber sensing systems.

Wavelength selection can be obtained by narrow bandpass and short or long wavelength pass optical filters. Bandpass filters have a maximum transmission (T) at a selected wavelength, and the bandpass is defined at the points where the transmission drops to $T/2$. Short and long pass filters are defined for cut-on or cutoff wavelengths, respectively, at which half the maximum transmission occurs. These filters are often employed in fluorescence measurements to reduce stray light and to act as prefilters for higher resolution interference filters. Monochomators produce high wavelength resolution (separation) by dispersion from a prism (in a transmissive mode) or a diffraction grating (dispersion by reflection). The important parameters of diffraction gratings are their size (which determines the effective aperture, or $F\text{-}number$), the number of lines ruled on the grating (typically 300 to 2,400 lines/mm, determining the resolution), and the blaze angle. Holographic gratings are produced at low cost by interference fringes obtained from laser beams projected onto photoresist layers deposited on a glass support. Compared to conventional mechanically ruled gratings, they usually have poorer rejection of stray light.

9.3.2 Fiber-Optic Components

Conventional passive components, such as polarizers, phase modulators, frequency shifters, and nonlinear components, have been developed as miniaturized all-fiber or integrated-optic designs, which are well suited to fiber sensor systems. Furthermore, the waveguide properties allow new functions and components [28,29]. As examples, a range of components are based on SELFOC® or graded refractive index (GRIN) lenses. The transmission property of GRIN lenses is that their magnitude is unity and that a desired focal length can be easily obtained by choosing their length. GRIN lenses can be easily coupled to optical fibers for use in chemical spectroscopy [30] and used as a quarter-pitch able to accommodate a larger beam width and reduce it to a smaller size, and vice versa. These GRIN rods can also serve as couplers, multiplexers, and other branching components [31].

Tapers are beam expanders for increasing or decreasing the fiber core size and numerical aperture and hence can serve as coupling elements between monomode and multimode fibers. A conical taper, with an elongated fiber and integrated spherical 10-μm-radius microlens, can produce a high coupling efficiency of 55% to a monomode fiber. Directional couplers, with a Y-branched or star geometry, divide the input light power between several fibers. They are fabricated from optical fibers by three different methods: etching, fusing, or mechanical polishing. These couplers can have polarization-maintaining properties or can separate the polarizations with low or high birefringence. Their main performance parameter properties (directivity, insertion loss, polarization dependence) have been summarized [28].

Methods for joining or connecting two optical fibers include permanent connections (splices) or demountable connectors [32]. Typical losses of a single connector are 0.5 to 1.5 dB due to misalignment of the two fibers, which can also affect the modal characteristics and hence the time dispersion.

9.3.2.1 Wavelength-Division Multiplexing

Sensing systems based on wavelength-division multiplexing (WDM) [33] employ multiple wavelengths for different channels (e.g., one wavelength for the signal and another for the reference) and require passive components for separating and combining several wavelengths. A typical system with fiber directional couplers is shown in Figure 9.8.

Fiber wavelength filters are made of doped fibers with selectively absorbing cores [34]. Other fiber wavelength filters have been realized based on the evanescent properties of the waveguide and the fiber mode dispersion [35]. For instance, the loss of a curved monomode fiber increases with wavelength (λ) if the part of the cladding near the core is modified to have an index (n_1) greater than the effective core index (n_c). The wavelength effect is due to the increasing difference ($n_1 - n_c$), since n_c varies

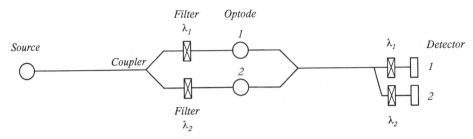

Figure 9.8   Principle of fiber WDM. A single light source is split by a coupler into two fibers. The two fibers are distinguished by two different wavelength filters, λ_1 and λ_2, which are detected by two photodetectors. Many other arrangements are possible.

with wavelength $f(\lambda)$. A second layer with an n_s index (equal to cladding index) has a guiding and coupling effect when $n_1 > n_c > n_s$ [29].

Evanescent field gratings of (p) period, fabricated on the polished surface of a fiber, result in a wavelength peak $\lambda_p = 2n_c \cdot p$ for the back-reflected signal into the fiber. The use of gratings on fibers on the polished support of the fiber or on a planar waveguide has produced many components such as filters, mirrors, switches, and modulators usable in an integrated sensor concept (see Chapter 10).

9.3.2.2 Spatial Multiplexing

Spatial multiplexing systems can be designed around optical switches to select between the different input and output channels. The switches are mechanical devices with mirrors and lenses or use electro-optic switching. Switches can also be realized with integrated optics in which the optical beam is deflected by piezoelectrical or acousto-optical effects (see Section 9.4). Spatial multiplexing is employed in the measurement of several optodes in commercial equipment, particularly in spectrophotometry.

In a typical system, two spatial switch elements allow both the source (white light) and the detector (e.g., a photodiode array) to be switched between the various optodes. Synchronization is necessary between the two switches. The wavelength selection is obtained by a grating (Figure 9.9(a)). Other systems employ combined spatial and wavelength multiplexing. In Figure 9.9(b), several sources (e.g., LEDs with different wavelengths) illuminate several optodes and a spatial switch directs the light to a single detector.

The Fresnel reflections that occur from fiber end faces and optical components constitute a major difficulty for spatial multiplexing systems. At present, these systems are not satisfactory for fluorescence and Raman spectrometry or for techniques in which the light levels are very low.

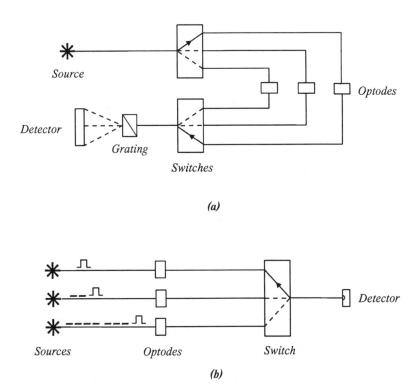

Figure 9.9   Spatial multiplexing of optodes for absorption measurements: (a) Using a single source and detector with a grating for multiwavelength detection (e.g., with a photodiode array). The two mechanical switches are synchronized to measure each optode in turn. (b) With several different sources and a single detector. Each source can be distinguished by electrical time multiplexing.

9.4 ACTIVE COMPONENTS

Active components have optical properties that may be changed by the effect of an external field [36].

9.4.1 Effects

The most important electro-optic effects employed in active components are the Pockels and the Kerr effects. In the Pockels effect, the change in refractive index of the material is directly proportional to the electric field (i.e., a linear electro-optic effect). Birefringent crystals of KH_2PO_4, $LiNbO_3$, and $LiTaO_3$ are often used in a planar configuration [37–39]. For the Kerr effect, the change in refractive index is proportional to the square of the electric field. Ceramic polycrystalline materials based on lead, lanthanum, zirconium, and titanium oxide compositions are used, and the birefringence is altered by selecting their composition.

An acoustic wave can generate a periodic mechanical strain in a material by the photoacoustic effect, resulting in a dynamic diffraction grating formed by the refractive index variations, which will diffract the optical wave. This diffraction is first-order for Bragg gratings and multiple-order for Raman-Nath systems. These acousto-optical effects are the basis for beam deflectors, Q-switching, multifrequency modulators, and components for signal processing (e.g., Fourier analysis). A typical integrated acousto-optic modulator can be fabricated in an X-cut $LiNbO_3$ waveguides with indiffused titanium [40].

The Faraday magneto-optic effect produces a rotation of the polarization that is independent of the direction of propagation of the light. This effect is applied to magnetic field sensing and is used for polarization rotation controllers [27].

9.4.2 Important Active Fiber Components

Polarization controllers transform an input polarization state into a desired output state. Polarization control is achieved for polarization-preserving (maintaining) fibers by lateral pressure on the fiber (squeezing the fiber) or by coiling or bending it to induce asymmetric radial strains and birefringence [41]. In addition, fiber polarizers have been fabricated by polishing the side of the fiber almost down to the core and then placing it against a birefringent crystal or adding a thin layer metal cladding to remove one of the polarization states. Other polarizers are made by altering the shape of the fiber so that one of the polarization states is below the cutoff (e.g., with special fiber structures such as the D-shaped or bow-tie fiber; see Chapter 7). Fibers do not maintain polarization over a long distance, however.

Modulators are components that act on the phase of the optical wave. They are used for phase modulation stabilization or heterodyne detection of the optical phase shift. One construction has a piezoelectric cylinder (e.g., of lead zirconate titanate (PZT)) wound with several turns of an optical fiber. Application of an electric field expands or contracts the cylinder, altering the length of the fiber, thereby introducing a phase shift. A second type comprises a piezoelectric plastic jacket that is coated on the fiber. In chemical sensing, these modulators are employed mainly in interferometric techniques and in gas detection by real-time correlation spectrometry [42]. Frequency shifters with birefringent fibers induce a change in the polarization state for a shift in frequency for heterodyne detection and switching. Other fiber frequency shifters include two-mode filters and phase modulators with a time-dependent sawtooth waveform (serrodyne frequency shifter).

Continuing development of components for chemical sensing will offer commercial integrated-optic devices made on planar waveguides [43]. Planar surface microrelief structures such as a laser diode-to-fiber connecting lens array [44] or a high-index TiO_2/SiO_2 mixed oxide film (n_f = 1.83) deposited on polycarbonate support (n_s = 1.57) [45] are well suited to low-cost mass replication.

References

[1] Modlin, D. N., and F. P. Milanovich, "Instrumentation for Fiber Optic Chemical Sensors," Chap. 6 in *Fiber Optic Chemical Sensors and Biosensors, Vol. 1*, O. S. Wolfbeis, ed., Boca Raton: CRC Press, 1991, pp. 237–302.

[2] Yurek, A. M., and A. Dandridge, "Optical Sources," Chap. 5 in *Optical Fiber Sensors*, J. P. Dakin, and B. Culshaw, eds., Norwood, MA: Artech House, 1988, pp. 151–187.

[3] Ungar, S., *Fibres Optiques, Theorie et Applications*, Paris: Dunod, 1989.

[4] Gillensen, K., and W. Schairen, *Light Emitting Diodes: Introduction*, Englewood Cliffs, NJ: Prentice Hall, 1987.

[5] Wilson, J., and J. B. F. Hawkes, *Lasers. Principles and Applications*, Englewoods Cliffs, NJ: Prentice Hall, 1987.

[6] Demtröder, W., *Laser Spectroscopy*, Berlin: Springer-Verlag, 1981.

[7] Lakowicz, J. R., and H. Szmacinski, "Fluorescence Lifetime Based Sensing of pH , Ca^{2+}, K^+ and Glucose," *Sensors and Actuators*, Vol. B 11, 1993, pp. 133–143.

[8] Dakin, J. P., and A. J. King, "Limitations of a Single Optical Fiber Fluorimeter System Due to Background Fluorescence," *IEE Proc.*, Vol. 131, Pt H., 4, 1984, pp. 273–275.

[9] Boisdé, G., W. Carvalho, P. Dumas, and V. Neuman, "Fluorescence Emission by Molecular Imputities in Irradiated and Non Irradiated Silica," *Proc. SPIE–Int. Soc. Opt. Eng.*, Vol. 506, 1984, pp. 196–201.

[10] Myrick, M. L., and S. M. Angel, "Elimination of Background in Fiber Optic Raman Measurements," *Appl. Spectrosc.*, Vol. 44, 1990, pp. 565–570.

[11] MacKenzie, H. A., G. B. Christison, P. Hodgson, and D. Blanc, "A Laser Photoacoustic Sensor for Analyte Detection in Aqueous Systems," *Sensors and Actuators*, Vol. B 11, 1993, pp. 213–220.

[12] Szmacinski, H., and J. R. Lakowicz, "Fluorescence Lifetime-Based Sensing and Imaging," *Sensors and Actuators*, Vol. B 29, 1995, pp. 16–24.

[13] Dereniak, E. L., and D. G. Crowe, *Optical Radiation Detectors*, New York: Wiley, 1984.

[14] Dennis, P. N. J., *Photodetectors: An Introduction to Current Technology*, New York: Plenum Press, 1986.

[15] Debney, B. T., and A. C. Carter, "Optical Detectors and Receivers," Chap. 4 in *Optical Fiber Sensors*, J. P. Dakin, and E. J. Culshaw, eds., Norwood, MA: Artech House, 1988, pp. 107–149.

[16] Sze, S. M., *Physics of Semiconductors Devices*, 2nd ed., New York: Wiley, 1981.

[17] Shur, M., *Physics of Semiconductor Devices*, Englewood Cliffs, NJ, Prentice Hall, 1990.

[18] Wolfbeis, O. S., G. Boisdé, and G. Gauglitz, "Optochemical Sensors," Chap. 12 in *Sensors: A Comprehensive Survey, Vol. 2*, W. Göpel, J. Hesse, and J. N. Zemel, eds., Weinheim: VCH, 1991, pp. 573–645.

[19] Andrews, D. L., *Perspectives in Modern Chemical Spectroscopy*, Berlin: Springer-Verlag, 1990.

[20] Smith, W. J., *Modern Optical Engineering*, New York: McGraw Hill, 1966.

[21] O'Shea, D. C., *Elements of Modern Optical Design*, New York: Wiley, 1986.

[22] Suematsu, Y., ed., *Optical Devices and Fibers*, Vol. 17, Amsterdam: OHM-North Holland, 1985.

[23] Briley, B. E., *An Introduction to Fiber Optics System Design*, Amsterdam: North Holland, 1988.

[24] Hentschel, C., *Fiber Optics Handbook*, Germany: Boeblingen, 1988.

[25] Horner, J. L., ed. *Optical Signal Processing*, San Diego: Academic Press, 1987.

[26] Vanderlugt, A., *Optical Signal Processing*, Wiley Interscience, 1992.

[27] Dakin, J. C., and B. Culshaw, eds., *Optical Fiber Sensors: Vol. I, Principles and Components*, 1988; *Vol. II, Systems and Applications*, Norwood, MA: Artech House, 1989.

[28] Digonnet, M. J. F, and B. Y. Kim, "Fiber Optic Components," Chap. 7 in *Optical Fiber Sensors: Vol. I, Principles and Components*, J. C. Dakin and B. Culshaw, eds., Norwood, MA: Artech House, 1988, pp. 209–248.

[29] Goure, J. P., "Composants à fibres optiques," *Ecole d'Eté, Systèmes Optiques IESC*, Les Ulis, Fr.: Editions de Physique, 1992, pp. 377–397.

[30] Landis, D. A., and C. J. Seliskar, "Fiber Optics /GRIN Lens Couples for Use in Chemical Spectroscopy," *Appl. Spectrosc.*, Vol. 49, 1995, pp. 547–555.

[31] Tomlinson, W. J., "GRIN-Rod Lenses in Optical Fiber Communication Systems," *Appl. Opt.*, Vol. 19, 1980, pp. 1127–1138.

[32] Miller, C. M., S. C. Mettler, and I. A. White, *Optical Fiber Splices and Connectors: Theory and Methods*, New York: Marcel Dekker, 1986.

[33] Senior, J. M., and S. D. Cusworth, "Wavelength Division Multiplexing in Optical Fibre Sensor Systems and Networks: A Review," *Opt. and Laser Technol.*, Vol. 22, 1990, pp. 113–126.

[34] Farries, M. C., J. F. Townsend, and S. B. Poole, "Very High Reflective Optical Fibre Filters," *Electron. Lett.*, Vol. 22, 1986, pp. 1126–1128.

[35] Parriaux, O., F. Bernoux, and G. Chartier, "Wavelength Selective Distributed Coupling Between Single Mode Fibers for Multiplexing," *J. Opt. Commun.*, Vol. 2, 1981, pp. 105–109.

[36] Carenco, A., "Composants actifs," *Ecole d'Eté, Systèmes Optiques IESC*, Les Ulis, Fr: Editions de Physique, 1992, pp. 93–143.

[37] Klein, R., and E. Voges, "Integrated Optic Ammonia Sensor," *Sensors and Actuators*, Vol. B 11, 1993, pp. 221–225.

[38] Stamm, C., and W. Lukosz, "Integrated Optical Difference Interferometer as Refractometer and Chemical Sensor," *Sensors and Actuators*, Vol. B 11, 1993, pp. 177–181.

[39] Hinkov, I., and V. Hinkov, "Two Layer Waveguiding Structure in $LiNbO_3$ With Birefringence Reversal for Refractive Index Sensors With Large Measurement Range," *IEEE J.Lightwave Technol.*, Vol. 11, 1993, pp. 554–559.

[40] Hinkov, I., and V. Hinkov, "Integrated Acousto-Optic Collinear TE-TM Converter for 0.8 µm Optical Wavelength Range," *Electron. Lett.* Vol. 27, 1991, pp. 1211–1213.

[41] Lefevre, H., *The Fiber-optic Gyroscope*, Norwood, MA: Artech House, 1993.

[42] Edwards, H. O., and J. P. Dakin, "Gas Sensors Using Correlation Spectroscopy Compatible With Fibre-Optic Operation," *Sensors and Actuators*, Vol. B 11, 1993, pp. 9–19.

[43] Kersten, R. T, "Integrated Optics for Sensors," Chap. 9 in *Optical Fiber Sensors: Vol. I, Principles and Components*, J. C. Dakin and B. Culshaw, eds., Norwood, MA: Artech House, 1988, pp. 277–317.

[44] Rossi, M., G. L. Bona, and R. E. Kunz, "Arrays of Anamorphic Phase-Matched Fresnel Elements for Diode-To-Fiber Coupling," *Appl. Opt.*, Vol. 34, 1995, pp. 2483–2488.

[45] Kempen, L. U., R. E. Kunz, and M. T. Gale, "Micromolded Structures for Integrated Optical Sensors," *Proc. SPIE-Int. Soc. Opt. Eng.*, Vol. 2639, 1995, pp. 278–285.

Chapter 10

Optodes and Sensing Cells

10.1 CLASSIFICATION OF OPTODES

The terms *optode* and *optrode* were first introduced simultaneously by the same authors [1,2] for CO_2 partial pressure (pCO_2) measurement by fluorescence detection. *Optrode* has been accepted in English-speaking countries [3], while *optode* is generally preferred in Europe. In this book, we have used *optode* for describing optical waveguide sensitive elements, since the concept of *optical way* (from Οπτικος in Greek) seems more appropriate for the majority of these sensor devices. Photoelectrode sensors (see Section 10.5) might be better described as optrodes.

Optical fiber sensors are often classified into two main classes: extrinsic and intrinsic types. This division can also be applied to optodes. In extrinsic optodes, the light is allowed to exit the optical fiber and is modulated in a separate zone before being relaunched back into a fiber. In intrinsic optodes, the light is modulated in response to the measurand while still being guided [4]. From a chemical standpoint, three possible configurations of increasing complexity have been considered [5], as shown in Figure 10.1.

Figure 10.2 summarizes different types of optodes. The principle of extrinsic optodes is derived from general optochemical sensor concepts (e.g., when source and receiver are placed near the measuring cell in a bulk optic system), for which waveguides are not necessary but offer advantages. In intrinsic optodes, planar waveguides or optical fibers form an integral part of the sensing scheme. In the case of a chemical-optical transduction (Figure 10.2(b,c,f,g)), an immobilized indicator (or label) and reagents are included in a solid or lipidic phase in a thin film or a membrane placed on the fiber and acting as a heterogeneous-phase optode (liquid-solid). When an intermediate analyte (B) is measured (Figure 10.2(c,g)), a multilayer system (e.g., with addition of a membrane between the sample analytes (A) and (B))is necessary. Examples of all these different types of optodes are given below.

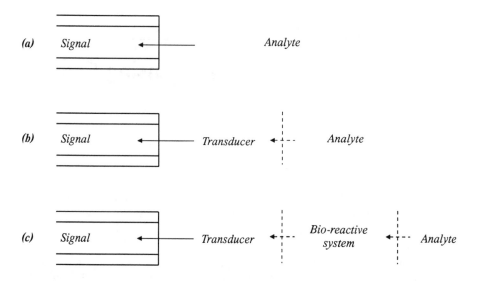

Figure 10.1   Classification of chemical fiber sensors: (a) direct measurement of the intrinsic optical properties of the analyte; (b) a chemical-optical transducer (such as an immobilized indicator) converts changes in the analyte into an optical signal; (c) when no indicator is known, the analyte is converted by a bioreactive system (e.g., an enzyme) into a compound that can be detected. (*Source:* [5]. Reprinted with permission.)

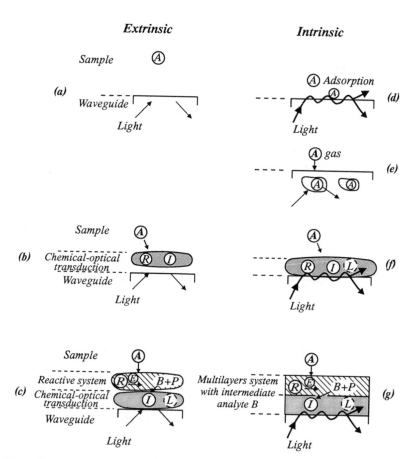

Figure 10.2 Classification of optodes into extrinsic (a–c) and intrinsic (d–g) sensor elements following Figure 10.1: (a) direct measurement; analyte A is in a homogeneous phase; (b) system with chemical-optical transduction using a reagent R and an indicator I; (c) use of an intermediate analyte, such as B or P by enzyme (E) reaction; (d) analyte adsorbed on the surface of the waveguide interacts directly with the optical evanescent wave; (e) analyte inside a porous optode; (f) a chemical-optical transduction obtained with the reagent R and indicator I or label L; and (g) an intermediate analyte is necessary for the measurement. The systems in (c,g) are multilayers with heterogeneous phases.

10.2 EXTRINSIC OPTODES

10.2.1 Sensing Cells

In most cases, the usual optical elements, such as lenses, mirrors, and glass or transparent plastic windows, can be used to obtain a light beam well adapted to fiber-optic probes or flow-through cells. For sensing cells with a long optical path length, the length limitation is about 50 cm for a tubular geometry where the light rays are almost parallel. For gas measurements, a White cell (with multiple internal reflections) equipped with fibers has been tried [6]. Long path cells including optical fibers, shown in Figure 10.3, have been conceived for measurements of trace uranium ions in solution with a typical optical path length of about 0.75m and a volume of 3 ml [7].

The small diameter of optical fibers allows cells having a very short optical path length (1 mm) and very low volume (1 to 3 µl) for flow injection analysis (FIA) or liquid chromatography. In FTIR, cells with two different path lengths have been proposed for obtaining high dynamic range [8]. Sometimes a membrane with immobilized indicator and reagent is included in the cell. Tubular cells containing several fibers with

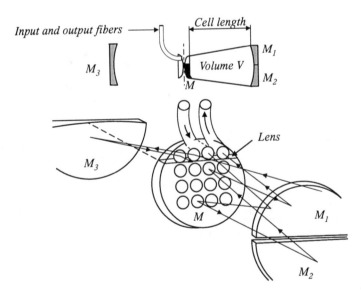

Figure 10.3   Example of an absorption cell with a long path length (White cell) connected to optical fibers for remote measurements. Multiple reflections between the three mirrors M, M_1, and M_2 produce successive images of the light from the input fiber. After several passages (in this case 16 with the mirror M_3), the light is focused into the output fiber. This increases the path length, which can be up to 1m for liquids and 30m for gases. (*After:* [7]).

immobilized compounds on their surface have been used to increase the sensitivity of the sensing element [9].

10.2.2 Simple Fiber Optodes

In the majority of extrinsic sensors, simple plain fibers (such as PCS or glass rods) are employed to transport the light in and out of the optode. The optode can be *passive*, such as a probe without reagent, or *reactive* with immobilized reagents.

Assuming that the numerical aperture of the source is equal or greater than that of the optical fiber (NA), the optical power guided by the fiber is proportional to the number of coupled bound modes P_m [10]:

$$P_m = \pi \cdot k_F \cdot I_\lambda \cdot A_f \cdot (NA)^2 \cdot \Delta\lambda/n_0^2 \qquad (10.1)$$

where I_λ is the radiance per unit wavelength, $\Delta\lambda$ the spectral width, A_f the cross section of the guiding area of the fiber, n_0 the refractive index of the medium surrounding the fiber, and k_F is the Fresnel reflection coefficient between the core (refractive index n_1) and the surrounding medium:

$$k_F = [(n_1 - n_0)/(n_1 + n_0)]^2 \qquad (10.2)$$

In fluorescence measurements, two main considerations determine the choice of the optode geometry and sensing arrangement. First, the interaction of the exciting light with the fiber produces interfering or parasitic light due to fluorescence from impurities and color centers in the fiber, as well as Raman and Brillouin scattering coming from the core or the cladding/buffer region. A single fiber, for both transmission of the exciting light and collection of the return signal, is the simplest arrangement and well adapted to remote measurement at long distances and on small samples, but the detectability is limited by noise in the background fluorescence signal. The net signal (S), which depends on the fluorescence emission collection efficiency, and the background noise (N) have been compared between a single-fiber configuration and double fibers (separate exciting and collection fibers) for remote sensing [11]. The double configuration provides improvement in the signal-to-noise ratio (S/N) by a factor of 300 and by a factor of 5 better for the detection limit. Indeed, measurements with the double-fiber arrangement are only limited by dark current noise.

Second, the optode collection efficiency depends on the numerical aperture of the fibers, their diameter, the angle (β) between the exciting and the collecting fibers, and the distance between the centers of these fibers.

The total fluorescence power (P_f) received by an optical fiber, for the configuration in Figure 10.4, is given by integrating over the whole sensing volume V:

$$dP_f = k_f \cdot P_0 \cdot P_y \cdot A_f \cdot (\Omega/4\pi) \cdot (S_c/S_f) \cdot dV \qquad (10.3)$$

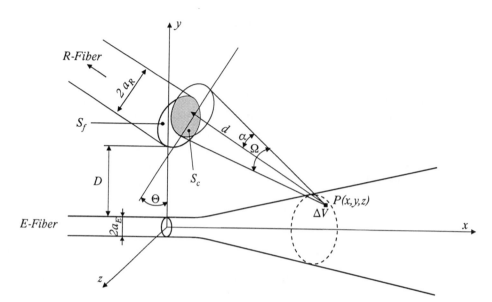

Figure 10.4 Model for the collection of fluorescent light in an optode by a receiving fiber R with end face area S_f. The exciting light comes from a fiber E. The fluorescence is produced in a volume ΔV in a solid angle Ω. The equivalent surface S_c is a function of the R-fiber numerical aperture. It is represented by the common surface (shaded) between S_f and the equivalent surface of the emitted light surface at a distance d. The total fluorescence power collected is given by (10.3).

where k_f is a parameter depending on the fluorescence for a given analyte, P_0 is the power of the excitation beam, P_y is a function characterizing the light power variations on an axis y perpendicular to the propagation direction x, A_f is the attenuation of the fluorescence in the medium, Ω is the solid angle $\Omega = 2\pi\,(1 - \cos \alpha)$, with the half-angle α of the cone at a distance d from the collecting fiber, and the term (S_c/S_f) is the ratio of the collected light area S_c on the end face area of the fiber S_f. A similar mathematical modeling of fluorescence collection efficiency has been applied to uranium trace detection by remote time-resolved fluorescence [12].

All these parameters are determined by the physical construction of the optode, the characteristics of the excitation beam, the diameter of the fibers, the distance D between them, their numerical aperture, and the measurement cell geometry and can be modeled to establish the best arrangement for single-fiber, multifiber, and coaxial fiber systems [13,14] as represented in Figure 10.5. The relative sensitivity of optodes with and without GRIN lenses or Winston cones (tapered cones) has been determined for luminescence and Raman scattering with a single- and double-fiber configuration [15]. The best performance for sampling over large volumes is obtained for a cell with the collecting fiber directly opposite the emitting fiber (*fiber face-to-face configuration*). The single-fiber configuration is best when the signal-to-noise

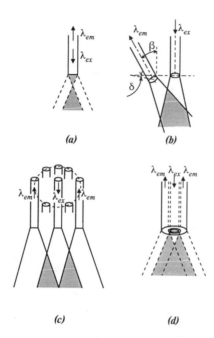

Figure 10.5 Extrinsic optodes using plain fibers without an immobilized reagent for measurement of fluorescence and Raman emission: (a) single fiber, (b) two fibers forming an angle β (from 0 to 90 degrees) in the vertical plane and δ in the horizontal plane, (c) multiple fibers in a circular arrangement, and (d) coaxial fibers. The hatched zone corresponds to the sensed volume.

ratio is expected to be large and for very small measured volumes. With dual fibers in an optode (used as in situ probe), an angle between excitation and collecting fibers gives a more sensitive device; the best efficiency is obtained for $15 > \beta > 30°$ [16].

Other geometries with special optics have been tried, such as a sphere [17], a capillary tube [18], and optics with a flower arrangement [19] or with an associated pierced mirror [20]. All these arrangements can normally be used for Raman spectrometry as well as fluorescence.

10.2.3 Optodes With Immobilized Compounds

A number of chemically reactive optodes have been described in Chapter 2 (Figures 2.6 to 2.9), Chapter 4 (Figures 4.20 to 4.22), Chapter 5 (Figures 5.3 and 5.4), Chapter 8 (Figure 8.12), and in Chapters 11 and 12 on applications. A typical extrinsic multilayer optode uses two membranes, one of which is selectively permeable to gas (e.g., oxygen), the other as a support for the reagent and enzyme (see Figures 5.4 and 11.6). A special configuration consists of a multilayer structure (see Figure 10.6) [21]. This configuration is convenient for a flow-through cell for analyte determination (e.g.,

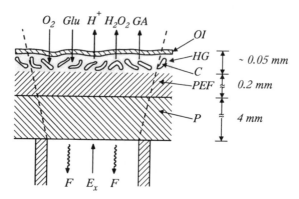

Figure 10.6 Cross section through the sensing layer of a fiber-optic glucose sensor. Abbreviations are as follows: PEF = polyester foil; C = cellulose particles with immobilized dye and enzyme; HG = hydrogel layer; and OI = optical isolation. Oxygen and glucose (Glu) diffuse to the enzyme, whereas protons, gluconic acid (GA), and H_2O_2 diffuse out of the sensing layer. Excitation (E_x) is from below and fluorescence (F) is collected in the same direction. (*After:* [21]).

glucose) by intermediate analyte measurement (e.g., pH detection) and for single-shot tests. The multilayer concept can be extended to intrinsic optodes in interferometry and evanescent spectroscopy.

10.3 INTRINSIC OPTODES

In intrinsic sensors, the fiber or waveguide is intrinsically the sensing element (e.g., in refractometry) or forms an essential part (support) of this element when a chemically sensitive layer is integrated on the support (e.g., in evanescence spectroscopy). Four broad classes of chemical sensors have been defined [22,23]: refractometric sensors, evanescent wave spectroscopic (EWS) sensors, coating-based sensors, and core-based sensors. This classification is used here with some minor amendments.

10.3.1 Refractometric Optodes

Various refractometric probes have been proposed in using optical fibers (Figure 10.7): straight fibers without their cladding [24], short lengths of declad PMMA fiber in a spiral form [25], fibers bent in a U-shape [26] or in a serpentine with reversed curvatures [27], tapered fibers [28], twisted fibers [29], and notches in fibers [30]. The geometry of the probe is often chosen to increase its sensitivity by increasing the higher order modes. Higher modes can also be increased by injecting the light at a higher angle of incidence into a multimode fiber [31]. Planar waveguides have the

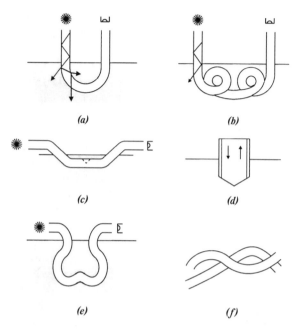

Figure 10.7   Sensors for refractive index measurement: (a) unclad U bend rod, (b) helicoidal rod, (c) slightly bent rod with notch (shown dotted), (d) fiber end with prism shape, (e) serpentine with reversed curvatures, and (f) twisted fibers.

advantage that they can be conceived of as integrated-optic refractometers (see Section 8.3).

10.3.2 Direct Evanescent Wave Spectroscopic Optodes

A direct EWS optode uses the optical evanescent field of guided modes in a waveguide to probe the vicinity of the interface between the core and the cladding. If a measurand molecule changes the optical properties of an absorbance- or fluorescence-based indicator dye, these changes can be measured by the evanescent field (see Section 8.7). The light absorbed in the cladding layer (cladding loss) can be analyzed by evanescence spectroscopy, and conversely the evanescent field can provide a means of coupling fluorescent light into the fiber.

Both fibers and planar waveguides are suitable for direct EWS (see also Section 10.4). For multimode fibers, normally the sample forms the cladding [32] and the measuring depth of the evanescent field can be controlled by the launch optics. The cladding power fraction (see Section 7.5.3.2) is dependent on the reciprocal of the V-parameter [33] over five orders of magnitude of analyte concentration, with only a small deviation proportional to $1/V^2$, for V numbers from 24 to 190. This function has been verified for different refractive indices of the core, cladding and measurand [34].

In direct EWS sensors, variations in the absorption of the sample induce direct changes in the waveguide transmission properties, and TIR or ATR methods can be applied. For instance, a declad section of a multimode plastic-clad fiber (PCS) has been used as an inline ATR cell, since a large number of reflections per unit length are possible [35]. Tunneling modes with spatial filtering were launched into a clad fiber in order to produce modes close to cutoff within the sensing region. Low-order modes can be blocked by an annular beam mask at launch and higher order modes are generated by tapering or bending the fiber.

If the evanescent absorbance of a dye solution varies linearly with the core length, the transmission loss increases with the square root of the analyte concentration. This shows why a major drawback of EWS sensors is *fouling* of the surface, since the surface effect is confined to the highest evanescent field region.

Generally, in direct EWS sensors, the measured absorbance spectra coincide with those obtained by conventional transmission methods. However, in multimode fibers the spectral dispersion can affect the V-parameter. The variation toward lower refractive index in a wavelength band caused by this dispersion induces more power to penetrate into the cladding region, and the fiber is therefore more sensitive to measurand absorbance in this band. In anomalous dispersion, with strong absorption bands, the absorbance peak may appear to be shifted to longer wavelengths.

A model of the cladding loss phenomenon has been given for TE modes and meridional rays of a multimode fiber [36]. The fraction of power in the cladding is typically much less than 1% for a strongly guiding multimode structure. Methods have been tried to enhance the sensitivity by increasing the higher order mode content and adjusting the mode distribution, for example, by using a long section of 0.1-mm fiber with a serpentine mode scrambler fiber entering the flow cell for conventional FTIR [37]. The mode scrambler generates an equilibrium distribution of modes and provides a twofold enhancement in performance.

Glasses derived from sol-gel can provide a potential solution to the weak sensitivity of EWS sensors by producing a porous overlayer [32]. This concept, used for monomode fibers, could be extended to multimode fibers and applied in the IR region.

The extension of EWS optodes to NIR and FTIR is the subject of extensive research, especially for gas sensing. Many probe types have been experimented with: PTFE-clad fluoride glass fiber probes [38], 50/125-µm multimode fibers with stripped cladding for methane measurement [39], ATR cells based on silver halide fibers in the 2- to 20-µm range [40], unclad chalcogenide-based fibers for detection of solvents (acetone, ethyl alcohol) and sulfuric acid [41], and sapphire fibers for monitoring the curing of epoxy resins [42]. The disadvantage of these methods is the long diffusion time of gases for clad fibers and the relatively weak sensitivity that requires the use of a long fiber length.

10.3.3 Intrinsic Coating-Based Optodes

In intrinsic coating-based (ICB) sensors, spectroscopy is performed on a coating placed near the fiber core, where the evanescent field is large. The analyte diffuses into the coating and changes the absorbance [43] or the fluorescence properties. Coatings can be produced by many methods. Fibers may be silanized and treated to make a fluorophore stick on both the cladding and the tip of the fiber [44]. Capillary glass or plastic tubes with chemically sensitive coatings on the inner surface offer an interesting alternative to fibers. The sample cavity has a small internal volume and acts as an optical waveguide (see Figure 7.9(b)). Input and output grating couplers can be added at each end of the capillary to make an integrated sensor. This tube optode has been tested for carbon dioxide measurements [45].

The combination of fluorescence and evanescence wave techniques (see Chapter 8) is common in fiber-optic biosensing. This is described in the book by Wise and Wingard [46], in which different aspects of fiber-optic ICB sensors are discussed: the theory [47], the chemistry and technology [48], and instrumentation [49].

The importance of fiber numerical aperture has been emphasized, both by semiquantitative [50] and quantitative [51] approaches. From these reported experimental results with a declad multimode fiber immersed in fluorescein solution, the excitation of the fluorophore by evanescent interaction and the detection of the fluorescence have been analyzed. The exit cone angle, which contains 90% of the transmitted optical power at the end of the fiber, is defined as $(\theta_0)_{max}$ and the launch numerical aperture is $NA_L = \sin(\theta_0)_{max}$.

The effects of the launch spot size and NA of the material ($NA_m = (n_1^2 - n_2^2)^{1/2}$) have also been determined. The quantitative predictions of the optical model are fully in agreement with the experimental data. If all modes have a high field intensity at the core-cladding interface, the total fluorescence signal (S_t) for an optical Lambertian source of radiance (I_0), a fiber length (L), a fiber radius (a), and the launch spot radius (r_{max}) is

$$S_t \propto I_0 \cdot L \cdot a \cdot (r_{max}/a)^2 \cdot (NA_L)^8 \cdot (NA_m)^{-4} \qquad (10.4)$$

A *button probe* has been developed for a compact portable fiber-optic probe, which supports and aligns the end of the fiber while maintaining the numerical aperture of the sensing element [49]. A ring of $2R$ diameter, slightly smaller than the fiber diameter ($2a$), contacts the fiber at distance Y from the proximal end of the fiber without interfering with the light input. A limiting value of the sensor numerical aperture is given by

$$NA = (a - R) \cdot n_2/Y \qquad (10.5)$$

When the *NA* is smaller than this limiting value, the fluorescent signal varies approximately as NA^9 [49]. A transparent seal can be integrated in the design and may also serve as a lens, which reduces the problems of the alignment ring and handling. An injection-molded polystyrene fiber seal has been designed for a high-NA optical system.

Theoretical modeling of light coupling from sources distributed throughout a cladding to guided modes in the fiber core has been treated in a general way [52]. The fluorescence collection efficiency from the cladding, originally thought to be a linearly increasing function of the V-parameter, is not well described by this parameter [53]. A better parameter is $K_a = 2\pi/\lambda$. However, in this model only the true bound modes in the fiber are considered, while leaky modes, which transport a great deal of energy in short-length fiber fluorosensors, are not taken into account. The main result shows that even for bulk-distributed sources, most of the guided energy is coupled into the fiber core from sources very close to the core surface.

Innovative systems for coating-based fiber-optic sensors are the use of the porous sol-gel technique [54], a transparent jacket with the cladding on its inside surface and air outside [55], and employing two gratings as a fiber Bragg coupler for directing the light out of the core and capturing the resultant fluorescence from the chemical reaction [56]. In another approach, an unjacketed optical fiber was inserted into a capillary tube (300-μm inner diameter) in which the propagating light was mode-filtered due to the change in the critical angle at the core/sensitive-cladding interface along the side of the fiber [57]. In this configuration, a column effect appears as in chromatography due to chemical compounds separating in the fiber cladding.

10.3.4 Core-Based Optodes

In the class of intrinsic core-based optodes, hollow fibers (Figure 10.8(a)) have been tried experimentally as liquid core fibers in spectrophotometry [58,59] and as waveguide capillary flow cells in fluorimetry [60]. However, few liquid solvents exist with a refractive index greater than glasses or silica fibers, such as CS_2 ($n = 1.628$) or toluene ($n = 1.494$). With fluoroethylenepropylene (FEP) tubing ($n = 1.338$), ethanol ($n = 1.359$) and some water-ethanol mixtures can be analyzed [61]. This method has been used to determine concentrations of 0.2 μg · g^{-1} of sulfur in steel [62]. Also, the concept of a regenerable reagent has been tried experimentally with ethylene glycol/water ($n = 1.38$) as a carrier for ammonia detection [63].

Porous fibers and/or cladding (Figure 10.8(b)) with multimodal characteristics are well adapted for gas measurements such as ammonia vapor [64] and carbon monoxide [65] because the diffusion times are relatively short. Of interest for these fibers is that porous polymers can be made by heterogeneous cross-linking polymerization, which can trap indicators and/or reagents in the porous matrix and exhibit both very high gas permeability and liquid impermeability. Also of interest is the use of silica gel products with a porous structure for applications in liquids [66].

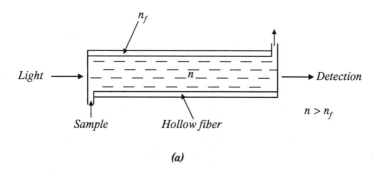

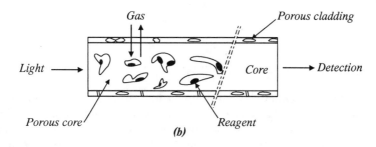

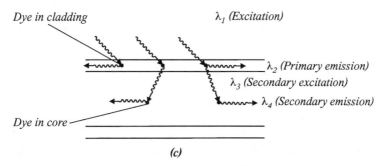

Figure 10.8 Intrinsic optodes: (a) hollow fiber with liquid sample; (b) porous fiber with adsorbed reagent (a porous cladding by itself with reagent can be used as shown in the right-hand side of the figure); and (c) a two-stage fluorescence coupling system (*After:* [67], with permission).

10.3.5 Hybrid ICB and Core-Based Optodes

A generic technique for two-stage fluorescence coupling [67] is shown in Figure 10.8(c). Fluorescent molecules present in the coating emit light after reaction with the analyte, and this fluorescent emission is absorbed by a second series of fluorescent molecules present in the core, which emit their own fluorescence in the core of the fiber. The fiber design maximizes the amount of the captured secondary fluorescence that is trapped in the fiber. The relatively small amount of emission from the coating coupled into the fiber core by evanescent field effects is also converted to secondary fluorescence. The two-stage process of coupling light from the cladding into the core is more efficient, 100 times more so than the simple evanescent fluorescent field approach. This technique is well adapted to distributed fluorescence sensing. Initial experiments with this technique had an oxygen-sensitive dye included in a permeable silicone coating placed on the outside of a plastic fiber with a fluorescent core.

10.4 PLANAR WAVEGUIDE DEVICES

The main advantages of planar waveguide optics for intrinsic chemical-optical transducers are their relatively large surface for interaction between the film coating and the gas or liquid phase chemical compounds to be detected, and also the possibility of integrating optical components (such as multiplexers, directional couplers, gratings) directly into the waveguide design, as well as active optoelectronic components (sources and detectors). Simple and practical geometries can be made conveniently at low cost in injection-molded transparent plastics coated by standard thin-film techniques, and are highly suitable for mounting in biochemical analyzers as disposable probes.

10.4.1 Fluorescent Capillary Fill Device

The fluorescent capillary fill device (FCFD) is a modification of an assay format suggested by Hirschfeld [68] in which the core of an optical fiber is placed inside a capillary tube, forming a cavity. The sample dissolves the labeled antigen that competes with the antigen to be determined for binding to antibodies immobilized on the fiber. Fluorescence is detected from the evanescent fraction of a wave traveling inside the optical fiber.

The elegance of FCFD (Figure 10.9) is based on two modifications: one wall rather than an extra fiber acts as the waveguide, and the sample is drawn up by capillary action. Also, FCFD is an attractive design for a disposable evanescent wave sensor. A planar structure with two reagent layers has been applied to fluoroimmunosensing using a label for interrogating an antibody-antigen reaction [69]. Two sheets of glass or plastic are held apart with a small gap of capillary dimensions (about 100 µm). The inner surfaces are coated with fluorescent labels and

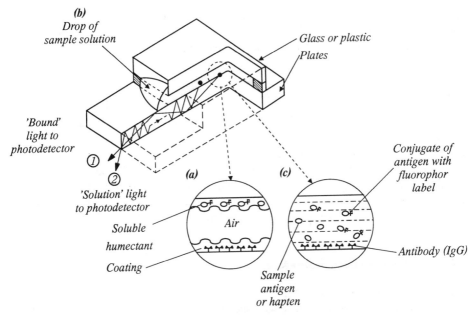

Figure 10.9 Construction of the fluorescence capillary fill device for competitive immunoassay: (a) The empty device has immobilized antibody on the lower coating and a reagent in the upper coating. (b) A drop of sample is sucked into the gap by capillary action and releases the immobilized reagents. (c) The filled device is represented in a competitive binding assay. (After: [69], with permission.)

all the reagents necessary for a competitive binding or a sandwich type immunoassay. Measurement with an optical evanescent wave allows discrimination between bound and free fluorescence, thus avoiding the time-consuming washing and/or separation steps of a normal immunoassay.

In order to account for sensor fluctuations (intensity variations of light source, detector, and loading with fluorescent label), internal references are provided by patterning: (1) a discrete region of reagents on the device surface to give a reference signal, and (2) a blank region in which the intrinsic fluorescence of the sample (blood) is measured [70,71].

Commercially available FCFD planar structures have been summarized recently and compared with optical-fiber systems [72]. The steps in the fabrication process for the sensor include the preparation of top and bottom plates (cleaning, surface activation, spin coating, and drying) and cell assembly (sandwich assembly, scribing and breaking, packaging for storage). These simple operations make the device easy to manufacture.

The inner walls of an FCFD may also be coated with a chromogenic or fluorogenic reagent, preferably in a form that allows rapid dissolution by the sample. Thus, a disposable sensor has been described for *para*-aminophenol (10 to 40 μm · mL^{-1}); determination is by visual inspection compared with known colors of standards [73].

10.4.2 Multianalyte Sensing

Another example of a disposable evanescent wave device has a number of narrow individual strip waveguides on a patterned planar sensor (Figure 10.10) [74]. The design includes multianalyte sensing, internal references, elimination of the washing step, and avoidance of the irreversibility problem of immunosensing (due to the high affinity constant of antibody-antigen reactions). An unambiguous determination is obtained by scanning a photomultiplier across the stripes with different chemical reactions. This multiple-site concept could also be applied to other sensors, such as the FCFD.

10.4.3 Guiding Monolayer Structures

The simplest structure of a chemical planar waveguide sensor consists of a sensitive chemical film coated on a narrow waveguide that is formed or engraved on a solid support. This support can be in plastic or glass, and optical fibers can be used for transfer-ring the light to the planar waveguide. Two configurations for interferometric measurement are shown in Figure 10.11: a structure with light reflection from a metallic mirror (Figure 10.11(a)) used for gas analysis [75] and a Mach-Zehnder

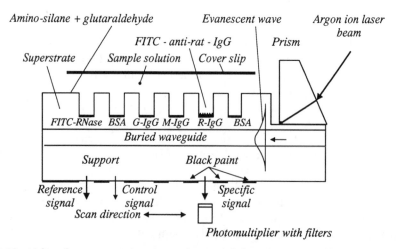

Figure 10.10   Multianalyte sensing with a patterned waveguide for measurement of bovine serum albumin (BSA). Abbreviations are as follows: G = goat, M = mouse, R = rat, IgG = immunoglobulin, and FITC = fluorescein isothiocyanate. (*After:* [74], with permission.)

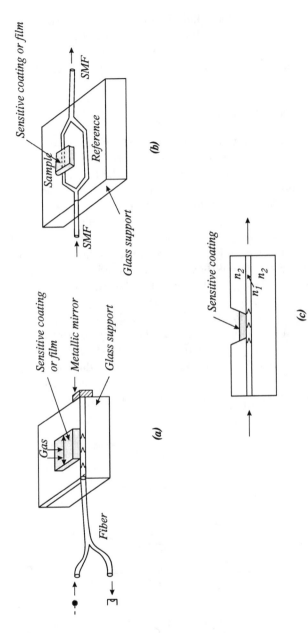

Figure 10.11 Guiding monolayer structures for chemical sensing: (a) cutaway view of a monolayer device sensitive to gas with a mirror for return of the light; (b) an integrated Mach-Zehnder interferometer structure coupled to a single-mode fiber (SMF); and (c) symmetrical planar optical waveguide with an exposed part for sensing.

setup (Figure 10.11(b)) in which the light is divided into two arms of the waveguide and the two beams interfere with each other at the output [76].

Other guiding monolayer structures have been exploited. One example is an embedded slab waveguide, part of which is exposed to allow contact with the chemical sample (see Figure 10.11(c)).

10.4.4 Guiding Multilayer Structures

A waveguide multilayer configuration (Figure 10.12) has been proposed for surface reaction sensing using the technique of TIRF spectroscopy [77]. The structure consists of three layers, namely, two layers of SiO_2 sandwiched around the core of silicon oxygen nitride. The modulation mechanism is based on radiationless energy transfer between donors and acceptors in a sensitive bilayer coating.

The guiding structure in case of SPR sensing is generally multilayer. The simplest structure for SPR consists of a thin metallic layer (thickness of several tens of nanometers) deposited on a waveguide, as shown in Kretschmann's arrangement (see Figure 8.14(b)). Other geometries have been proposed such as the one shown in Figure 10.13 [77,78]. The main guiding structure has the $SiO_2/SiON$ or Si_3N_4/SiO_2 configuration. A short segment of 100 μm can be modified by a surface plasmon multilayered guiding structure in which evaporated silver is the reactive surface placed between a ZrO_2 layer, as the active layer, and a TiO_2 layer to prevent optical coupling between the silver and SiO_2 layers. One or more layers that are sensitive to the measurand are then coated on this second multilayer guiding structure.

These structures can be fabricated with conventional techniques such as silicon-based technology [79,80] for integrated optics [81] and solid-state chemical

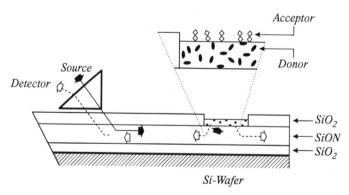

Figure 10.12   Example of a multilayer planar waveguide structure with bilayer coating for TIRF measurement (*After:* [77], with permission). The sensor is based on the modulation of the quantum efficiency of luminescence centers (donors) by radiationless energy transfer to quenching centers (acceptors) immobilized on top of the waveguide. This structure has been tested as a glucose sensor with affinity binding [97].

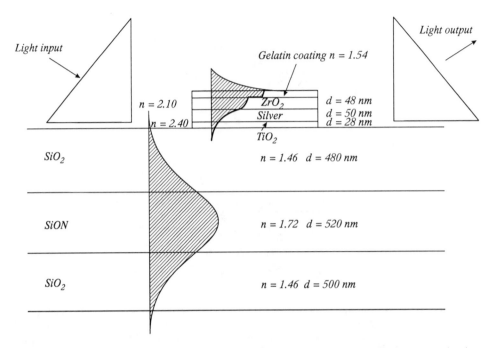

Figure 10.13 A surface plasmon waveguide sensor with a planar multilayer guiding structure. d is the thickness of each layer, and n is the refractive index. (*After:* [78], with permission.)

sensors [82], including titanium in-diffusion in LiNbO$_3$ [83], field-assisted ion exchange in glass [74], and the low-temperature process for producing metal indium phosphate waveguides [84].

10.4.5 Integrated Optical Sensors

Planar waveguides (see Section 7.3) are the basic component for integrated optical sensors (IOSs) [85]. The term *integrated* covers various sensors either as a partially integrated or as a fully integrated form of the IOS concept.

As an example, the integrated optical polarimetric interferometer [86] can be considered as a partially integrated sensor. Indeed, the integrated part of the sensor performs only the chemical-optical transduction. When polarized light is coupled into the end of a multilayer planar waveguide, coherently excited TE$_0$ and TM$_0$ modes are produced; the sensor measures the time-dependent variation of the phase between these modes under the influence of the measurand. In the configuration illustrated in Figure 7.6, the grating is incorporated in the structure and performs an essential function (spectral dispersion).

Totally integrated optical sensors have been proposed and realized [87] based

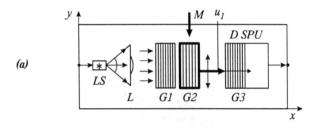

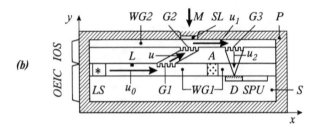

Figure 10.14 A totally integrated optical sensor based on a GREFIN sensing waveguide coupled to an optoelectronic integrated circuit (OEIC): (a) top view and (b) cross section. A light source (LS) illuminates a grating coupler (G1) via a lens (L) by the guided wave (u_0) in the waveguide (WG1). The collimated wave u emerging from G1 is then coupled into the GREFIN waveguide (WG2) by grating coupler G2. Optical changes occur from the measurand (M) in the sensing layer (SL). An absorber (A) suppresses the stray light. A grating coupler (G3) focuses the wave u_1 into the beam u_2 imaged on the detector array (D) and processed by an integrated signal processing unit (SPU). (*After:* [88], with permission)

on an IOS deposited on the top of an optoelectronic integrated circuit [88]. These combine the gradient of effective refractive index (GREFIN) approach with a dual-waveguide scheme (Figure 10.14). In these sensors the nonuniformity of the waveguide is produced by the GREFIN, in which the thickness of the guiding layer varies spatially. The guided photons propagate along a curved trajectory (photon deflection). Light is coupled into the waveguide by a grating in a porous Ta_2O_3 film.

10.5 PHOTOELECTRODE AND OPTOELECTROCHEMICAL SENSORS

Photoelectrode and optoelectrochemical-sensitive elements [89] combining electrical and optical sensing techniques might be best described as *optrodes*, by optical analogy to the term *electrode*. Figure 10.15 illustrates the essential differences between the concepts of an optode and an optrode.

Examples of photoelectrode or optrode sensors are electrically generated

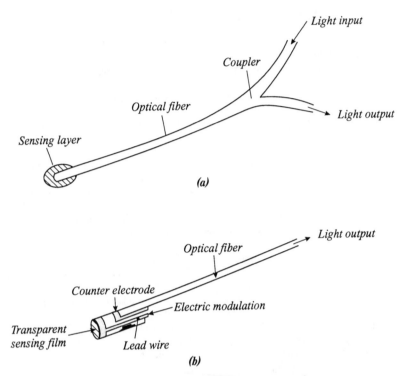

Figure 10.15 Schematic representation of (a) an optode and (b) an optrode.

luminescence [90–92], electrically modulated sensors [93], and electrochemical generation of an indicator [94]. Also, planar optical waveguides can be electrochemically modulated and used to measure fluorescence or plasmon resonance [95,96].

References

[1] Lübbers, D. W., and N. Opitz, "Die pCO$_2$/pO$_2$ Optode," *Z. Naturforsch.*, Vol. 30 C, 1975, pp. 532–533.

[2] Opitz, N., and D. W. Lübbers, "A New Fast Responding Optical Method to Measure pCO$_2$ in Gases and Solutions," *Pflugers Archiv. Eur. J. of Physiol.*, Vol. S 355, 1975, PR120.

[3] Borman, S. A., "Optrodes," *Anal.Chem.*, Vol. 53, 1981, pp. 1616A–1618A.

[4] Culshaw, B., "Basic Concepts of Optical Fiber Sensors," Chap. 2 in *Optical Fiber Sensors, Vol. 1*, J. P. Dakin, and B. Culshaw, eds., Norwood, MA: Artech House, 1988, pp. 9–24.

[5] Wolfbeis, O. S., ed., *Fiber Optic Chemical Sensors and Biosensors, Vol. 1*, Boca Raton: CRC Press, 1991, p. 4.

[6] Inaba, H., T. Kobayasi, M. Hirama, and M. Hanza, "Optical Fibre Network System for Air-Pollution

Monitoring Over a Wide Area by Optical Absorption Method," *Electron. Lett.*, Vol. 15, 1979, pp. 749–751.

[7] Boisdé, G., and A. Boissier, "Photometer With Concave Mirror and Field Optics," U.S. Pat. 4,188,126 and 4,225,232, 1980.

[8] Hirschfeld, T., "Lens and Wedge Absorption Cells for FTIR Spectroscopy," *Appl. Spectrosc.*, Vol. 39, 1985, pp. 426–430.

[9] Kar, S., and M. A. Arnold, "Cylindrical Sensor Geometry for Absorbance Based Fiber Optic Ammonia Sensors," *Talanta*, Vol. 41, 1994, pp. 1051–1058.

[10] Modlin, D. N., and F. P. Milanovich, "Instrumentation for Fiber Optic Chemical Sensors," Chap. 6 in *Fiber Optic Chemical Sensors and Biosensors, Vol. 1*, O. S. Wolfbeis, ed., Boca Raton: CRC Press, 1991, pp. 237–302.

[11] Louch, J., and J. D. Ingle, "Experimental Comparison of Single and Double Fiber Configurations for Remote Fiber Optic Fluorescence Sensing," *Anal. Chem.*, Vol. 60, 1988, pp. 2537–2540.

[12] Moulin C., S. Rougeault, D. Hamon, and P. Mauchien, "Uranium Determination by Remote Time-Resolved Laser-Induced Fluorescence," *Appl. Spectrosc.*, Vol. 47, 1993, pp. 2007–2012.

[13] Plaza, P., N. Q. Dao, M. Jouan, H. Fevrier, and H. Saisse, "Simulation et Optimisation de Capteurs à Fibres Emboitées pour Spectrométrie Raman," *Analusis*, Vol. 15, 1987, pp. 504–507.

[14] Boisdé, G., G. Cardin, G. Mazé, and M. Poulain, "Conduit optique coaxial et procédé de fabrication de ce conduit," Fr. Pat. BF 2,683,638, 1991.

[15] Myrick, M. L., S. M. Angel, and R. Desiderio, "Comparison of Some Fiber Optic Configurations for Measurement of Luminescence and Raman Scattering," *Appl. Opt.*, Vol. 29, 1990, pp. 1333–1344.

[16] Plaza, P., N. Q. Dao, M. Jouan, H. Fevrier, and H. Saisse, "Simulation et Optimisation des Capteurs à Fibres Optiques Adjacentes," *Appl. Opt.*, Vol. 25, 1986, pp. 3448–3454.

[17] Milanovich, F. P., and T. Hirschfeld, "Process, Product, and Waste Stream Monitoring With Fiber Optic," *Adv. Instrum.*, Vol. 38, 1983, pp. 404–418.

[18] MacCreery, R. L., M. Fleichmann, and P. J. Hendra, "Fiber Optic Probe for Raman Spectrometry," *Anal. Chem.*, Vol. 55, 1983, pp. 146–147.

[19] Dao, N. Q., and M. Jouan, "The Raman Laser Fiber Optics (RLFO) Method and Its Applications," *Sensors and Actuators*, Vol. B 11, 1993, pp. 147–160.

[20] Boisdé, G., B. Kirsch, P. Mauchien, and S. Rougeault, "Nouvelle Optode Passive pour la Spectrofluorimétrie et la Spectrométrie Raman," *Proc. Opto 88*. Paris: ESI Pub., 1988, pp. 294–299.

[21] Trettnak, W., M. J. P. Leiner, and O. S. Wolfbeis, "Fibre Optic Glucose Sensor With a pH Optrode as the Transducer," *Biosensors*, Vol. 4, 1989, pp. 15–26.

[22] Lieberman, R. A., "Intrinsic Fiber Optic Chemical Sensors," Chap. 5 in *Fiber Optic Chemical Sensors and Biosensors, Vol. 1*, O. S. Wolfbeis, ed., Boca Raton: CRC Press, 1991 pp. 193–235.

[23] Lieberman, R. A., "Recent Progress in Intrinsic Fiber Optic Chemical Sensing," *Sensors and Actuators*, Vol. B 11, 1993, pp. 43–55.

[24] Cole, C. F., R. A. Sims, and A. J. Adams, "An On-line Process Fiber Optic Refractometer for Measuring Edible Oil Hydrogenation," *Proc. Arkansas Acad. Sci.*, Vol. 44, 1990, pp. 28–29.

[25] Brossia, C. E., and S. C. Wu, "Low Cost in Soil Organic Contaminant Sensor," *Proc. SPIE–Int. Soc. Opt. Eng.*, Vol. 1368, 1990, pp. 115–120.

[26] Takeo, T., and H. Hattori, "Application of a Fiber Optic Refractometer for Monitoring Skin Condition," *Proc. SPIE–Int. Soc. Opt. Eng.*, Vol. 1587, 1992, pp. 284–287.

[27] Harmer, A. L., "Optical Fiber Refractometer Using Attenuation of Modes," First OFS, Inst. Electr. Eng. London, *IEEE*, Vol. 221, 1983, pp. 104–108.

[28] Kumar, K., T. B. V. Subhrahmanyam, A. D. Sharma, K. Thyagarajan, B. P. Pal, and I. C. Goyal, "Novel Refractometer Using a Tapered Optical Fiber," *Electron. Lett.*, Vol. 20, 1984, pp. 534–535.

[29] Smella, E., and J. J. Santiago-Aviles, "A Versatile Twisted Optical Fiber Sensor," *Sensors and Actuators*, Vol. 13, 1988, pp. 117–129. -

[30] Harmer, A. L., "Chemical and Biochemical Instrumentation With Optical Sensors," *Tutorial course,* *SPIE T53,* Boston, O-E/fiber lase, 1988.

[31] Archenault, M., H. Gagnaire, J. P. Goure, and N. Jaffrezic-Renault, "A Simple Intrinsic Refractometer," *Sensors and Actuators,* Vol. B5, 1991, pp. 173–179.

[32] MacCraith, B. D., "Enhanced Evanescent Wave Sensors Based on Sol-Gel Derived Porous Glass Coatings," *Sensors and Actuators,* Vol. B 11, 1993, pp. 29–34.

[33] Paul, P. H., and G. Kychakoff, "Fiber Optic Evanescent Field Absorption Sensor," *Appl. Phys. Lett.,* Vol. 51, 1987, pp. 12–14.

[34] Safaai-Jazi, A., C. K. Jen, and G. W. Farnell, "Optical Fiber Sensor on Differential Spectroscopic Absorption," *Appl. Opt.,* Vol. 24, 1985, pp. 2341–2345.

[35] Ruddy, V., B. MacCraith, and J. A. Murphy, "Evanescent Wave Absorption Spectroscopy Using Multimode Fibers," *J. Appl. Phys.,* Vol. 67, 1990, pp. 6070–6075.

[36] Ruddy, V., "An Effective Attenuation Coefficient for Evanescent Wave Spectroscopy Using Multimode Fiber," *Fiber Integrated Opt.,* Vol. 9, 1990, pp. 142–150.

[37] Colin, T. B., K. H. Yang, and W. C. Stwalley, "The Effect of Mode Distribution Evanescent Field Intensity: Applications in Optical Fiber Sensors," *Appl. Spectrosc.,* Vol. 45, 1991, pp. 1201–1295.

[38] Ruddy, V., and S. MacCabe, "Detection of Propane by IR-ATR in a Teflon-Clad Fluoride Glass Optical Fiber," *Appl. Spectrosc.,* Vol. 44, 1990, pp. 1561–1463.

[39] Tai, H., H. Tanaka, and T. Yoshino, "Fiber Optic Evanescent Wave Methane Gas Sensor Using Optical Absorption for the 3.392 µm Line of a He-Ne Laser," *Opt. Lett.,* Vol. 12, 1987, pp. 437–439.

[40] Simhoney, S., E. M. Kosower, and A. Katzir, "Fourier Transform Infrared Spectra of Organic Compounds in Solution and as Thin Layers Obtained by Using an Attenuated Total Internal Reflectance Fiber-Optic Cell," *Anal. Chem.,* Vol. 60, 1988, pp. 1908–1910.

[41] Heo, J., M. Rodrigues, S. J. Saggesse, and G. H. Sigel, "Remote Fiber Optic Chemical Sensing Using Evanescent Wave Interactions in Chalcogenide Glass Fibers," *Appl. Opt.,* Vol. 30, 1991, pp. 3944–3951.

[42] Druy, M. A., L. Elandjian, W. A. Stevenson, R. D. Driver, G. M. Leskowitz, and L. E. Curtiss, "Fourier Transform Infrared (FTIR) Fiber Optic Monitoring of Composites During Cure in an Autoclave," *Proc. SPIE–Int. Soc. Opt. Eng.,* Vol. 1170, 1988, pp. 150–159.

[43] Giuliani, J. F., H. Wohltjen, and N. L. Jarvis, "Reversible Optical Waveguide Sensor for Ammonia Vapors," *Opt. Lett.,* Vol. 8, 1983, pp. 54–56.

[44] Hurum, D. C., R. Von Wandruszka, and A. E. Grey, "Production of Fluorescent Quartz Fibers for an Optical Sensor," *Anal. Lett.,* Vol. 24, 1991, pp. 905–911.

[45] Weigl, B. H., and O. S. Wolfbeis, "Capillary Optical Sensors," *Anal. Chem.,* Vol. 66, 1994, pp. 3323–3327.

[46] Wise, D. L., and L. B. Wingard, eds., *Biosensors With Fiberoptics,* Clifton, NJ: Humana Press, 1991.

[47] Love, W. F., L. J. Button, and R. E. Slovacek, "Optical Characteristics of Fiberoptic Evanescent Wave Sensors," in *Biosensors With Fiberoptics,* D. L. Wise and L. B. Wingard, eds., Clifton, NJ: Humana Press, 1991, pp. 139–180.

[48] Thompson, R.B., and F. S. Ligler, "Chemistry and Technology of Evanescent Wave Biosensors," in *Biosensors With Fiberoptics,* D. L. Wise and L. B. Wingard, eds., Clifton, NJ: Humana Press, 1991, pp. 111–136.

[49] Lackie, S. J., T. R. Glass, and M. J. Block," Instrumentation for Cylindrical Waveguide Evanescent Fluorosensors," in *Biosensors With Fiberoptics,* D. L. Wise and L. B. Wingard, eds., Clifton, NJ: Humana Press, 1991, pp. 225–251.

[50] Glass, T. R., S. Lackie, and T. Hirschfeld, "Effect of Numerical Aperture on Signal Level in Cylindrical Waveguide Evanescent Fluorosensors," *Appl. Opt.,* Vol. 26, 1987, pp. 2181–2187.

[51] Love, W. F., and L. J. Button, "Optical Characteristics of Fiber Optic Evanescent Wave Sensors," *Proc. SPIE–Int. Soc. Opt. Eng.,* Vol. 990, 1988, pp. 175–180.

[52] Egalon, C. O., and R. S. Rogowski, "Model of a Thin Film Optical Fiber Fluorosensor," *Proc. SPIE–Int. Soc. Opt. Eng.*, Vol. 1368, 1990, pp. 134–150.

[53] Marcuse, D., "Launching Light Into Fiber Cores From Sources Located in the Cladding," *J. of Lightwave Technol.*, Vol. 6, 1988, pp. 1273–1279.

[54] MacCraith, B. D., V. Ruddy, C. Potter, B. O'Kelly, and J. F. McGilp, "Optical Waveguide Sensor Using Evanescent Wave Excitation of Fluorescent Dye in Sol-Gel Glass," *Electron. Lett.*, Vol. 27, 1991, pp. 1247–1248.

[55] Lal, S., and M. C. Yappert, "Development, Characterization and Application of a Double Waveguide Evanescent Sensor," *Appl. Spectrosc.*, Vol. 45, 1991, pp. 1607–1612.

[56] Meltz, G., W. W. Morey, and J. R. Dunphy, "Fiber Bragg Grating Chemical Sensor," *Proc. SPIE–Int. Soc. Opt. Eng.*, Vol. 1587, 1992, pp. 350–361.

[57] Synovec, R. E., A. W. Sulya, L. W. Burgess, M. D. Foster, and C. A. Bruckner, "Fiber Optic Based Mode-Filtered Light Detection for Small Volume Chemical Analysis," *Anal. Chem.*, Vol. 67, 1995, pp. 473–481.

[58] Fujiwara, K., and K. Fuwa, "Liquid Core Optical Fiber Total Reflection Cell as a Colorimetric Detector for Flow Injection Analysis," *Anal. Chem.*, Vol. 57, 1985, pp. 1012–1016.

[59] Wei, W., H. Qushe, W. Tao, F. Minzhao, L. Yuanmin, and R. Gouxia, "Absorbance Study of Liquid Ion Optical Fibers in Spectrophotometry," *Anal. Chem.*, Vol. 64, 1992, pp. 22–25.

[60] Fujiwara, K., J. B. Simeonsson, B. W. Smith, and J. D. Wineforder, "Waveguide Capillary Flow Cell for Fluorimetry," *Anal. Chem.*, Vol. 60, 1988, pp. 1065–1068.

[61] Tsunoda, K., A. Nomura, J. Yamada, and S. Nishi, "The Use of Poly(Tetrafluoroethylene-co-Hexafluoropropylene) Tubing as a Waveguide Capillary Cell for Liquid Absorption Spectrometry," *Appl. Spectrosc.*, Vol. 44, 1990, pp. 163–165.

[62] Chiba, K., I. Inamoto, K. I. Tsunoda, and H. Akawai, "Sensitive Determination of Sulfur in Steel by Spectrophotometry With a Waveguide Capillary Cell," *Analyst*, Vol. 119, 1994, pp. 709–712.

[63] Hong, K., and L. W. Burgess, "Liquid Core Waveguide for Chemical Sensing," *Proc. SPIE–Int. Soc. Opt. Eng.*, Vol. 2293, 1994, pp. 71–79.

[64] Shahriari, M. R., Q. Zhou, and G. H. Sigel, "Porous Optical Fibers for High Sensitivity Ammonia Vapor Sensors," *Opt. Lett.*, Vol. 13, 1988, pp. 407–409.

[65] Zhou, Q., and G. H. Sigel, "Porous Polymer Fiber for Carbon Monoxide Detection," *Proc. SPIE–Int. Soc. Opt. Eng.* Vol. 1172, 1989, pp. 157–161.

[66] Shahriari, M. R., and J. Y. Ding, "Doped Sol-Gel Films for Fiber Optic Chemical Sensors," Chap. 13 in *Sol-Gel Optics: Processing and Applications*, L. C. Klein, ed., Boston: Kluwer, 1994.

[67] Lieberman, R. A., and K. E. Brown, "Intrinsic Fiber Optic Chemical Sensor Based on Two-Stage Fluorescence Coupling," *Proc. SPIE–Int. Soc. Opt. Eng.*, Vol. 990, 1988, pp. 104–108.

[68] Hirschfeld, T. B., "Fluorescent Immunoassay Employing Optical Fiber in Capillary Tube," U.S. Pat. 4,447,546, and 4,582,809, 1984.

[69] Badley, R. A., R. A. L. Drake, I. A. Shanks, A. M. Smith, and P. R. Stephenson, "Optical Biosensors for Immunoassays: The Fluorescent Capillary Fill Device," *Phil. Trans. R. Soc. London, Ser. B*, Vol. 316, 1987, pp. 143–160.

[70] Deacon, J. F., A. M. Thomson, A. L. Page, J. E. Stops, P. R. Roberts, S. C. Whiteley, J. W. Attridge, C. A. Love, G. A. Robinson, and G. P. Davidson, "An Assay for Human Chorionic Gonadatropihin Using the Capillary Fill Immunosensor," *Biosens. Bioelectron.*, Vol. 6, 1991, pp. 193–199.

[71] Robinson, G. A., J. W. Attridge, J. K. Deacon, and S. C. Whiteley, "The Fluorescent Capillary Fill Device," *Sensors and Actuators*, Vol. B 11, 1993, pp. 235–238.

[72] Robinson, G. A., "The Commercial Development of Planar Optical Biosensors," *Sensors and Actuators*, Vol. B 29, 1995, pp. 31–36.

[73] Fogg, A. G., A. K. Deisingh, and R. Pirzad, "Disposable Capillary Fill Device for the Colorimetric Determination of para-Aminophenol as Indophenol Blue," *Anal. Lett.*, Vol. 25, 1992, pp. 937–946.

[74] Zhou, Y., J. V. Magill, R.M. De la Rue, P. J. R. Laybourn, and W. Cushley, "Evanescent Fluorescence Immunoassays Performed With a Disposable Ion-Exchanged Patterned Waveguide," *Sensors and Actuators*, Vol. B 11, 1993, pp. 245–250.

[75] Klein, R., and E. Voges, "Integrated Optic Ammonia Sensor," *Sensors and Actuators*, Vol. B 11, 1993, pp. 221–225.

[76] Gauglitz, G., and J. Ingenhoff, "Integrated Optical Sensor for Halogenated and Non-halogenated Hydrocarbons," *Sensors and Actuators*, Vol. B 11, 1993, pp. 207–212.

[77] Kreuwel, H. J. M., "Planar Waveguides Sensors for the Chemical Domain," Thesis, University Twente, Netherlands, 1988.

[78] Kreuwel, H. J. M., P. V. Lambeck, J. V. Gent, T. J. A. Popma, "Surface Plasmon Dispersion and Luminescence Quenching Applied to Planar Waveguide Sensors for the Measurement of Chemical Concentrations," *Proc. SPIE–Int. Soc. Opt. Eng.*, Vol. 798, 1987, pp. 218–224.

[79] Valette, S., S. Renard, J. P. Jado, P. Gidon, and C. Erbeia, "Silicon Based Integrated Optics Technology for Optical Sensor Applications. *Sensors and Actuators*," Vol. A 21/23, 1990, pp. 1087–1091.

[80] Levy, R. A., ed., *Novel Silicon Based Technologies*, Netherlands: Kluwer, 1991.

[81] Kersten, R. T., "Integrated Optics for Sensors," Chap. 9 in *Optical Fiber Sensors*, J. P. Dakin, and B. Culshaw, eds., Norwood, MA: Artech House, 1991, pp. 277–317.

[82] Madou, M. J., and S. R. Morrison, *Chemical Sensing With Solid State Devices*, Boston: Academic Press, 1989.

[83] Nishizawa, K., E. Sudo, M. Yoshida, and T. Yamasaki, "High Sensitivity Waveguide-Type Hydrogen Sensor," *Proc. OFS Conf. Tokyo*, Japan, 1986, pp. 131–134.

[84] Sloper, A. N., J. K. Deacon, and M. T. Flanagan, "A Planar Indium Phosphate Monomode Waveguide Evanescent Field Immunosensor," *Sensors and Actuators*, Vol. B 1, 1990, pp. 589–591.

[85] Lukosz, W., "Integrated Optical Chemical and Direct Biochemical Sensors," *Sensors and Actuators*, Vol. B 29, 1995, pp. 37–50.

[86] Stamm, C., and W. Lukosz, "Integrated Optical Difference Interferometer as Refractometer and Chemical Sensor," *Sensors and Actuators*, Vol. B 11, 1993, pp. 177–181.

[87] Kunz, R. E., "Totally Integrated Optical Measuring Sensors," *Proc. SPIE–Int. Soc. Opt. Eng.*, Vol. 1587, 1992, pp. 98–113.

[88] Kunz, R. E., "Gradient Effective Index Waveguide Sensors," *Sensors and Actuators*, Vol. B 11, 1993, pp. 167–176.

[89] Aizawa, M., M. Tanaka, and Y. Ikariyama, "A Fiber Optic Electrode for Optoelectrochemical Biosensors," Chap. 9 in *Chemical Sensors and Instrumentation*, R. W. Murray, R. E. Dessy, W. R. Heineman, J. Janata, and W. R. Seitz, eds., ACS Symp. Series, Vol. 403, 1989, pp. 129–138.

[90] Van Dyke, D. A., and H. Y. Cheng, "Electrochemical Manipulation of Fluorescence and Chemiluminescence Signals at Fiber Optic Probes," *Anal. Chem.*, Vol. 61, 1989, pp. 633–636.

[91] Egashira, N., H. Kimasako, and K. Ohga, "Fabrication of a Fiber Optic Based Electrochemiluminescence Sensor and Its Application to the Determination of Oxalate," *J. Anal. Sci.*, Vol. 6, 1990, pp. 903–904.

[92] Kuhn, L. S., A. Weber, and S. G. Weber, "Microring Electrode/Optical Waveguide: Electrochemical Characterization and Application to Electrogenerated Luminescence," *Anal. Chem.*, Vol. 62, 1990, pp. 1631–1636.

[93] Gunasingham, H., and C. H. Tan, "Electrochemically Modulated Optrode for Glucose," *Biosens. Bioelectron.*, Vol. 7, 1992, pp. 353–359.

[94] Gunasingham, H., C. H. Tan, and K. L. Seow, "Fiber Optic Glucose Sensor With Electrochemical Generation of Indicator Reagent," *Anal. Chem.*, Vol. 62, 1990, pp. 755–759.

[95] Piraud, C., E. K. Mwarania, J. Yao, K. O'Dwyer, D. J. Schiffrin, and J. S. Wilkinson, "Optochemical Transduction on Planar Optical Waveguides," *J. Lightwave Technol.*, Vol. 10, 1992, pp. 693–699.

[96] Poznyck, S. K., and A. I. Kulak, "An Electroluminescence Optical Sensor System Based on TiO_2 Film

Electrodes for Continuous Measurements of H_2O_2 Concentration in Solution," *Sensors and Actuators,* Vol. B 22, 1994, pp. 97–100.

[97] Hesselink, G. L. J., H. J. M. Kreuwel, P. V. Lambeck, H. J. Van de Bovenkamp, J. F. J. Engbersen, D. N. Reinhout, and T. J. A. Popma, "Glucose Sensor Based on the LUQUEN Principle," *Sensors and Actuators,* Vol. A 7, 1992, pp. 363–366.

$P_{art\ IV:}$

Applications

All sensors can be described using three aspects derived from general considerations [1], as shown in Figure 11.1 for optical chemical sensors. This method of presentation includes the optical and chemical theories (described in Part I and Chapter 6, Part III), the detection principles (in Chapters 7, 8, and 9, Part III), the chemical and optical technologies (in Part II and Chapter 10, Part III), the measured parameters, and the application fields. These two last aspects, concerning their performance and use, are described in Part IV (Chapters 11 and 12). Chapter 11 is devoted to biosensors and Chapter 12 to sensors for environmental, process control, and gas sensing applications.

Many examples of optical chemical sensors and biosensors and their applications (mainly up to 1989) may also be found in two books [2,3] and in more recent *Opt(r)ode* conferences (published in Sensors and Actuators B 11 and B 29).

This part offers a simple introduction to applications. It is not intended to provide a comprehensive coverage, since the number and types of sensors are continually increasing and many application fields are yet to be exploited commercially. The objective of these two chapters is to summarize major developments in recent years, to discuss the most important measured parameters, and to examine the main trends in this field without being exhaustive.

Figure 11.2 shows in a general way the classes of the most commonly measured parameters and their application fields. However, the requirements for individual sensor performance (e.g., sensitivity, measurement range, stability, and temperature) can differ widely depending on the specific application. In addition, some measurements, such as pH and ion determination, are to be found in almost all applications.

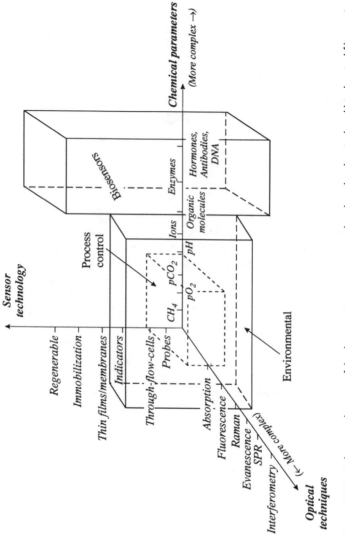

Figure 11.1   Schematic diagram of the three main parameters used to describe chemical and biochemical fiber-optic sensors. Three examples are given.

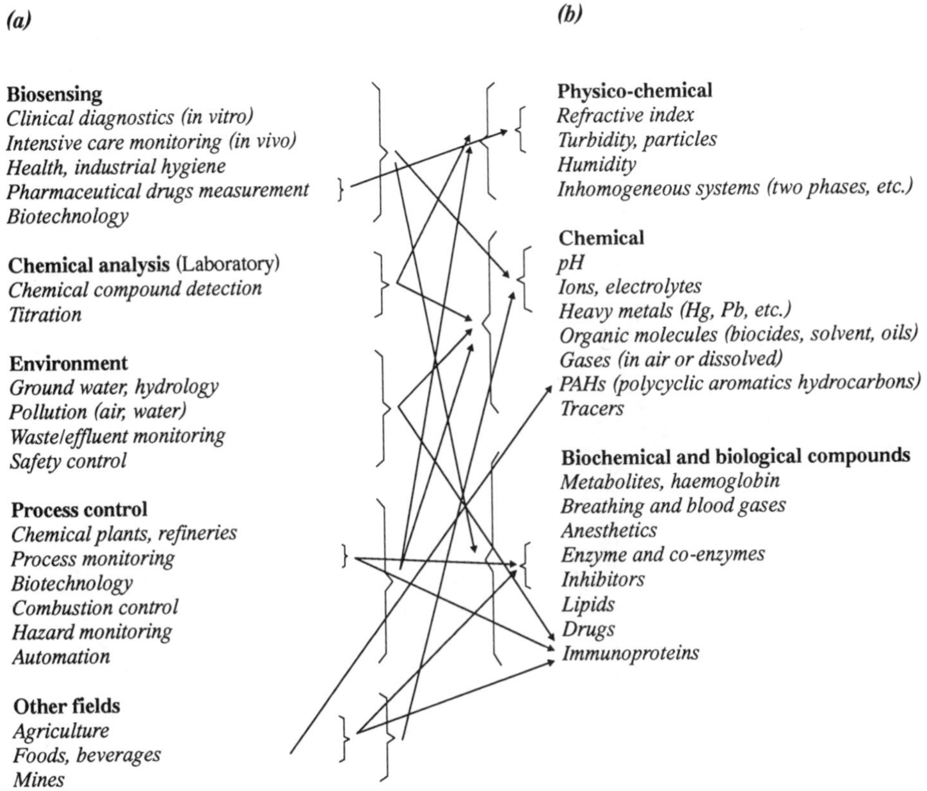

(a)

Biosensing
Clinical diagnostics (in vitro)
Intensive care monitoring (in vivo)
Health, industrial hygiene
Pharmaceutical drugs measurement
Biotechnology

Chemical analysis (Laboratory)
Chemical compound detection
Titration

Environment
Ground water, hydrology
Pollution (air, water)
Waste/effluent monitoring
Safety control

Process control
Chemical plants, refineries
Process monitoring
Biotechnology
Combustion control
Hazard monitoring
Automation

Other fields
Agriculture
Foods, beverages
Mines

(b)

Physico-chemical
Refractive index
Turbidity, particles
Humidity
Inhomogeneous systems (two phases, etc.)

Chemical
pH
Ions, electrolytes
Heavy metals (Hg, Pb, etc.)
Organic molecules (biocides, solvent, oils)
Gases (in air or dissolved)
PAHs (polycyclic aromatics hydrocarbons)
Tracers

Biochemical and biological compounds
Metabolites, haemoglobin
Breathing and blood gases
Anesthetics
Enzyme and co-enzymes
Inhibitors
Lipids
Drugs
Immunoproteins

Figure 11.2 (a) Applications and (b) measured physico-chemical, chemical, and biochemical parameters of optical waveguide sensors. The arrows show the main relationships between columns (a) and (b).

Chapter 11

Biochemical Sensors and Biosensors

11.1 INTRODUCTION

Chemical and biochemical sensors can measure, respectively, chemical quantities (e.g., ion, glucose) or biochemical quantities (e.g., reduced nicotinamide adenine dinucleotide (NADH)) in a biological medium (e.g., whole blood, serum, urine, tissue). Biosensors incorporate a biochemical compound (enzyme, antibody) in the sensing element. Enzyme-based sensors and immunosensors based on immunological techniques (immunoassays) form the two main classes of biosensors.

These sensors are applied mainly in medicine either for clinical analysis (e.g., medical diagnosis) and in vitro discrete measurements or for in vivo critical care monitoring with extracorporeal or corporeal supervision (e.g., pH, pO_2, pCO_2 measurement) and surgery (anesthetics). Other applications include biotechnology, drug and food monitoring, and experiments on laboratory animals.

Typical examples of these sensors have been described [4,5], particularly for in vivo monitoring [6,7].

11.2 CLINICAL DIAGNOSTICS

Whole blood is composed of plasma (serum and fibrinogen) and cellular elements (erythrocytes, leucocytes, platelets). Table 11.1 gives the concentrations of the most common chemical and biochemical substances in blood. Data are similar for serum.

Figure 11.3 shows the typical concentration range of some species in clinical chemistry, biochemistry, and immunoassay techniques.

In the period from 1970 to 1980, some of these compounds started to be measured for clinical diagnostics by optical techniques, especially by absorption and

Table 11.1
Common Chemical and Biochemical Substances in Blood and
Their Concentrations

Compounds	In mg · L^{-1}	In mM
Inorganic		
Ammonia	0.4–1.12	$2.2–6.2 \times 10^{-2}$
Bicarbonates	$1.3–1.8 \times 10^{3}$	21–29
Calcium	90–102	2.2–2.6
Chloride	$3.4–3.7 \times 10^{3}$	96–105
Iron	0.65–1.8	$1.2–3.2 \times 10^{-2}$
Phosphore (inorganic)	24–45	0.8–1.45
Potassium	120–180	3.1–4.7
Sodium	$3–3.4 \times 10^{3}$	132–148
Organic		
Albumin	$40–50 \times 10^{3}$	
Bilirubin (total)	2–12	$3.4–21 \times 10^{-3}$
Cholesterol (total)	$1.5–2.5 \times 10^{3}$	3.9–6.5
Creatinine	5–13	$4.4–11.5 \times 10^{-2}$
Glucose	$7.5–11.2 \times 10^{2}$	4.2–6.2
Hemoglobin	$1.15–1.7 \times 10^{5}$	7–10.5
Lipids (total)	$3.5–8 \times 10^{3}$	
Protides	$6.5–8 \times 10^{4}$	
Triglycerides	$0.5–1.4 \times 10^{3}$	0.56–1.6
Urea	$1.5–4.8 \times 10^{2}$	2.5–8
Uric acid	34–80	0.2–0.48

fluorescence with autoanalyzers in continuous flow analyzers (e.g., from Technicon Corp.) and discrete multichannel analyzers equipped with optical fibers as light guides (e.g., DIMA™, Commissariat Energie Atomique, France). Today, many methods in clinical chemistry [8] use flow injection analysis, a general technique based on "the repetitive action of a well defined zone of analyte and reagent on a target (sensor) situated in a continuous unsegmented carrier stream" [9]. Figure 11.4 is an example of a manifold adapted for ammonia and urea (via urease) determinations [10]. FIA systems can also be used in biotechnology, coupled to optodes sensitive to pH, oxygen, H_2O_2, carbon dioxide, or NADH in conjunction with enzymatic reactions [11].

11.2.1 Electrolytes

The main blood electrolytes considered here are sodium (Na^+), potassium (K^+), lithium (Li^+), calcium (Ca^{2+}), and chloride (Cl^-) ions. Ion determination requires an accurate knowledge of the selectivity coefficients (see Section 4.3.6) because several ions with almost identical properties are present simultaneously in blood or serum. The

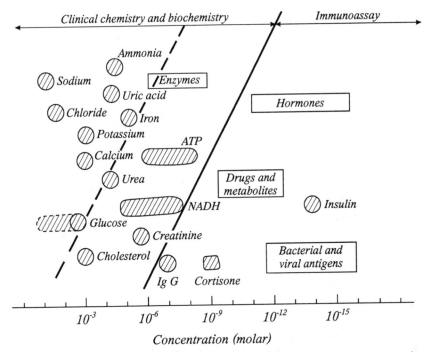

Figure 11.3 Relative concentration range of some species in clinical chemistry, biochemistry, and immunoassay technique. The range in blood or serum is represented by a circle. The dashed line divides approximately clinical and biochemistry areas, and immunoassay is to the right of the solid line.

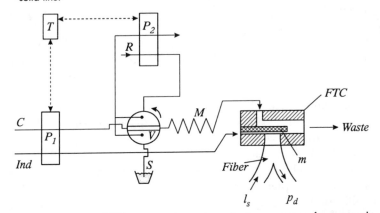

Figure 11.4 Flow injection manifold for NH_3 and urea determination consisting of two peristaltic pumps (P_1 and P_2), controlled by a timer (T), that pump the reagent (R), carrier (C), and indicator (Ind) through supply tubes. A two-port valve (V) alternatively sends the sample (S) and reagent (R) through a mixing coil (M) to the flow-through cell (FTC). This has a gas-permeable membrane (m) separating the measuring chamber with a bifurcated optical fiber, connected to a light source (l_s) and photodetector (p_d). (*After:* [10], with permission.)

optical selectivity coefficient ($K_{ij}^{opt} = K_i/K_j$) between the studied ion (i) and interfering ion (j) is determined from the total optical signal (S) and individual signals (S_i for i and S_j for j compounds) with

$$S = S_i + (K_{ij}^{opt}) \cdot S_j \tag{11.1}$$

Similarly, for several interfering compounds this becomes

$$S = S_i + \Sigma\{(K_{ij}^{opt}) \cdot S_j\} \tag{11.2}$$

This coefficient, given generally in logarithmic form, can be obtained in a simplified form by graphical means (Figure 11.5) from the optical signals (S_i and S_j) of the individual compounds for the same activity ($a_i = a_j$). A simplified equation for the absorbance measurement in a reaction involving ion pair formation has been given [12] when the total ionophore (a_{ti}) and anionic dye (a_{td}) activities are equal:

$$K_{ij}^{opt} = \{S_i(k \cdot \varepsilon \cdot d \cdot a - S_j)^2\}/\{S_j(k \cdot \varepsilon \cdot d \cdot a - S_i)^2\} \tag{11.3}$$

where ε is the molar absorptivity of the dye (in its associated form), d the effective length of the light path in the membrane, k a factor for the efficiency of light detection, and S the absorbances from the theoretical curves of cations i and j with an activity ($a = a_{ti} = a_{td}$).

Centimolar K^+ concentrations have been determined by absorption using crown ether [13], but the K^+/Na^+ selectivity ratio (about 1:10) is too weak for clinical purposes when sodium is present. The crown ether can be covalently linked to a chromophore as demonstrated in serum [14].

Promising sensing schemes are possible with ion carriers, such as valinomycin. One method has applied LB film techniques to PSDs in the 0.01- to 10-mM range, with a logarithmic selectivity coefficient of -2.5 to -3.5 in an acidic bilayer with valinomycin and rhodamine B-C_{18} ester [15]. In a similar sensing scheme, using a hexalayer membrane with arachidic acid (a tetralayer with valinomycin and a bilayer with rhodamine dye), this coefficient (about -4) has been increased by the use of a reference optode [16]. However, in this case, one of the advantages of optodes over electrodes, which require a reference cell, is lost.

New sensing schemes with ion exchange or coextraction of neutral ion carriers in a membrane (see Chapter 4) are also encouraging. In an absorbance K^+ sensor, a novel lipophilic anionic dye, synthesized from 2-nitrophenyloctyl ether (NPOE) and sensitive to pH, was investigated experimentally with dibenzo-18-crown-6 and valinomycin as ion-selective neutral ionophores [12]. Another transduction is based on two different dyes; one absorbent compound acts as a proton carrier and the second is a stable and pH-independent fluorophore whose excitation and emission overlap the

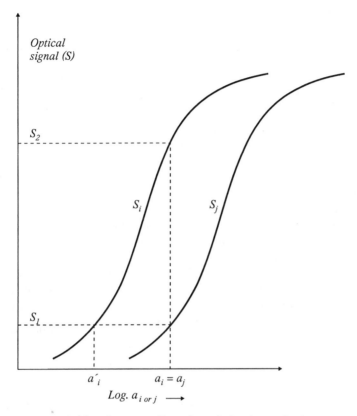

Figure 11.5 Determination of the selectivity coefficient from idealized curves for the variation of the optical signal (S) as a function of the logarithmic activity (a) of the two components, the measured ion (i) and the interfering ion (j). The selectivity coefficient can be determined after calibration from the S_2 and S_1 signals by the ratio (a'_i/a_i), where a'_i is defined by the S_1 signal on the S_i curve and a_i by $a_i = a_j$ on the S_j curve. The S_i and S_j curves can have different shapes, which implies that the determined value is only relative to a given medium.

absorption band of the absorber dye [17]. The fluorescence intensity is modulated by an absorber with the so-called inner filter effect (see Section 8.6). This highly sensitive sensor is fully reversible over the range of concentrations 1 μM to 10 mM. It can be used only with dilute serum and a buffer to adjust the pH, and also requires changing the characteristics of the blood to avoid extracting carriers and plasticizer from the membrane.

Studies in whole blood at 40%, after washing to remove endogenous potassium, have been made by fluorescence with a precision of 1% without optical interference from red blood cells [18]. MEDPIN (see Section 4.3.3.2), a hydrophobic indonaphthol absorbance indicator, forms a stable ternary complex with the valinomycin-K^+ ion pair in a PVC membrane. Energy transfer between a fluorescent dye and the anionic form of MEDPIN permits detection of K^+ ions in human blood. However, the sensitivity of potassium decreases with the presence of Na^+. Measurement of potassium, almost independent of pH, has been obtained by a scheme in which the dye is an ion pair with a lipophilic anion [19].

Potassium sensors from PSDs are based on a commercially available Ames Seralyzer solid-state strip with a coextraction mechanism (see Section 4.6) of the valinomycin (ion carrier) and lipophilic erythrosin B as a colored counter-anion in a PVC membrane [20]. In another approach, the Reflotron test is based on an ion-exchange mechanism with valinomycin and MEDPIN as protonated dye contained in the membrane [21,22].

The absorbance determination of potassium ions has been reported in diluted human plasma with a selective optode membrane [23,24]. This sensor is based on the use of one ionophore, one neutral carrier chromoionophore, anionic sites in the membrane, and an ion exchange reaction in the liquid phase. The lifetime of the system with diluted serum samples in an FTC determined by the partition coefficient (P) at the membrane/solution interface can attain 30 days if a minimal lipophilicity ($\log P_{TLC}$ superior to 10) of the chromoionophore is obtained [25]. The lipophilicity value (P_{TLC}) is established from the reversed-phase thin-layer chromatography (TLC) method by means of the relation

$$P_{TLC} = (V_m/V_s) \cdot (1/R_f - 1) \tag{11.4}$$

where V_m is the volume of the mobile phase, V_s a volume representative of the stationary phase, and R_f a retention coefficient defined as the ratio of the distances moved between the compound and the solvent front. Also, quantitative lipophilicity can be indirectly accessible from previously measured exchange constants on thin membranes in liquid-liquid extraction experiments [26].

Sodium and calcium determinations have been demonstrated experimentally in similar systems [24]. In particular, a calcium optode based on the absorption of an ion exchange was realized by combining a conventional Ca^{2+} neutral selective carrier and a new H^+ selective chromoionophore of the phenoxazine type together with anionic sites in a plasticized PVC membrane [27]. The selectivity coefficients are similar to the corresponding potentiometric coefficients. Also, a highly lipophilic ionophore (with a water/1-octanol partition coefficient P up to 10^{17}) of the phenylenedioxydiacetamide type (called ETH 4120) has been synthesized for Na^+ measurement in blood [28]. This ionophore is doped in bis(1-butylpentyl)adipate (BBPA) acting as a solvent polymeric membrane.

Other sensing schemes for Ca^{2+} use a natural carboxylic polyether antibiotic as a complexing agent with calcium (also magnesium) [29], hydrophobically C-12 derived calcein as a reagent [30] or a LB-membrane potential [31], and a fluorescence measurement.

A novel neutral ionophore of the diazocrown ether type and a lipophilic anionic dye (LAD) of the diphenylamine type have been synthesized for calcium determination in human serum [32]. In a reflectance flow cell, the optode shows an excellent Ca^{2+} selectivity (10^{-4}) against Na^+, K^+, and Mg^{2+} for normal Ca^{2+} serum concentrations (2.5 mM).

Table 11.2 summarizes examples of results of selectivity coefficients for Na^+, K^+, Li^+ [33], and Ca^{2+} measurements from different authors.

For anion sensing, a reversible chloride sensor incorporating trioctyl tin chloride (TOTC of the type R_3SnCl, where R is an alkyl or aryl compound) has been developed for human plasma diagnosis [34]. The selectivity of the indicator in combination with TOTC in BBPA plasticized in PVC matrices is 1, 0.3, -0.7, -1.8, and -2.8 with interference from Br^-, I^-, ClO_4^-, NO_3^-, and SO_4^{2-} ions, respectively. The tricyclo-hexyltin chloride (TCTC)-based optode using LAD as dye exhibits the best response characteristics for chloride ions according to a coextraction model [35]. The selectivity coefficients are 0.6, 1.0, -3.7, -3.1, and -3.7 with respect to the same anions.

All these examples show the increasing interest of membrane optodes in clinical diagnostics.

11.2.2 Glucose Biosensor

The optical glucose sensor is a typical and important example of a biosensor, and many approaches have been tried. Glucose, a metabolite, is one of the main analytes determined by discrete measurement or continuous monitoring in clinical diagnostics (blood, serum, urine), and also in biotechnology and the food industry. Table 11.3

Table 11.2

Logarithmic Selectivity Coefficients (Log K_{ij}^{opt}) of Cation-Selective Optodes

	Li^+	Na^+	K^+	Ca^{2+}	NH_4^+	Mg^{2+}	References
			Interfering ion (j)				
Primary Ion (i)							
Na^+	-1.1	0	-1.2	-1.2		-2.5	[24]
K^+	< -4.5	-4.0	0	-4.5	-3.0		[12]
K^+		-4.0	0				[16]
K^+	-3.7	-3.5	0	-3.7		-4.0	[24]
Li^+	0	-4.3	-4.0	-4.0		-4.0	[33]
Ca^{2+}	-3.1	-3.6	-3.8	0		-4.1	[24]
Ca^{2+}	-3.1	-3.6	-3.8	0		-4.1	[27]
Ca^{2+}	-4	-4.5	-4.1	0		-4.1	[32]

Note: i is the primary ion being measured, and j is the interfering ion.

Table 11.3
Typical Sensing Schemes for Glucose Determination

Sensor Type	Intermediate Analyte	Characteristics	Measured Parameter	Optical Technique	References
Affinity binding		One label	Intensity	Fluorescence	[36]
Affinity binding		Two labels	Intensity	Energy transfer (F)	[37]
Affinity binding		Two labels	Lifetime	Energy transfer (F)	[38]
Affinity binding		Two labels	Quenching	Energy transfer (F)	[39]
Enzyme-based	Oxygen	Decacyclene as indicator	Quenching	Fluorescence	[40]
Enzyme-based	Oxygen	Ruthenium compound (I)	Quenching	Fluorescence	[43]
Enzyme-based	Oxygen	Pyrene butyric acid (I)	Quenching	Fluorescence	[41]
Enzyme-based	Oxygen		Dynamic quenching	Fluorescence	[42]
Enzyme-based	Oxygen	Metalloporphyrins (I)	Lifetime	Phosphorescence	[44]
Enzyme-based	pH		Intensity	Fluorescence	[45]
Enzyme-based	Oxygen + pH	Spectral overlap of dyes	Intensity	Energy transfer (F)	[46]
Enzyme-based	H_2O_2	Peroxalate reaction	Intensity	Chemiluminescence	[50]
Enzyme-based	H_2O_2	Luminol reaction	Intensity	Chemiluminescence	[51]
Enzyme-based	H_2O_2	Two enzymes with HPA	Intensity	Fluorescence	[52]
Enzyme-based	H_2O_2	Luminol and HRP	Intensity	Chemiluminescence	[48]
Enzyme-based	NAD	Coenzyme reaction	Intensity	Fluorescence	[55]
Enzyme-based	NADP	Coenzyme reaction	Intensity	Fluorescence	[56,57]
Enzyme-based	GOx	Flavins as coenzymes	Intensity	Intrinsic fluorescence	[54]
Enzyme-based		Redox reaction with TTF	Intensity	Reflectance	[53]
Enzyme-based		Direct	Refractive index	Interferometry	[58]
Spectroscopic		Noninvasive	Intensity	Reflectance, absorption	[60]
Spectroscopic			Intensity	Photoacoustic	[59]

Note: Abbreviations are as follows: F = fluorescence; NAD = nicotinamide adenine dinucleotide; NADP = phosphored-NAD; GOx = glucose oxidase; I = indicator; HPA = hydroxyphenyl acetic acid; HRP = horseradish peroxidase; TTF = tetrathiafulvalene.

shows the main approaches of optical biosensing methods that have been realized with fiber and waveguide optics, especially with competitive binding reactions (affinity sensors), with enzymatic systems (enzyme-based sensors), and with simple IR spectroscopic measurements.

The first optical glucose affinity biosensor [36] employed a dextran label with a fluorescent marker (F^*) in a cavity within the sensing zone of the light (see Figure 2.6). The basis of the sensor is a competitive binding reaction of glucose (Glu) and dextran (Dex) with concanavalin A (Con A):

$$Glu + Con\ A \rightleftharpoons Glu - Con\ A \tag{11.5}$$

$$F^* - Dex + Con\ A \rightleftharpoons F^* - Dex - Con\ A \tag{11.6}$$

The quantity of free or unbound dextran is determined by the strength of the fluorescent signal and hence the quantity of bound glucose by inference. In later developments of this technique, two labels (F^* and T^*) are used for an energy transfer interaction. The reaction (11.5) competes with

$$F^* - \text{Dex} + T^* - \text{Con A} \rightleftharpoons F^* - \text{Dex} - T^* - \text{Con A} \qquad (11.7)$$

This energy transfer can be measured by either the fluorescence intensity [37], the lifetime [38], or the quenching [39].

Enzyme-based biosensors are schematized as in Figure 11.6 or the simple

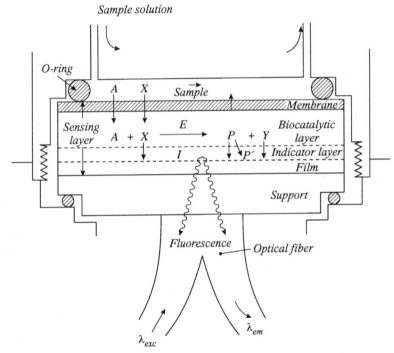

Figure 11.6 An enzyme-based biosensor with FTC. The production of an intermediate analyte (P and Y) or the consumption of an intermediate analyte (X) after diffusing through a membrane, which has an immobilized enzyme (E) attached, is detected in the presence of an indicator (I) by absorbance, chemiluminescence, fluorescence, quenching fluorescence, or a change in pH (by producing P′ derived from P). The input (λ_{ex}) and output (λ_{em}) light is brought to and from the cell by a bifurcated optical fiber or waveguide. In the example of the glucose sensor according to (11.8), A is glucose, E is glucose oxidase, X is O_2, P is glucolactone giving rise to P′ gluconic acid, and Y is H_2O_2.

example in Figure 3.5. Different reactions for glucose measurement are possible. The most common is based on the oxidation of glucose in the presence of glucose oxidase (GOx):

$$\text{Glu} + O_2 \overset{\text{GOx}}{\rightleftharpoons} \text{Gluconolactone} + H_2O_2 \qquad (11.8)$$

Gluconolactone is transformed into gluconic acid, which induces a change in pH. In this scheme O_2, pH, H_2O_2, and GOx act as intermediate analytes (chemical-chemical transduction).

The consumed oxygen quenches the indicator and the pO_2 decreases across the enzyme layer in proportion to the concentration of oxidized glucose. The glucose fluorescent calibration curves can be represented approximately by

$$F_0/F = 1 + K_{sv} \cdot pO_2 + K_g \cdot \text{Glu} \qquad (11.9)$$

where F and F_0 are the fluorescence intensities in the presence and absence of glucose, respectively, K_{sv} is the quenching constant, and K_g is the sensitivity coefficient for glucose. This method has been developed for the 0.1- to 20-mM glucose range [40] and the 1- to 150-kPa pO_2 range [41]. Further developments for glucose analysis via oxygen determination include the extension to higher concentration ranges up to 200 mM at a pH 6.9 [42], the use of indicators such as those based on ruthenium, which are well adapted to urine determination [43], and a combined measurement of the luminescence signal and lifetime [44].

A fully reversible biosensor (see Figure 10.6) has been produced with pH as an intermediate analyte but for a limited range of glucose concentrations [45]. An improvement has been proposed to use two indicators, each responding to two different intermediate analytes that take part in the enzymatic reaction, such as pH and O_2. Measurement is based on the fluorescence energy transfer between these indicators (e.g., the reaction pair GOx-fluorescein and the GOx-Ruthenium complex) when their absorption spectra overlap [46].

In a similar way, H_2O_2 is frequently used as the basis for the chemical transducer [47,48]. The advantage is that chemiluminescent or bioluminescent reactions are possible [49], such as with bis-(2,4,6,-trichlorophenol)-oxalate [50], or with luminol (see (4.13)) in the presence of a surfactant for enhancing the optical signal [51]. A sensor has been realized with two enzymes (HRP and GOx) immobilized on the inside surface of a nylon tube; the first catalyzes the production of H_2O_2, the second the reaction of H_2O_2 with HPA in the presence of a fluorescent agent, 6,6',8 hydroxy-(1,1'-diphenyl)-3,3' diacetic acid (DBDA). This concept produced a linear and very stable response for glucose measurement in serum calculated from Michaeli-Menten kinetics for soluble and immobilized enzymes [52].

Another sensor based on GOx (which can be considered an *optrode*) generates electrochemically a redox mediator dye of glucose, tetrathiafulvalene (TTF) acting as an indicator reagent [53]. The final reaction is independent of oxygen concentration:

$$\text{Glucose} \overset{\text{GOx + TTF}}{\rightleftharpoons} \text{Gluconolactone} + 2H^+ + 2e^- \tag{11.10}$$

The detection limit is about 0.2 mM, and the most common compounds (e.g., uric acid, vitamin C) do not interfere with this measurement technique, in contrast to amperometric detection methods.

The FAD (flavine adenine dinucleotide) part of GOx is an intrinsic green fluorescent coenzyme and can form the basis for analytical measurements. The signal response is fully reversible in the presence of oxygen and the enzyme itself acts as a chemical-chemical transducer. The concentrations of glucose can be detected in a narrow range, typically 1.5 to 2 mM [54].

A coenzyme, such as NADH or phosphored-NADH (NADPH), can serve as an intermediate analyte (λ_{exc} = 350 nm, λ_{em} = 450 nm). The reaction is catalyzed by glucose deshydrogenase (GDH) or phosphored-GDH (GDPH):

$$\text{Glu} + \text{NAD(P)}^+ \overset{\text{GD(P)H}}{\rightleftharpoons} \text{Gluconolactone} + \text{NAD(P)H} + H^+ \tag{11.11}$$

A reproducible probe having a good linearity in the 1.1- to 11.0-mM glucose range has been demonstrated [55]. These systems may be used without loss of enzyme and coenzymes in an FIA system with a miniaturized fluorimeter [56] and can be extended to fructose determination [57].

Direct monitoring of glucose concentration, without an intermediate analyte, has been tried with an integrated-optic sensor [58]. A Mach-Zehnder interferometer measures directly the change in refractive index due to glucose. However, this refractive index variation is only 1.3334 to 1.3337 for a variation in the sugar concentration of 0.1% to 0.3% (typical range for diabetes). A laser photoacoustic method appears a better choice for glucose monitoring in blood and human plasma [59], making use of the proportional change with glucose concentration in the 1,180 nm and 1,700 nm absorption bands, due to C-H vibrational harmonics. Noninvasive systems of glucose determination for home use by diabetics are being actively pursued based on direct measurement of the reflection and transmission spectra in the NIR [60].

Glucose sensing is important for many applications and has considerable potential for future markets. The different techniques discussed here for glucose determination demonstrate some of the many possibilities for optics as sensing schemes and the importance of an intermediate analyte in biosensing.

11.2.3 NH₃/NH₄⁺ System

Ammonia sensors have been developed in the 5- to 20-µM range with absorption [61] and fluorescence techniques [62]. A continuous flow system has tested a renewable reagent sensor (see Figure 2.8) with a pH-bromothymol indicator from concentrations of 1.5 to 800 µM with an accuracy of 1% [63]. However, these ammonia sensors based on pH effects are sensitive to uncharged amines and acidic gases (e.g., CO_2), and are not specific.

The equilibrium between gas and solution is given by the equation

$$NH_3 \text{ (gas)} + H^+ \text{ (sol)} \rightleftharpoons NH_4^+ \text{ (sol)} \qquad (11.12)$$

A membrane that is permeable to gas (and not to protons) can be added to separate the sample solution from the internal cavity where the indicator is immobilized on an optical fiber or waveguide as shown in Figure 11.7 [2].

In early experiments, the determination of ammonia in blood and serum with an irreversible ninhydrin reaction was monitored for clinical determination in the 0 to 40-mg · L⁻¹ range by total internal reflection on a cylindrical waveguide coated with a suitable polymer [64].

Using fluorescence and nonactin as a receptor/carrier, a lipophilic Nile blue dye in a PVC membrane can detect ammonium ions in the 0.03- to 10-mM range at neutral pH [65]. This sensor is compatible with solid-state electronics (LED and photodiodes) and can be applied to biochemical reactions where the ammonium ion is the intermediate analyte. Further development has led to a new absorbance sensor based on a lipophilized pH indicator immobilized in a silicone elastomer (permeable

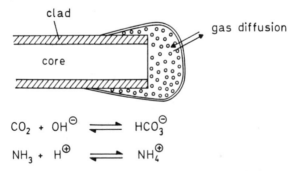

Figure 11.7 A miniaturized fiber-optic sensor for NH_3 or CO_2 gas detection. The equilibria (see (11.12) and (11.18) to (11.20)) occur inside the enclosed cell. A buffer to control the internal pH and an indicator are trapped in a polymer emulsion that coats the end of the fiber. The membrane is permeable to gas and impermeable to protons. The membrane can also enhance the signal by reflecting the measuring light and absorbing ambient light, preventing it from entering the cell. (*Source:* [2]. Reprinted with permission.)

to gas and not permeable to protons) in the concentration range 0.017 to 17 ppm with a detection limit of 12 ppb [66].

A fluorescence ammonia sensor has also been tried in untreated serum for clinical purposes [67]. At a pH 7.5, the detection of gaseous ammonia is only in the range of 0.3 to 2.7 μM. The key to the fiber-optic approach and achieving submolar detection limits (0.09 μM, better than for electrochemical sensors) is to increase the concentration of the indicator dye, and hence the ammonia input in the solution in the sensing cavity. A cylindrical geometry (a tube) permeable to gas, in which the pH-sensitive dye is enclosed, improves the detection limit to about 20 nM [68].

11.2.4 Urea and Creatinine Sensors

The concentration of urea in blood is an important parameter in clinical chemistry. The sensing scheme is based on the hydrolysis of urea in the presence of the enzyme urease:

$$\text{H}_2\text{N}-(\text{C}=\text{O})-\text{NH}_2 + 3\,\text{H}_2\text{O} \overset{\text{Urease}}{\rightleftharpoons} 2\,\text{NH}_4^+ + \text{HCO}_3^- + \text{OH}^- \quad (11.13)$$

Consequently the pH, ammonium ions (and hence ammonia), urease activity, and bicarbonate ions are key elements as intermediate analytes in the determination of urea and creatinine.

In a urea biosensor, the change of pH can be determined colorimetrically [69] by using the FIA technique after separation of the ammonia by diffusion [70] or in a gas-diffusion flow injection system [71]. The enzyme urease, which is covalently immobilized onto an alkyl amine monolayer acting as the membrane, together with a pH-sensitive fluorescent compound, has been immobilized onto the surface of an optical fiber for urea determination [72]. The fluorescence response decreases linearly as the pH increases. The detection limit is about 40 μM. Another urea sensor, based on the pH shift due to the enzymatic reaction, has been tested experimentally in simulated dialysis [73].

Urea determination from the assessment of ammonia gas has also been reported for absorption [74] and fluorescence measurements [75]. The fluorescence sensor responds significantly more quickly than a similar absorbance biosensor, and its performance was demonstrated on dilute serum samples.

Absorption measurements in solution of NH_4^+ [76] or NH_3 with neutral ionophores [77] has been extended to urea determination through fluorescence ($\lambda_{ex} = 550$ nm, $\lambda_{em} = 630$ nm), with the urease immobilized on a nylon membrane [78]. This measurement is based on selective recognition and binding of ammonium ions by nonactin acting as an ammonium carrier and transporting the ions through a PVC membrane. A proton is released from a lipophilic Nile blue type dye acting as a pH indicator in the membrane. Figure 11.8 shows the cross section of the

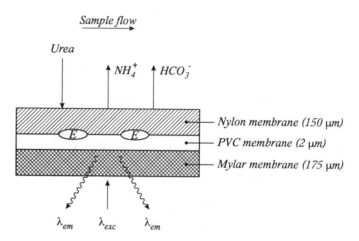

Figure 11.8   Cross section of the sensing layer of a urea biosensor showing the successive layers, with E the enzyme. The arrows indicate the diffusion processes and the directions of the exciting (λ_{ex}) and emitted (λ_{em}) light. (*After:* [78].)

sensing layer. The sensor is fully reversible and very stable. It responds to ammonium ion concentration in the range 30 μM to 10 mM and to urea with a detection limit between 0.03 and 0.6 mM for near neutral pH.

In the same way, creatinine can be measured by an enzymatic reaction with creatinine iminohydrolase; this generates ammonia, which is separated from the matrix by gas diffusion and affects a pH-sensitive indicator measured by reflectance [79]. A better method employs the properties of the ammonium fluorosensor already described [65]. The detection range obtained is from 0.03 to 1.0 mM.

11.2.5 Cholesterol and Other Biosensors

Free cholesterol, when bound to esters, can be liberated by cholesterol esterase (ChE), and oxidized by O_2 under the action of cholesterol oxidase (ChOx):

$$\text{Cholesterol ester} + H_2O \overset{\text{ChE}}{\rightleftharpoons} \text{acid gras} + \text{Cholesterol} \qquad (11.14)$$

$$\text{Cholesterol} + O_2 \overset{\text{ChOx}}{\rightleftharpoons} \text{Cholest-4-en-3-one} + H_2O_2 \qquad (11.15)$$

At a pH of 7.25, the oxygen consumption measured by the fluorescence quenching of an oxygen sensitive dye has been used in a cholesterol biosensor in which the cholesterol oxidase was immobilized covalently on a nylon membrane [80]. The analytical range was 0.2 to 3 mM. The drawback of this sensor is its slow time response, about 10 minutes. Another fiber-optic biosensor with cholesterol esterase and cholesterol oxidase coimmobilized on a membrane with chemically activated carboxylic groups has been developed for the determination of both total and free cholesterol [81]. The concentration of the analyte was correlated with the signal intensity of a colored oxidized dye formed by reaction with H_2O_2. The reaction time is less than 2 minutes, and the sensor is designed for monitoring with a linear range of 0.5 to 5 mM and 1 to 8 mM for free and total cholesterol, respectively.

A further example of a biosensor also makes use of the ammonium ion, present at physiological pH values, acting as an intermediate agent. The ammonium ion provides selective recognition of the 1-propanolol enantiomer (optical isomer) phenylethylamine and norephedrine by reversible interaction of biogenic amines with an optically active dibutyl tartrate as the carrier in a PVC membrane [82]. The selectivity coefficient depends on the type of tartrate receptor (carrier) and varies between 0 and 0.5, the best value being obtained for norephedrine (0.5). In addition, the measurement of 3-hydroxybutyrate, an important compound in the metabolic disorder of diabetes and alcoholic ketosis, has been measured in serum by using $NAD^+/NADH$ as an intermediate analyte in an FIA system coupled to a chemiluminometer [83].

Other biosensors, such as lactate, penicillins, and drug sensors, are categorized under biotechnology (Section 11.4).

11.3 IN VIVO MONITORING

Both invasive (extracorporeal or corporeal) and noninvasive monitoring can be performed with optical-fiber sensors. This section describes some of the different approaches, particularly for measurement of oxygen content (a vital parameter), pH, and CO_2 in the blood. Intrinsic indicators, such as hemoglobin, cytochromes, NADH, and FAD coenzymes, are present in the human body and can be used to assess the state of tissue O_2 supply.

11.3.1 Oximetry

Most blood oxygen is carried by hemoglobin. The amount of oxygen bound to hemoglobin, as HbO_2 in erythrocytes, relative to the maximum O_2 carrying capacity of hemoglobin, defines the oxygen saturation (OS) of blood, expressed as a percentage. Typically values of OS are 95% and 75% in arterial and venous blood, respectively. The optical measurement of OS is based on the difference between the optical spectra of deoxygenated (Hb) and oxygenated (HbO_2) hemoglobin. The simplest method consists of measuring the OS as a linear function of the difference between the absorption (or reflectance) coefficients at the isobestic point (where Hb and HbO_2 have the same absorption values, around 830 nm) and at the peak absorption of HbO_2 (around 650 nm).

In reflectance:

$$OS = \% \; HbO_2 = a - b \cdot (R_{iso}/R_{peak}) \qquad (11.16)$$

where R_{iso} and R_{peak} are the reflectances at the isobestic point and at the maximum for HbO_2, and a and b are empirical parameters considered as second-degree polynomials as a function of Hb concentration [2]. A large amount of work has been devoted to oximetry over the last 30 years since the first invasive oximeter was invented [84]. Many types of fiber oximeter are commercially available.

More comprehensive assessments, including cytochromes [85,86], can be obtained by measuring the whole blood spectrum. Intravascular oximetry is a clinically accepted technique but remains limited when the oxygen variations are large (e.g., coming from respiratory gas mixtures under anesthesia).

Recent research includes pulse oximetry [87–89], a method that is attracting increasing attention for noninvasive monitoring (see Section 11.3.7). A NIR spectrometer using frequency domain and eight light sources (LEDs) located at different distances from a detector fiber has also been the subject of experiments [90].

11.3.2 NADH as Indicator

The redox potential of the $NAD^+/NADH$ system, at a pH 7, reflects the redox status of tissues. When the system is excited around 340 nm, a wavelength near that of a pulsed N_2 laser (337 nm), only the NADH fluoresces ($\lambda_{em} = 460 \pm 20$ nm) and can serve to monitor the tissue metabolism. Using a fiber-optic catheter, some experimental measurements of the NADH to NAD^+ ratio have been taken in the beating heart. A second dye laser at 585 or 805 nm (isobestic points of hemoglobin) was employed as a reference in order to compensate for blood disturbances [91]. In a similar way, further work has monitored brain and other tissue functions [92]. In biological tissues, the fluorescence decay time response of NADH is not constant and oscillations of the response signal are observed with differences in results for different tissues [93]. It has been suggested that this be used in vivo to differentiate between the NADH in cytoplasm and in isolated cells. Tests in a guinea pig heart during ischemia have also been reported [94].

11.3.3 Principles of Blood Gas and pH Measurements

For patients undergoing surgery or in intensive care, it is necessary to monitor important body functions, which includes CO_2, O_2, and pH, with both extracorporeal and corporeal invasive techniques.

11.3.3.1 pH

The principles of pH determination have been outlined already in Section 2.7. Extensive work has been done in this area, summarized in several surveys [95–97]. The variation of pH in the blood is normally small, between 6.9 and 7.4, and the choice of a suitable chromophore (see Chapter 4) is determined by selecting its pK'_a (around 7.2) and longer measurement wavelengths (e.g., above 600 nm) in order to allow the use of solid-state light sources [98,99].

For absorbance measurements, (2.63) gives the relation between the pH and the optical signal. For fluorescence measurements, a similar relation is obtained with

$$pH = pK'_a - \log\left[(F_{max}/F_{I^-}) - 1\right] \tag{11.17}$$

where F_{max} is the fluorescence intensity of the undissociated indicator (HI) completely in the phenolate form, and F_{I^-} is the intensity of the I^- form at a given pH.

Frequently HPTS (8-hydroxy-1,3,6,-pyrene trisulfonic acid), a fluorophore with a pK'_a of 7.3, is chosen. However, it is highly sensitive to ionic strength. This fluorophore is covalently bonded to a cellulose membrane. It is excited at 410 nm (in the acid form) and 460 nm (in the basic form), and it has a single emission band (at 520 nm). The pH is measured from the ratio of the fluorescence intensities of the two excited forms (see Figure 4.5). A temperature increase of 5°C can change the pK'_a by around +0.2 units and thus it is important to measure or compensate for temperature changes.

11.3.3.2 $pCO_2/HCO_3^-/CO_3^{2-}$ Systems

In solution, CO_2 is present in three forms, as carbonic acid (H_2CO_3), bicarbonate (HCO_3^-), and carbonate (CO_3^{2-}) ions described by the following equilibria:

$$CO_2(sol) + H_2O \rightleftharpoons H_2CO_3(sol) \tag{11.18}$$

with the K_h (hydration) constant and pK_h (-log K_h) = 2.58;

$$H_2CO_3 \rightleftharpoons HCO_3^- + H^+ \tag{11.19}$$

with the dissociation constant K_1 (pK_1 = 6.4);

$$HCO_3^- \rightleftharpoons CO_3^{2-} + H^+ \tag{11.20}$$

with the dissociation constant K_2 (pK_2 = 10.2). The equation for relating the total CO_2

concentration (C) (including both unhydrated CO_2 and hydrated H_2CO_3 forms) to the hydrogen ion concentration $[H^+]$ is

$$K_1 \cdot C = ([H^+]^2 + N \cdot [H^+] - K_w)/(1 + K_2/[H^+]) \qquad (11.21)$$

where K_w is the dissociation constant of the water and N the hydrogen carbonate concentration.

The concept of the CO_2 optode was invented by Lübbers and Opitz [100] and the theory of optical CO_2 sensing was subsequently developed by Zhujun and Seitz [101]. In the basic sensor, carbon dioxide diffuses across a membrane and changes the pH, which is detected by means of the fluorescent dye HPTS. In equilibrium the partial pressure in the sample is equal to that in the sensor reservoir, which contains hydrogen carbonate. The optical signal is normally translated into pCO_2 (partial pressure of CO_2) in the usual 10- to 120-torr (mm Hg) range (7.5 to 16 kPa) or, in special cases, in the 4- to 150-torr range (0.5 to 20 kPa) for biological systems.

11.3.3.3 Oxygen and pO₂ Measurements

The dissolved oxygen is normally given as pO_2 over the 0 to 200-torr range. The most common technique for in vivo fiber-optic sensing, first used by Peterson et al. [102], is fluorescence (or phosphorescence) quenching that obeys the Stern-Volmer equation:

$$F_0/F = 1 + K_{sv} \cdot pO_2 \qquad (11.22)$$

This relates pO_2 to the fluorescence intensities in the absence (F_0) and the presence (F) of oxygen. K_{sv} is the dynamic quenching Stern-Volmer constant (between 1 and 2,000 M^{-1}).

This equation can also be written in terms of the decay time of the fluorescence, τ_0 in the absence and τ in the presence of oxygen, respectively:

$$\tau_0/\tau = 1 + K_{sv} \cdot pO_2 \qquad (11.23)$$

with τ_0 typically 5×10^7 to 10^9 sec.

Numerous fluorophores, such as polycyclic aromatic hydrocarbons, are quenched by oxygen, and a large variety of sensors use quenching fluorescence, decay time, or energy transfer [2]. The fluorophores based on metallo-organic compounds such as ruthenium complexes have the advantages of long lifetimes (0.2 to 5.0 μs) and a convenient excitation wavelength (450 to 490 nm) that is suitable for an argon ion laser (488 nm) or other common excitation sources. Recent developments have incorporated ruthenium (II) as bipyridyl or tris-diphenyl complexes in sol-gel porous coatings [103,104] or encapsulated zeolite supercages [105], as diimin complexes photoexcited by LEDs [106], or as silicone soluble complexes [107]. However, the Stern-Volmer function is not linear over the whole pO_2 range, and a new model, which considers the interaction of a fluorophore in a nonuniform environment, has been tested for defining the matrix polymer effect [108]. In a miniature reversible oxygen sensor for which only a sample volume of 100 fL is required, a detection limit of 10 to 17 mol has been achieved [109]. However, the sensor must be optimized because of leaching of the ruthenium complexes from the acrylamide matrix. Also, phosphorescent probes based on platinum-porphyrin derivatives have been suggested for biosensing [110].

In addition, some preliminary experiments have tried to determine oxygen by absorbance measurements in the presence of viologen [111] or by reflectance with bis-(histidinato)-cobalt (II) immobilized on a silicone layer [112].

Two analytes can be measured simultaneously in the same sensor, either with two layers having different emission wavelengths [113], as shown in Figure 11.9, or by multiparametric analysis (e.g., by measuring both intensity and lifetime) of a single indicator.

11.3.4 Extracorporeal Blood Gas Sensors

CDI-3M Health Care (Irvine, CA) has developed an extracorporeal blood gas monitor, which employs the initial Lübbers and Opitz's concept for the pH sensor with HPTS as the fluorophore. For pCO_2 measurement, this same fluorophore is dissolved in a bicarbonate buffer encapsulated in a silicone layer acting as a membrane. For pO_2, a modified decacyclene ($\lambda_{ex} = 385$ nm, $\lambda_{em} = 515$ nm) is incorporated with another fluorophore, insensitive to oxygen and acting as a reference, into a silicone membrane permeable to oxygen. The three indicators are put in a disposable optode built into

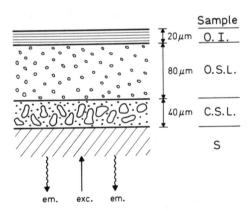

Figure 11.9   Cross section of the fluorescent sensing layer of a sensor that measures both oxygen and carbon dioxide gas. The exciting light (λ_{ex}) stimulates emission (λ_{em}) from both the carbon dioxide-sensitive layer (CSL), giving a green fluorescence, and the oxygen-sensitive layer (OSL), giving a red fluorescence. S is the transparent support and OI an optical isolation. (*Source:* [113]. Reprinted with permission.)

a disposable FTC in an extracorporeal blood circuit. Two membranes, permeable to gas, separate the blood stream from the sensing elements. A thermistor is added for temperature correction (between 10 and 45°C). A thin layer of black PTFE acts as optical isolation and decreases the sensitivity to halothanes (anesthetics). The optical fluorescent signals are translated into readings of pH (6.5 to 8.0), pCO_2 (1 to 120 torr, 1 to 16 kPa), and pO_2 (20 to 500 torr, 2 to 67 kPa). An independent evaluation of the performance of this sensor has been published [114]. More than 10,000 disposable probes are produced monthly.

Other extracorporeal blood gas systems are available; for example, from Bio-medial Sensors (England) in which pH and pCO_2 are determined by a reflectance technique. Disposable FTCs for in vitro blood analysis have also been developed by AVL-List (Austria) for pH, pCO_2, and pO_2 determination [115].

11.3.5 Invasive Techniques

11.3.5.1 Intravascular Sensors

A miniaturized version (0.62 mm in diameter) of the sensor described in the previous section has also been developed by CDI (3M-Health) for incorporation in a fiber-optic arterial catheter to monitor pH, pCO_2, and pO_2 simultaneously [116]. The probe (Figure 11.10) has three sensing elements placed at the tip of three 140-μm fibers and a thermocouple included in the same envelope. The pH optode is based on HPTS fluorophore immobilized on aminoethyl cellulose, the pCO_2 optode has HPTS in a bicarbonate buffer encapsulated in silicone matrix, and the pO_2 optode has decacyclene solubilized in a silicone rubber. The precision in bovine blood was 0.03 pH units, 2 torr for pCO_2, and 4 torr for pO_2 [117]. However, a *wall effect* occurs when the fiber tip is in contact with the arterial wall, decreasing the values of the measured tissue oxygen. Also, thrombus formation around the sensor must be diminished with

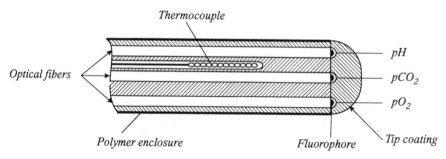

Figure 11.10 Optode for simultaneous in vivo monitoring of pH, pO_2, and pCO_2 in human blood. (*After:* [116], with permission.)

anticoagulanting agents (e.g., coating the sensor with immobilized heparin) or employing an oscillatory blood flow [118].

Other intravascular sensors have been developed for arterial blood gases and pH, such as the SensiCath device [119] and the Puritan Bennett Corporation system. Also, an integrated device with multiple sensing elements on a single fiber has been described [120].

Various important operating factors must be considered for in vivo sensing: sterilization, biocompatibility, blood compatibility, shape of the sensor, temperature compensation, and methods of calibration [121,122]. Initial evaluation and tests are generally carried out by arterial measurements in sheep, swine [123], and dogs [124]. Full tests are then made on human subjects in the operating room [125].

11.3.5.2 Gastroenterology

Bilirubin and pH are the two main in vivo parameters for gastroesophageal reflux control. pH requires a broad range, from 1 to 8 units. This can be achieved with a chromophore having several suitable pK'_as or with more than one dye (generally two to cover this range). Fluorescein and eosine compounds have been proposed as fluorophores for coimmobilization (see Section 5.4.3) [126], and HPTS has been extended to the 1- to 3-pH-units range [127]. A combination of two dyes has been tried in optical absorption sensors: one dye is bromophenol blue and the other has two pK'_as, either metacresol purple [128] or thymol blue [129]. Figure 11.11 shows an optical probe for in vivo monitoring.

Figure 11.12 compares an electrode and a fiber-optic sensor for in vivo measurements. The fiber device is less sensitive to minor movements of the stomach tube.

Bile is always present in enterogastric refluxes. An absorption measurement

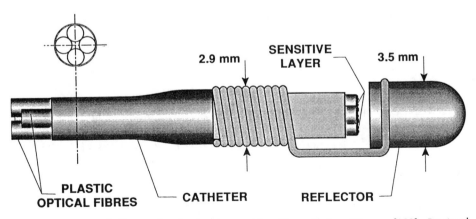

Figure 11.11 Optical fiber probe for in vivo gastric pH monitoring. (*Source:* [129]. Reprinted with permission.)

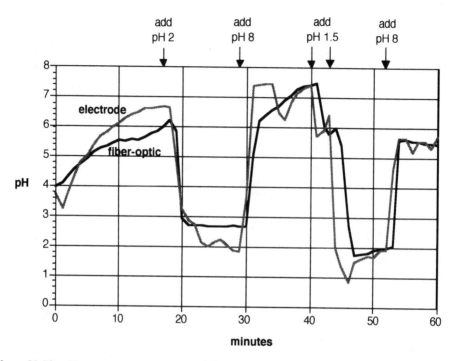

Figure 11.12 pH measurement comparison of the in vivo performance of a fiber-optic sensor and an electrode sensor in an animal stomach. Unlike the electrode, the fiber-optic sensor is not sensitive to minor movements of the stomach tube. (*Source:* [128]. Reprinted with permission.)

using two LEDs at 465 and 570 nm (as a reference wavelength) has been exploited for in vivo measurements on many patients [130]. The Bilitec 2000 is a commercial device manufactured by Prodotec (Italy). Today, bile and pH are routinely determined by in vitro and in vivo measurements [131].

11.3.5.3 Respiratory Analysis

Oxygen analyzers are important in critical care, artificial respiration, and breathing analysis, either for intermittent measurements or preferably for continuous monitoring. Response times are generally fast, on the order of 20 to 30 ms [132,133].

11.3.6 Transcutaneous Techniques

Transcutaneous measurement of pCO_2 and pO_2 is based on the gas permeability of human skin and is employed when the measured organs are not directly accessible (for example, the brain). From a wavelength-by-wavelength comparison of the HbO_2

spectra obtained by reflectance from the skin and by absorption with a cuvette, a transformation function can be used for evaluating the reflectance spectrum of skin tissue by nonlinear multicomponent analysis [6]:

$$\cosh[y(\lambda)] = [a(\lambda) + s(\lambda)]/s(\lambda) \tag{11.24}$$

where $y(\lambda)$ is the tissue reflection spectrum, $a(\lambda)$ the absorption coefficient, and $s(\lambda)$ the scattering coefficient at a wavelength λ. The ratio $a(\lambda)/s(\lambda)$ can be determined from the Kubelka-Munk reflectance function. This multicomponent analysis has been performed for calibration using ten components: three cytochromes in oxidized and reduced forms, oxygenated and deoxygenated myoglobin, and two scattering functions. The spectrum can be rapidly scanned by rotating interference filters (at 1,900 rpm) to minimize problems of patient movements [134].

The heterogeneity of tissue oxygen supply can be characterized by macrohistograms or microhistograms or topograms (spatial histogram) of the different values of the local HbO_2 saturation [135]. This heterogeneity varies as a function of time during the day-night rhythm as observed for the brain [136] and for the vegetative (sympathic and parasympathic) nervous system [137].

Also, O_2-flux optodes, which measure a pO_2 difference between two sites of a membrane, can be constructed with two different indicators (or optodes), one on the upper side and the other on the lower side of the membrane. This system is usable for testing the vitality of organs for transplantation and giving a continuous noninvasive indirect image of the microcirculation [6]. Recently, frequency domain fluorometry (see Chapter 8) has been applied for measuring the O_2-flux versus time in human skin [138].

11.3.7 Noninvasive Techniques

Skin tissue is quite transparent in the NIR region between 650 and 1,100 nm. In earlobe, fingertip, and toe oximeters, the oxygen saturation is calculated from special algorithms, because blood contains not only HbO_2 but also Hb and derivatives of Hb compounds. The measurement is done by phase-sensitive detection with either heartbeat or a squeezer device, but it is influenced by blood flow and scattering from tissues.

Reflectance spectrometry with a CCD detector of 1,024 pixels and plastic optical fibers has been proposed for noninvasive measurements [139]. Time-resolved spectrophotometry with an ultrashort laser pulse ($\Delta t = 4$ ps, $\lambda = 783$ nm) has been applied through the head of a newborn infant [140] and to the human calf muscle [141]. The emerging distribution of the light varies according to the transmit time within the tissues. In this scattering medium, a *differential path length factor* is defined in order to correct the photometric data.

Noninvasive measurements with NIR spectrometry have been proposed for other analytes such as glucose, cholesterol, and drugs, taking advantage of a large number of new laser systems [142]. For example, reflectance measurements in the NIR with interference filters have been applied to cholesterol determination in serum [143]. In addition, single-mode fibers, which define a very small probe volume at the fiber tip (below 10 μm^3), could have advantages for the noninvasive investigation of biological tissues for determining optical parameters, such as refractive index, transmission properties, and tissue thickness [144].

11.4 BIOTECHNOLOGY

Biotechnology can be regarded as the exploitation of biological organisms, systems, and processes coming from living matter [145,146]. Four groups of processes can be found:

1. Microbiological processes that categorize the biochemistry and interactions of microorganisms and define the parameters for bioreactions (e.g., biomass plants). For instance, all living cells synthesize ATP, which is a key compound in the determination of these parameters.
2. Enzyme applications that examine metabolic reactions in the presence of these biocatalytic compounds.
3. Genetic techniques that study the recombination processes of DNA, a universal compound of living cells. The addition of new molecules to produce genetically altered DNA and insertion into a host for reproduction opens up many applications in health, agriculture, immunology, and the pharmaceutical industries.
4. Process control of natural or synthetic products (e.g., certain foodstuffs, methane, and alcohol production) related to living matter.

Numerous sensors are used in biotechnology [147,148]. Table 11.4 outlines the main areas of biotechnology and the various chemical parameters measured by optochemical sensing. Overviews of this field have been provided by several workers [149–151].

Three main types of fiber-optic chemical sensors can be used for monitoring production processes as shown in Figure 11.13:

1. Peripheral sensors for autoanalyzers, especially for FIA, in which the sample is brought to the sensor from the process. They form the major part of sensors in biotechnology. Optodes and FTCs are used for liquid and gaseous samples. A filtration module serves as a microorganism barrier.
2. FTCs integrated into process bypass systems (e.g., for biomass, NADH, or ATP content).
3. In situ optodes (on-line sensors). The sensor is placed directly in the reaction chamber or in the sample.

Adverse operating conditions (particularly sterilization, deposition of proteins on the probe surface, and microbial growth effects) increase the problems associated with

Table 11.4
Chemical Fiber-Optic Biosensors in Biotechnology

Processes	Measurement Parameters and Chemical Compounds	Examples of Products
(1) *Bioreactions* (Fermentation plants, biomass reactors)	Turbidity, biomass pH, oxygen, CO_2, NH_3/NH_4^+, glucose, urea, ethanol, lactate, pyruvate, glutamate,	Microorganisms
	Nucleic acids (DNA), coenzymes (NAD(P)H, ATP)	Proteins
(2) *Enzyme Reactions*	Turbidity, refractive index, biomass, pH, ionic strength, redox, pO_2, pCO_2	
	Enzyme activity, NAD(P)H, ATP	Enzymes, inhibitors
(3) *Genetic Techniques*	Immunosensors, enzyme activity	Reproduction of cells, genetically altered DNA
(4) *Process Control*		
• Pharmacy	Immunosensors, products	Penicillin, aspirin, vitamins, hormones, drugs, etc.
• Industrial products	CO_2, products	CH_4, methanol, ethanol, etc.
• Agrifoods	Glucose, alcohols, pH, humidity in solids, etc.	Foods, beer, wines, etc.
• Agriculture	Nutrients, immunosensors	Plant culture

the construction and utilization of these sensors, particularly for on-line and permanently installed sensors. Important examples of each of these types are given below.

11.4.1 Biomass, NADH, and ATP

Biomass content and cell density can be determined by nephelometry and scattering [152] or by total absorbance measurements directly in situ. However, absorbance values are very high (typical optical densities are 20 for a 1-cm path length). The major reason for using optical fibers is to construct optical cells with a very small path length (1 mm), which can be sterilized.

A better alternative to the usual dry biomass determination is to assess the concentration of compounds such as DNA, NAD(P)H, and ATP in cell cultures to estimate biomass concentration [153–155]. DNA assessment is examined in Section 11.5. NAD(P)H measurement can be related to turbidity in bioreactors for obtaining information about biomass and the metabolic state of baculoviruses during cultivation [156]. Measurements have been made with a modular fiber-optic fluorometer and a tungsten-halogen lamp; emission occurs at 360 nm for NADH and scattering is measured at wavelengths longer than 700 nm. This device can also

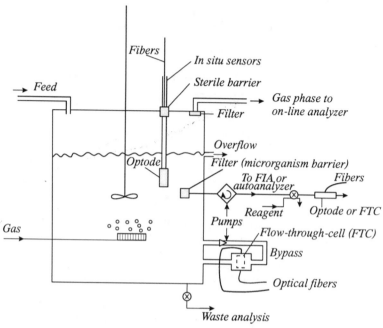

Figure 11.13   Schematic diagram showing the main categories of optical sensors in a complete sensing system for monitoring a fermentation process.

monitor bacteria confined at the fiber tip and changes in the NADH pool of enzyme complexes inside the cells during enzymatic reactions. NAD(P)H and F_{420} (a fluorescent coenzyme) have also been used for monitoring the methanogenic fermentation of glucose [157].

Chemoluminescence and bioluminescence are suitable for optical-fiber sensing in batch production or continuous-flow systems [158,159]. A typical example is the determination of both NAD(P)H and ATP in the same sensor from the coimmobilization of two enzymatic compounds on a polyamide membrane interrogated by bifurcated fiber optics [160]. NADH is measured via the light emission as the result of a multistep reaction catalyzed by a bacterial luciferase. Furthermore, a complex mechanism of light emission is catalyzed by the firefly luciferase, an enzyme specific to ATP. The reactions are shown in Figure 11.14. Measurements are possible within a large range, about 10^3 for NADH and 10^5 for ATP. The detection limits are as low as 0.25 pM for ATP and 5 pM for NADH, illustrating the potential of bioluminescent techniques. In a recent NAD(P)H sensor, poly(vinyl)alcohol (PVA)–Nafion is used as the matrix and the coreactants (flavine mononucleotide (FMN) and decanal) are immobilized in this reagent biosensor [161].

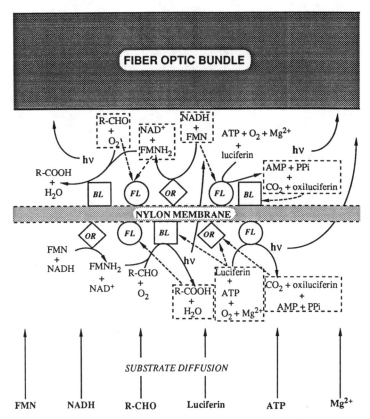

Figure 11.14 Multistep reactions occuring between the substrates and the analytes for measurement of NADH and ATP in a bioluminescent enzymatic sensor with a polyamide membrane placed at the tip of a fiber-optic bundle. FL = firefly luciferase; BL = bacterial luciferase; OR = oxidoreductase; FMN = flavine mononucleotide; $FMNH_2$ = reduced FMN; ATP = adenosine triphosphate; AMP = adenosine monophosphate; NAD^+ = nicotinamide dinucleotide. (*Source:* [160]. Reprinted with permission.)

11.4.2 pH, pO_2, and pCO_2 Measurements

Measurement of these three parameters is essential for bioreactor monitoring, but few practical demonstrations have been reported. The most significant is a sterilizable (by steam) pH sensor operating in the range of 6 to 10 pH units with an accuracy of ±0.1 units [162]. A pH-sensitive dye is covalently immobilized on a cellulose triacetate membrane glued onto a polyester support. The instrumentation uses plastic optical fibers, an LED light source, and a *pin* photodetector. This experiment has been extended to a triple sensor for pH, pO_2, and pCO_2 [163]. As with the pH sensor, CO_2 is measured by the absorbance from the color change of a pH-sensitive dye. Oxygen is determined by fluorescence quenching using a ruthenium complex immobilized on silica gel particles dispersed in transparent silicone. CO_2 and O_2 sensors are sterilizable with hydrogen peroxide and ethanol. This project is still under development with the aim of engineering reliable sensors for use in bioreactors in orbital stations. Work on monitoring these three fermentation parameters includes the calibration in the culture medium [164] and is in progress for other parameters, such as nutrients [165].

11.4.3 Enzyme Activity

Determination of enzyme activity, often applied in immunoassay and clinical chemistry, is also essential during enzyme production and fermentation control. Initial experiments with fiber optics have demonstrated the feasibility of such measurements for very small sample volumes [166] and have quantified the optical efficiency of the enzyme substrate immobilized onto membranes [167] and the fiber surface [168].

Dehydrogenase activity has been assessed by luminescence [169] and amylase activity by fluorescence energy transfer [170]. Urease activity, from the enzymatic hydrolysis of urea, has been measured with fiber optics and FIA techniques [171]; the fluorescence signal varies linearly with urease activity in the range 1- to 30-$U \cdot mL^{-1}$ range. These measurements have applications in the understanding of enzyme inhibition and detection of toxic substances [172].

11.4.4 Enzyme Optodes with Intermediate Analytes

In fermentation processes, sensing schemes with optodes for enzymatic reactions generally use an intermediate analyte such as H_2O_2, $NAD(P)H$, NH_3/NH_4^+, O_2, CO_2, or pH. Glucose, urea, ethanol, lactate, pyruvate, glutamate, and pharmaceutical products such as penicillin, vitamins, and aspirin are the main products measured in bioreactors with fiber-optic and FIA techniques. Glucose biosensors have been mentioned in Section 11.2.2. A specific application of continuous monitoring in wine and fruit juices has been based on continuous monitoring of the oxidation by glucose oxidase with an oxygen optode [173].

Ethanol biosensors operate by fluorescence oxygen quenching after oxidation by alcohol oxidase and the addition of a catalase for decomposing the hydrogen peroxide [174] or measurement of H_2O_2 by chemiluminescence in the presence of luminol [175]. However, the most frequent scheme for bioreactors is the reaction

$$\text{Ethanol} + NAD^+ \rightleftharpoons \text{Acetaldehyde} + NADH + H^+ \tag{11.25}$$

in the presence of alcohol dehydrogenase as an enzyme. Acetaldehyde is consumed by semicarbazide, giving semicarbazone. The regeneration reaction of NADH is obtained by the addition of pyruvate in the presence of lactate dehydrogenase:

$$\text{Pyruvate} + \text{NADH} + \text{H}^+ \rightleftharpoons \text{Lactate} + \text{NAD}^+ \qquad (11.26)$$

The same reaction has been proposed for pyruvate and lactate determination with pyruvate formation at a pH of about 7.4 and lactate creation at about pH 8.6 [176]. In bioreactors, an on-line ethanol optode has been demonstrated with a macromolecular polyethylene glycol–NADH coenzyme [177]. Figure 11.15 shows a comparison of on-line and off-line ethanol data [150]. This scheme, in the presence of NADH, has been extended to the determination of lactate, pyruvate, phenylpyruvate, glucose, D-mannitol, and formate concentrations [178].

It should be noted that lactate can be measured from an oxidation reaction in the presence of lactate monooxydase (LMOx), a fluorescent flavoprotein enzyme [179] or of L-lactate oxidase (LOx), as in the on-line determination of lactic acid during kefir fermentation [180]. Production of H_2O_2 by oxidation of lactate by LOx can serve as a cosubstrate for the luminol reaction catalyzed by peroxidase (POx). Compartmentalization of the enzyme barrier by stacking a POx membrane on a LOx

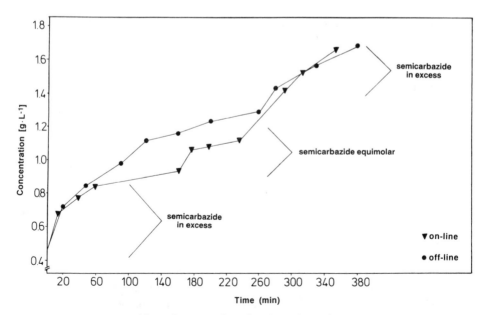

Figure 11.15   Comparison of the performance of an ethanol optode in a fermentation process: (▼) on-line and (●) off-line ethanol data. (*Source:* [150]. Reprinted with permission.)

membrane enhances the response signal. In this configuration, the signal is 22 times higher than the response obtained with two enzymes randomly coimmobilized on one membrane [161,181].

Glutamate, a common amino acid in foods and pharmaceutical preparations, can be measured fluorimetrically by coupling glutamate dehydrogenase with an oxidoreductase and bacterial luciferase system in a scheme with NAD^+ as intermediate analyte [176]. Two other methods are based on the detection of oxygen consumption in the presence of glutamate oxidase or the generation of carbon dioxide under the action of glutamate decarboxylase [182]. Oxygen transduction gave better results than CO_2 transduction. Glutamatinase and glutamate oxidase can also act in sequence on glutamine and produce H_2O_2 measured by chemiluminescence in animal cell cultures [183].

The detection of penicillin is normally done with a pH-sensitive dye after biocatalytic conversion in penicilloate in the presence of penicillase. Many methods of immobilization have been described [184–187], particularly a method of interaction between the glycoprotein avidin and the vitamin biotin [188]. A penicillin G amidase sensor has demonstrated a linear fluorescent signal up to $28 \text{ g} \cdot \text{L}^{-1}$ under real fermentation conditions [150]. Using a multivariate analysis, this penicillin measurement can be made independent of fluctuations of pH or buffer capacity of the sample [189].

11.4.5 Other Methods and Techniques

Three other sensor types in biotechnology should be mentioned. The first is based on (bio-)chemical nonenzymatic reactions; the second, discussed in process control (Chapter 12), employs direct spectroscopic measurement and applies especially to foods, agronomical, and pharmaceutical products; the third on gas measurements (mainly NH_3, O_2, CO_2) is not specific to biotechnology (see Sections 11.2.3 and 11.3.3).

Notable examples of the first type include the pH measurement of milk in the range of 6.2 to 7.2 ± 0.2 pH units [190] and the determination of aspirin [191] or penicillin V or G [192] by the technique of coextraction in a PVC membrane mounted in a FTC.

11.5 IMMUNOSENSORS

Immunosensors are employed in many areas: in biomedical applications for critical care of patients and detection of drugs or toxins in blood and serum, environmental applications (contaminants), biotechnology (antibody production), agriculture (pesticides, herbicides), and in the food industry (biological parameters). Excellent reviews on fiber-optic immunosensors cover initial work up to 1988 [193–195]. This early period was mainly devoted to establishing the basic concepts of immunosensing, characterizing the binding of proteins (competitive binding, theoretical modeling of structures) onto optical supports (planar waveguides and fibers), and experimenting with the techniques of total internal reflection, evanescent spectroscopy, and SPR [196]. Many problems still remain, such as improving the accuracy of sensors, their sensitivity, detection limits, reversibility, regenerability, biocompatibility and anticontamination properties; also, simpler, faster, and inexpensive instrumentation is needed. These are all areas being studied by many laboratories worldwide.

This section briefly reviews major developments within this field, particularly some of the characteristics of immunosensors, regenerable and reusable sensors, new polymers and dyes, and advances in devices and technology, and provides some appropriate examples.

11.5.1 Immunosensor Characteristics

The sensitivity of immunosensors depends on many factors, such as the ability of the immobilized antibody or antigen to maintain activity, the load (the amount) of immobilized compound, and increasing the incubation time (compensated by a longer analysis time). The importance of the immobilization procedure [197] has also been mentioned in Chapter 5. Additional factors, such as temperature, pH, and physical parameters (e.g., flow rate), influence the recognition reaction of the antibody-antigen complex at the solid-liquid interface. For instance, in a flow immunosensor, the nonequilibrium conditions change the kinetics of immobilized antibody binding, hence the time-varying signal [198,199].

On an optical fiber, 10^{11} antigens can be bound to a probe with an active surface of $0.57\,cm^2$ when the binding sites are saturated [200]. The fluorescent signal threshold can detect 10^9 labeled antigens (Mw = 100,000) and subnanomolar concentrations can be measured over two orders of magnitude. With longer incubations, the limit of detection (LOD) of the complex antibody BPT (Benzo[α]pyrene tetraol, a product of a DNA-adduct form for early cancer diagnosis) is lowered and the absolute LOD attainable is 40 attomoles (4×10^{-17} moles) [194]. The LOD also depends on the measurement technique. For example, a comparison has been made of the LOD of three optical sensing devices based on the perturbation of guided optical waves: by internal reflection with an input grating coupler or a differential interferometer, and by a metal dielectric interface using SPR [201]. For HBsAg (hepatitis B surface antigen) in human serum, the SPR and input grating coupler techniques have a similar sensitivity (2×10^{-5} in the effective change of refractive index) and the differential interferometer technique has at least one order of magnitude more sensitivity.

11.5.2 Reversible and Regenerable Immunosensors

A reversible fiber-optic immunosensor has been developed for the continuous monitoring of phenytoin, an anticonvulsant drug [202] (see Figure 11.16). Fluorescence

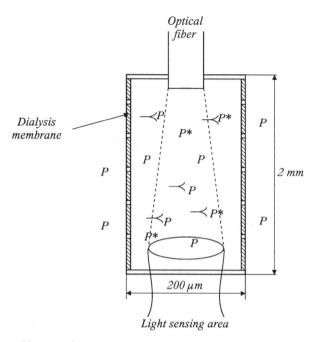

Figure 11.16   Reversible sensor for measurement of phenytoin based on quenching fluorescence, energy transfer, and competitive binding. P = unlabeled phenytoin; P* = phenytoin labeled with b-phycoerythrin (b-PE) compound. The antibody is labeled with Texas red. The unlabeled phenytoin crosses the dialysis membrane and competes with b-PE for antibody binding sites. The change in the fluorescent signal is reversibly proportional to the analyte P concentration. (*After:* [202], with permission.)

energy transfer with competitive binding is a system in which reagents are not consumed in the reaction and in which no physical separation of free and bound antibody-hapten complexes is necessary. The response time is controlled by the dissociation rate constant $(4 \times 10^{-3} \ \text{sec}^{-1}$ for the first-order reaction of labeled analyte-antibody complex) and is independent of the association rate constant [203]. High-affinity antibodies give response times on the order of minutes. The dynamic range of the sensor (1 to 20 μM) is well adapted to clinical purposes.

Regenerable biosensors have been described for the direct measurement of BPT in nanomolar concentrations with a reproducibility of 10% [204]. The optode is a special capillary tube delivery system in which regeneration occurs by delivering new reagents into the sensing chamber without removing the sensor from the sample. Two designs have been experimented with. For the first design, a permeable membrane traps a liquid antibody solution in the sensing chamber. For the second, the sensing chamber is a hollowed stainless-steel frit that retains the antibodies bound to silica beads. Aspiration produces fast delivery of the analyte into the sensing chamber. However, the system is complex and the sensor is not reversible.

An SPR sensor can be regenerated by using an acid glycine buffer at a pH 2.8 [205], or by simply immersing the sensing tip of the fiber in a chaotropic medium in order to disrupt the antigen-antibody complex without affecting subsequent measurements. This approach has been well developed for antibody fragments of human serum albumin (HSA) as analyte model [206]. $F_{(ab')}$ antibody fragments are immobilized by silylation onto quartz plates at the distal end of a fiber-optic probe. Then the substrate is washed with HSA solution to block $F_{(ab')}$ active sites prior to fluorescent labeling. When the immobilized anti-HSA $F_{(ab')}$ binds to HSA, the fluorescence of the dansyl label increases. The fluorescence is measured in a fixed time. Phosphoric acid is employed as a regeneration reagent; the sensor can be recycled over 50 times and can be stored for up to four months.

This approach has been extended to the assessment of haptens; different regeneration reagents (e.g., H_3PO_4, NaSCN, sodium lauryl sulfate, urea) affect the reaction time, the number of regeneration cycles, and the storage time [207].

Fouling of the sensor, as a result of nonspecific adsorption of contaminants from the sample on the surface, blocks receptor sites and hence leads to a reduction in signal of the sensor. Refunctionalization of fluoropolymers (PTFE, FEP) has been explored as a novel substrate for covalent stabilization of antibodies on human IgG and antihuman IgG [208]. These improvements have led to a greater stability, a decrease in the fouling, and longer storage times (one year) for the immunosensor, while maintaining the same detection limit and dynamic range as previous probes (Figure 11.17).

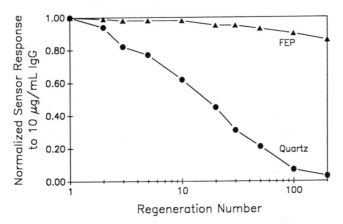

Figure 11.17 Comparison between FEP (fluoroethylenepropylene) and quartz-based immunoprobes as a function of the number of regeneration cycles. (*Source:* [208]. Reprinted with permission.)

11.5.3 Progress in Immunosensing

Current developments in immunosensors are being made in several directions, as illustrated by the following examples:

1. Theoretical and practical studies of planar waveguides and their design to optimize the optical signal for scattering measurements [209], with grating couplers [210–212], and with patterned ion-exchange waveguides [213]. Thus, for thin films, the sensitivity of a grating-coupled evanescent wave step-index slab waveguide can be given by a single normalized expression [214].
2. New instrumentation techniques, such as measuring the fluorescent lifetime with multifrequency phase modulation (Figure 11.18) for determining DNA adducts of

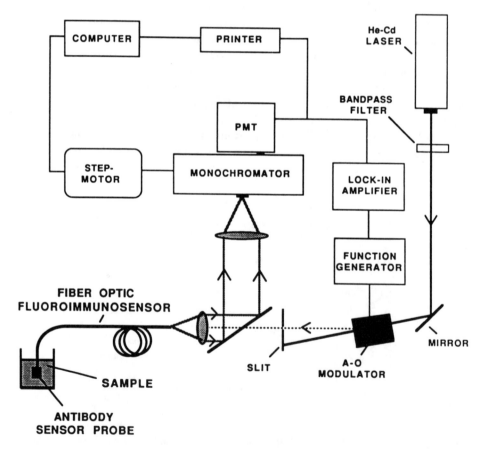

Figure 11.18 Schematic diagram of a phase-resolved fiber-optic fluoroimmunosensor. (*Source:* [215]. Reprinted with permission.)

benzo[α]pyrene in human placenta samples [215,216], an integrated-optic differential interferometer [217]), and an integrated-optic acoustic sensor [218] for assessing antigen-antibody interactions.

3. Feasibility studies of evanescent fluorescence with optical fibers, for example, by detection of skew rays [219] or with a single-mode tapered optical fiber [220]. Conical optodes molded in low-cost plastics have also been proposed in order to increase the signal-to-noise ratio for evanescent wave immunoassay [221,222].

4. Practical investigation of the chemical parameters of labels in the NIR spectral region [223–225], of new immobilization techniques such as sol-gel [226], and of the hydrophobic characteristics of surface and ligand in avidin-biotin interactions [227].

11.5.4 Examples of Optical Immunosensing

Table 11.5 summarizes some important compounds of blood, drugs, toxins, and hormones that have been determined by optical immunosensors.

In biosensing, another important species is DNA. In toxicology and cancer research, the detection of traces of polycyclic aromatic compounds, particularly benzo[α]pyrene, a metabolite that can bind covalently to DNA, has already been reported [194]. These aromatic compounds have a structure composed of several rings and can compete with known fluorescent agents (acridine orange, ethidium bromide, and proflavin) attached to double-stranded DNA [237]. This assay has been demonstrated by the technique of fluorescent polarization spectroscopy [237] and the evanescent wave technique [238]. By specific coupling of a suitable dye to a link group at the 3'-terminus of DNA, the influence of nucleic acid-dye interactions on the fluorescence lifetime and quantum yield of the coupled dye has been investigated by time-resolved fluorescence spectroscopy [239]. The difference in the quenching efficiency for the bases adenine, cytosine, thymine, and guanine (see Chapter 3) has been determined in mononucleotides and oligonucleotides in order to characterize DNA.

A new type of DNA gene probe is based on label detection by SERS, using a silver-coated alumina substrate [240]. Hybridization, which involves the joining of a nucleic acid strand with its corresponding mirror image, is a technique for identifying DNA sequences. A hybridized DNA probe can be used to detect DNA targets (e.g., gene sequences, bacteria, viral DNA fragments). Evanescent wave techniques are possible for human genome detection in real time using DNA hybridization on optical waveguides [241]. A covalent attachment of DNA to a silicon nitride (SiON) surface without an intermediate polymer has been tested experimentally with [242]. The adsorption capacity of SiON is not less than 200 pg · mm^{-2}, and the hybridization efficiency can be close to 100%. This level seems sufficient for enabling DNA detection by interferometry or SPR techniques with an integrated-optic concept.

A further example is single-stranded DNA (sDNA) as a selective recognition

Table 11.5
Examples of Optical Immunosensing With Fiber-Optic Probes

Usual name	Range	Label	Technique	References
Blood Compounds				
Ferritin	$0.01–1~\mu g \cdot mL^{-1}$	FITC	Evanescent fluorescence in flow cell	[228]
Albumin	$0.5–10~\mu g \cdot mL^{-1}$	Dansyl	Fluorescence from quartz plates	[206]
Albumin	$0.1–100~\mu g \cdot mL^{-1}$		Chemiluminescence catalyst immunoassay	[229]
Lipoproteins	$1–50~\mu g \cdot mL^{-1}$	TRITC	Total internal reflection fluorescence	[230]
Prostate antigen	$0.5–100~ng \cdot mL^{-1}$	APC	Fluorescent capillary fill device	[231]
Drugs and Toxins				
Phenytoin, Digoxin, and Theophyllin	25–500 nM	Dansyl	Fluorescence in regenerable immunosensor	[208]
Phenytoin	5–300 nM	Texas red/PE	Fluorescence energy transfer	[202]
Phenytoin and Lidocain	$0.02–20~mg \cdot mL^{-1}$	FITC/PE	Fluorescence on plates	[232]
Theophyllin	30–350 nM	CF	Fluorescence liposome flow injection assay	[233]
C. Botulinum	$5–200~ng \cdot mL^{-1}$	TRITC	Fiber-optic fluorimetry	[234]
Ricin	About $ng \cdot mL^{-1}$		Evanescent wave detection	[235]
Hormones				
Thyroxine	LOD = $0.1~\mu M$		Fluorescence quenching of TBG in flow cell	[236]

Note: Abbreviations are as follows: LOD = limit of detection; FITC = fluorescein isothiocyanate, TRITC = tetramethylrhodamine-isothiocyanate, APC = allophycocyanin; PE = phycoerythrin; CF = carboxyfluorescein, TBG = thyroxine binding globulin.

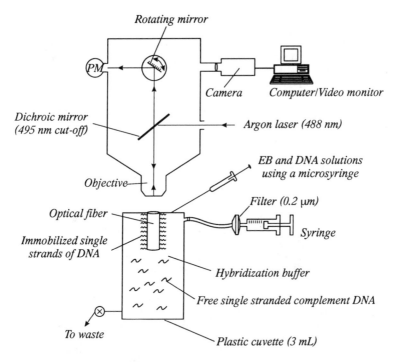

Figure 11.19 Schematic diagram of the equipment for the fluorimetric detection of DNA hybridization using optical fibers with immobilized DNA as the immunosensing layer. Abbreviations are as follows: PM = photomultiplier; EB = ethidium bromide. (*After:* [243], with permission.)

agent, which is covalently immobilized onto an optical fiber [243]. It can be hybridized with a complementary DNA (cDNA) to form double-stranded DNA (dDNA). This event is detected by total internal reflection after addition of the fluorescent sDNA stain, ethidium bromide. The device is shown in Figure 11.19. The detection limit of the sensor is about 86 ng · mL^{-1} of sDNA. The sensor retains the same activity after three months' storage.

References

[1] Göpel, W., J. Hesse, and J. N. Zemel, eds., *Sensors: A Comprehensive Survey*, Vols. 2 and 3, Weinheim: VCH, 1991.

[2] Wolfbeis, O. S., ed., *Fiber Optic Chemical Sensors and Biosensors*, Vols. 1 and 2, Boca Raton: CRC Press, 1991.

[3] Wise, D. L., and L. B. Wingard, eds., *Biosensors With Fiberoptics*, Clifton, NJ: Humana Press, 1991.

[4] Arnold, M. A., "Fiber Optic Based Biocatalytic Biosensors," Chap. 20 in *Chemical Sensors and*

Microinstrumentation, R. W. Murray, R. E. Dessy, W. R. Heineman, J. Janata, and W. R. Seitz, eds., Washington, ACS Symp. Series, Vol. 403, 1989, pp. 303–317.

[5] Trettnak, W., M. Hofer, and O. S. Wolfbeis, "Applications of Optochemical Sensors for Measuring Environmental and Biochemical Quantities," Chap. 18 in *Sensors: A Comprehensive Survey, Vol. 3*, W. Göpel, J. Hesse, and J. N. Zemel, eds., Weinheim: VCH, 1991, pp. 931–967.

[6] Lübbers, D. W., "Chemical in Vivo Monitoring by Optical Sensors in Medecine," *Sensors and Actuators*, Vol. B11, 1993, pp. 253–262.

[7] Baldini, F., and A. G. Mignani, "In-Vivo Biomedical Monitoring by Fiber Optic Systems," *Proc. SPIE–Int. Soc. Opt. Eng.*, Vol. 2293, 1994, pp. 80–89.

[8] Pesce, A. J., and L. A. Kaplan, eds., *Methods in Clinical Chemistry*, St-Louis: Mosby, 1987.

[9] Ruzicka, J., and E. H. Hansen, *Flow Injection Analysis*, 2nd ed., New York: Wiley, 1989.

[10] Ruzicka, J., and E. H. Hansen, "Optosensing at Active Surfaces: A New Detection Principle in Flow Injection Analysis," *Anal. Chim. Acta*, Vol. 173, 1985, pp. 3–21.

[11] Wolfbeis, O. S., "Optical Sensors in Flow Injection Analysis," *J. Mol. Struct.*, Vol. 292, 1992, pp. 133–140.

[12] Susuki, K., H. Ohzora, K. Tohda, K. Mizayaki, K. Watanabe, H. Inoue, and T. Shirai, "Fibre Optic Potassium Ion Sensor Based on a Neutral Ionophore and a Novel Lipophilic Anionic Dye," *Anal. Chim. Acta*, Vol. 237, 1990, pp. 155–164.

[13] Al-Amir, S. M. S., D. C. Ashworth, and R. Narayanaswamy, "Synthesis and Characterization of Some Chromogenic Crown Ethers as Potential Optical Sensors for Potassium Ions," *Talanta*, Vol. 36, 1989, pp. 645–650.

[14] Garcia, R. P., F. A. Moreno, M. F. Diaz Garcia, A. Sanz Mendel, and R. Narayanaswamy, "Serum Analysis for Potassium Ions Using a Fiber Optic Sensor," *Clin. Chim. Acta*, Vol. 207, 1992, pp. 31–40.

[15] Wolfbeis, O. S., and B. P. H. Schaffar, "Optical Sensors: An Ion Selective Optrode for Potassium," *Anal. Chim. Acta*, Vol. 198, 1987, pp. 1–12.

[16] Schaffar, B. P. H., O. S. Wolfbeis, and A. Leitner, "Effect of Langmuir-Blodgett Layer Composition on the Response of Ion-Selective Optrodes for Potassium, Based on the Fluorimetric Measurement of Membrane Potential," *Analyst*, Vol. 113, 1988, pp. 693–697.

[17] He, H., H. Li, G. Mohr, B. Kovacs, T. Werner, and O. S. Wolfbeis, "Novel Type of Ion-Selective Fluorosensor Based on the Inner Filter Effect: An Optrode for Potassium," *Anal. Chem.*, Vol. 65, 1993, pp. 123–127.

[18] Roe, J. N., F. C. Szoka, and A. S. Verkman, "Fibre Optic Sensor for the Detection of Potassium Using Fluorescence Energy Transfer," *Analyst*, Vol. 115, 1990, pp. 353–358.

[19] Wolfbeis, O. S., "Fluorescence Based Ion Sensing Using Potential Sensitive Dyes," *Sensors and Actuators*, Vol. B 29, 1995, pp. 140–147.

[20] Charlton, S. C., R. L. Fleming, and A. Zipp, "Solid Phase Colorimetric Determination of Potassium," *Clin. Chem.*, Vol. 28, 1982, pp. 1857–1861.

[21] Vogel, P., D. Thym, M. Fritz, and D. Mosoiu, "Test Carrier for the Determination of Ions in Biological Fluids," Ger. Offen Pat. 4,015,590, 1991.

[22] Ng, R. H., K. M. Sparks, and B. E. Statland, "Colorimetric Determination of Potassium in Plasma and Serum by Reflectance Photometry With a Dry Chemistry Reagent," *Clin. Chem.*, Vol. 38, 1992, pp. 1371–1372.

[23] Wang, K., K. Seiler, W. E. Morf, U. E. Spichiger, W. Simon, E. Lindner, and E. Pungor, "Characterization of Potassium Selective Optode Membranes Based on Neutral Ionophores and Application in Human Blood Plasma," *Anal. Sci.*, Vol. 6, 1990, pp. 715–720.

[24] Spichiger, U. E., K. Seiler, K. Wang, G. Suter, W. E. Morf, and W. Simon, "Optical Quantification of Sodium, Potassium, and Calcium Ions in Diluted Human Plasma Based on Ion-Selective Membranes," *Proc. SPIE–Int. Soc. Opt. Eng.*, Vol. 1510, 1991, pp. 118–130.

[25] Dinten, O., U. E. Spichiger, N. Chaniotakis, P. Gehrig, B. Rusterholz, W. E. Morf, and W. Simon,

"Lifetime of Neutral Carrier Based Liquid Membranes in Aqueous Samples and Blood and the Lipophilicity of Membrane Components," *Anal. Chem.*, Vol. 63, 1991, pp. 596–603.

[26] Bakker, E., and E. Pretsch, "Lipophilicity of Tetraphenylborate Derivatives as Anionic Sites in Neutral Carrier-Based Solvent Polymeric Membranes and Lifetime of Corresponding Ion-Selective Electrochemical and Optical Sensors," *Anal. Chim. Acta*, Vol. 309, 1995, pp. 7–17.

[27] Morf, W. E., K. Seiler, B. Rusterholz, and W. Simon, "Design of a Calcium Selective Optode Membrane Based on Neutral Ionophores," *Anal. Chem.*, Vol. 62, 1990, pp. 738–742.

[28] Gehrig, P., B. Rusterholz, and W. Simon, "Very Lipophilic Sodium Selective Ionophore for Chemical Sensors of High Lifetime," *Anal. Chim. Acta*, Vol. 233, 1990, pp. 295–298.

[29] Suzuki, K., K. Tohda, H. Ohzora, S. Nishihama, H. Inoue, and T. Shirai, "Fiber Optic Magnesium and Calcium Ion Sensor Based on Natural Carboxylic Poyether Antibiotic," *Anal. Chem.*, Vol. 61, 1989, pp. 382–384.

[30] Wang, Y., J. M. Baten, S. P. McMaughan, and D. R. Bobbitt, "Optical Fiber Based Sensor for Calcium Using Hydrophobically Associated Calcein and Laser-Induced Fluorescence Detection," *Microchem. J.*, Vol. 50, 1994, pp. 385–396.

[31] Schaffar, B. P. H., and O. S. Wolfbeis, "A Calcium Selective Optrode Based on Fluorimetric Measurement of Membrane Potential," *Anal. Chim. Acta*, Vol. 217, 1989, pp. 1–9.

[32] Hisamoto, H., K. Watanabe, E. Nakagawa, D. Siswanta, Y. Shichi, and K. Susuki, "Flow-Through Type Calcium Ion Selective Optodes Based on Novel Neutral Ionophores and a Lipophilic Anionic Dye," *Anal. Chim. Acta*, Vol. 299, 1994, pp. 179–187.

[33] Watanabe, K., E. Nakagawa, H. Yamada, H. Hisamoto, and K. Susuki, "Lithium Ion Selective Optical Sensor Based on a Novel Neutral Ionophore and a Lipophilic Anionic Dye," *Anal. Chem.*, Vol. 65, 1993, pp. 2704–2710.

[34] Tan, S. S. S., P. C. Hauser, K. Wang, K. Fluri, K. Seiler, B. Rusterholz, G. Suter, M. Krüttli, U. Spichiger, and W. Simon, "Reversible Optical Sensing Membrane for the Determination of Chloride in Serum," *Anal. Chim. Acta*, Vol. 255, 1991, pp. 35–44.

[35] Hisamoto, H., K. Watanabe, H. Oka, E. Nakagawa, U. E. Spichiger, and K. Suzuki, "Flow Through Type Chloride Ion Selective Optodes Based on Lipophilic Organometallic Chloride Adducts and a Lipophilic Anionic Dye," *Anal. Sci.*, Vol. 10, 1994, pp. 615–622.

[36] Schultz, J. S., S. Mansouri, and I. J. Goldstein, "Affinity Sensor: A New Technique for Developing Implantable Sensors for Glucose and Other Metabolites," *Diabetes Care*, Vol. 5, 1982, pp. 245–253.

[37] Meadows, D., and J. S. Schultz, "Fiber Optic Biosensors Based on Fluorescence Energy Transfer," *Talanta*, Vol. 35, 1988, pp. 145–150.

[38] Lakowicz, J. R., and H. Szmacinski, "Fluorescence Lifetime Based Sensing of pH , Ca^{2+}, K^+ and Glucose," *Sensors and Actuators*, Vol. B 11, 1993, pp. 133–143.

[39] Meadows, D. L., and J. S. Schultz, "Design, Manufacture and Characterization of an Optical Fiber Glucose Affinity Sensor Based on an Homogeneous Fluorescence Energy Transfer Assay System," *Anal. Chim. Acta*, Vol. 280, 1993, pp. 21–30.

[40] Trettnak, W., M. J. P. Leiner, and O. S. Wolfbeis, "Fibre Optic Glucose Biosensor With an Oxygen Optrode as the Transducer," *Analyst*, Vol. 113, 1988, pp. 1519–1523.

[41] Opitz, N., and D. W. Lübbers, "Electrochromic Dyes, Enzymes Reactions and Hormone Protein Interactions in Fluorescence Optic Sensor (Optode) Technology," *Talanta*, Vol. 35, 1988, pp. 123–127.

[42] Schaffar, B. P. H., and O. S. Wolfbeis, "A Fast Reponding Fiber Optic Glucose Biosensor Based on an Oxygen Optrode," *Biosens. Bioelectron.*, Vol. 5, 1990, pp. 137–148.

[43] Moreno-Bondi, M. C., O. S. Wolfbeis, M. J. P. Leiner, and B. P. H. Schaffar, "Oxygen Optrode for Use in a Fiber Optic Glucose Biosensor," *Anal. Chem.*, Vol. 62, 1990, pp. 2377–2380.

[44] Papkovsky, D. B., J. Olah, and I. N. Kurochkin, "Fibre Optic Lifetime Based Enzyme Biosensor," *Sensors and Actuators*, Vol. B 11, 1993, pp. 525–530.

[45] Trettnak, W., M. J. P. Leiner, and O. S. Wolfbeis, "Fiber Optic Glucose With a pH Optrode as the Transducer," *Biosensors*, Vol. 4, 1989, pp. 15–26.

[46] Sharma, A., and O. Juptner, "Investigations on Mixed Indicator Sensing Scheme (MISS) for Developing Optical Biosensors," *Proc. SPIE–Int. Soc. Opt. Eng.*, Vol. 2131, 1994, pp. 607–613.

[47] Arnold, M. A., X. Zhou, and R. S. Petsch, "Gas Sensing Internal Enzyme Fiber Optic Biosensor for Hydrogen Peroxide," *Talanta*, Vol. 41, 1994, pp. 783–787.

[48] Zhou, X., and M. A. Arnold, "Internal Enzyme Fiber Optic Biosensors for Hydrogen Peroxide and Glucose," *Anal. Chim. Acta*, Vol. 304, 1995, pp. 147–156.

[49] Blum, L. J., and P. R. Coulet, eds., *Biosensor Principles and Applications*, New York: Marcel Dekker, 1991.

[50] Abdel-Latif, M. S., and G. G. Guilbault, "Fiber Optic Sensor for the Determination of Glucose Using Micellar Enhanced Luminescence of Peroxalate Reaction," *Anal. Chem.*, Vol. 60, 1988, pp. 2671–2674.

[51] Abdel-Latif, M. S., and G. G. Guilbault, "Peroxide Optrode Based on Micellar Mediated Chemiluminescence Reaction of Luminol," *Anal. Chim. Acta*, Vol. 221, 1989, pp. 11–17.

[52] Genovesi, L., H. Pedersen, and G. H. Sigel, "The Development of a Generic Hydrogen Peroxide Sensor With Application to the Detection of Glucose," *Proc. SPIE–Int. Soc. Opt. Eng.*, Vol. 990, 1988, pp. 22–28.

[53] Gunasingham, H., C-H. Tan, and J. K. L. Seow, "Fiber Optic Glucose Sensor With Electrochemical Generation of Indicator Reagent," *Anal. Chem.*, Vol. 62, 1990, pp. 755–759.

[54] Trettnak, W., and O. S. Wolfbeis, "Fully Reversible Fibre Optic Glucose Biosensor Based on the Intrinsic Fluorescence of Glucose Oxidase," *Anal. Chim. Acta*, Vol. 221, 1989, pp. 195–203.

[55] Narayanaswamy, R., and F. Sevilla, "An Optical Probe for the Determination of Glucose Based on Immobilized Glucose Deshydrogenase," *Anal. Lett.*, Vol. 21, 1988, pp. 1165–1175.

[56] Schelp, C., T. Scheper, F. Bückmann, and K. F. Reardon, "Two Fibre Optic Sensors With Confined Enzymes and Coenzymes: Development and Applications," *Anal. Chim. Acta*, Vol. 255, 1991, pp. 223–229.

[57] Lee, S. J., M. Saleemuddin, T. Scheper, H. Loos, and H. Sahm, "A Fluorimetric Fiber Optic Biosensor for Dual Analysis of Glucose and Fructose Using Glucose-Fructose-Oxidoreductase Isolated From Zymomonas Mobilis," *J. Biotechnol.*, Vol. 36, 1994, pp. 39–44.

[58] Liu, Y., P. Hering, and M. O. Scully, "An Integrated Optical Sensor for Measuring Glucose Concentration," *Appl. Phys.*, Vol. B 54, 1992, pp. 18–23.

[59] MacKenzie, H. A., G. B. Christison, P. Hodgson, and D. Blanc, "A Laser Photoacoustic Sensor for Analyte Detection in Aqueous Sytems," *Sensors and Actuators*, Vol. B 11, 1993, pp. 213–220.

[60] Schlager, K. J., "Non Invasive Near Infrared Measurement of Blood Analyte Concentrations," U.S. Pat., 4,882,492, 1992.

[61] Rhines, T. D., and M. A. Arnold, "Simplex Optimization of a Fiber Optic Ammonia Sensor Based on Multiple Indicators," *Anal. Chem.*, Vol. 60, 1988, pp. 76–81.

[62] Wolfbeis, O. S., and H. E. Posch, "Fiber Optic Fluorescing Sensor for Ammonia," *Anal. Chim. Acta*, Vol. 185, 1986, pp. 321–327.

[63] Berman, R. J., and L. W. Burgess, "Renewable Reagent Fiber Optic Based Ammonia Sensor," *Proc. SPIE–Int. Soc. Opt. Eng.*, Vol. 1172, 1989, pp. 206–214.

[64] Smock, P. L., T. A. Ororfino, G. W. Wooten, and W. S. Spencer, "Vapor Phase Determination of Blood Ammonia by Optical Waveguide Technique," *Anal. Chem.*, Vol. 51, 1979, pp. 505–508.

[65] Wolfbeis, O. S., and H. Li, "LED-Compatible Fluorosensor for Ammonium and Its Application to Biosensing," *Proc. SPIE–Int. Soc. Opt. Eng.*, Vol. 1587, 1991, pp. 48–58.

[66] Werner, T., I. Klimant, and O. S. Wolfbeis, "Novel Matrix for Ammonia Sensing Based on Immobilized Indicator Ion Pairs," *Analyst*, Vol. 120, 1995, pp. 1627–1631.

[67] Rhines, T. D., and M. A. Arnold, "Determination of Ammonia in Untreated Serum With a Fiber Optic Ammonia Gas Sensor," *Anal. Chim. Acta*, Vol. 231, 1990, pp. 231–235.

[68] Kar, S., and M. A. Arnold, "Cylindrical Sensor Geometry for Absorbance Based Fiber Optic Ammonia Sensors," *Talanta*, Vol. 41, 1994, pp. 1051–1058.

[69] Polster, J., W. Höbel, A. Papperger, and H-L. Schmidt, "Fundamentals of Enzyme Substrate Determinations by Fiber Optics Spectroscopy," *Proc. SPIE–Int. Soc. Opt. Eng.*, Vol. 1172, 1989, pp. 273–286.

[70] Yerian, T.D., G. D. Christian, and J. Ruzicka, "Flow Injection Analysis as a Diagnostic Technique for Development and Testing of Chemical Sensors," *Anal. Chim. Acta*, Vol. 204, 1988, pp. 7–28.

[71] Spinks, T. L., and G. E. Pacey," Utilization of Adsorption Immobilized Urease in Gas Diffusion Flow System," *Anal. Chim. Acta*, Vol. 237, 1990, pp. 503–508.

[72] Brennan, J. D., R. S. Brown, A. D. Manna, K. M. R. Kallury, P. A. Piunno, and U. J. Krull, "Covalent Immobilization of Amphiphilic Monolayers Containing Urease Onto Optical Fibers for Fluorimetric Detection of Urea," *Sensors and Actuators*, Vol. B 11, 1993, pp. 109–119.

[73] Mueller, C., F. Schubert, and T. Scheper, "Multicomponent Fiber Optical Biosensor for Use in Hemodialysis Monitoring," *Proc. SPIE–Int. Soc. Opt. Eng.*, Vol. 2131, 1994, pp. 555–562.

[74] Xie, X., A. A. Suleiman, and G. G. Guilbault, "A Urea Fiber Optic Biosensor Based on Absorption Measurement," *Anal. Lett.*, Vol. 23, 1990, pp. 2143–2153.

[75] Rhines, T. D., and M. A. Arnold, "Fiber Optic Biosensor for Urea Based on Sensing of Ammonia Gas," *Anal. Chim. Acta*, Vol. 227, 1989, pp. 387–396.

[76] Seiler, K., W. E. Morf, B. Rusterholz, and W. Simon, "Design and Characterization of a Novel Ammonium Ion Selective Optical Sensor Based on Neutral Ionophores," *Anal. Sci.*, Vol. 5, 1989, pp. 557–561.

[77] Osawa, S., P. C. Hauser, K. Seiler, S. S. Tan, W. E. Morf, and W. Simon, "Ammonia Gas Selective Optical Sensors Based on Neutral Ionophores," *Anal. Chem.*, Vol. 63, 1991, pp. 640–644.

[78] Wolfbeis, O. S., and H. Li, "Fluorescence Optical Urea Biosensor With an Ammonium Optrode as Transducer," *Biosens. Bioelectron.*, Vol. 8, 1993, pp. 161–166.

[79] Jeppesen, M. T., and E. H. Hansen, "Determination of Creatinine in Undiluted Blood Serum by Enzymatic Flow Injection Analysis With Optosensing," *Anal. Chim. Acta*, Vol. 214, 1988, pp. 147–159.

[80] Trettnak, W., and O. S. Wolfbeis, "A Fiber Optic Cholesterol Biosensor With an Oxygen Optrode as the Transducer," *Anal. Biochem.*, Vol. 184, 1989, pp. 124–127.

[81] Krug, A., A. A. Suleiman, and G. G. Guilbault, "Enzyme Based Fiber Optic Device for the Determination of Total and Free Cholesterol," *Anal. Chim. Acta*, Vol. 256, 1992, pp. 263–268.

[82] He, H., G. Uray, and O. S. Wolfbeis, "Enianto-Selective Optodes," *Anal. Chim. Acta*, Vol. 246, 1991, pp. 251–257.

[83] Kiba, N., H. Koemado, J. Inagaki, and M. Furusawa, "Determination of 3-Hydroxybutyrate in Serum by Flow Injection Analysis Using a Co-immobilized 3-Hydroxybutyrate Deshydrogenase/NADH Oxidase Reactor and a Chemiluminometer," *Anal. Chim. Acta*, Vol. 298, 1994, pp. 129–133.

[84] Kapany, N. S., and N. Silbertrust, "Fiber Optics Spectrophotometer for in-Vivo Oximetry," *Nature*, Vol. 208, 1964, pp. 138–145.

[85] Cope, M., P. Van der Zee, M. Essenpreis, S. R. Arridge, and D. T. Delpy, "Data Analysis Methods for Near Infra-Red Spectroscopy of Tissue: Proble in Determining the Relative Cytochrome aa3 Concentration," *Proc. SPIE–Int. Soc. Opt. Eng.*, Vol. 1431, 1991, pp. 251–262.

[86] Miyaki, H., S. Nioka, A. Zaman, D. S. Smith, and B. Chance, "The Detection of Cytochrome in Oxidase Heme Iron and Copper Absorption in the Blood-Perfused and Blood-Free Brain in Normoxia and Hypoxia," *Anal. Biochem.*, Vol. 192, 1991, pp. 149–155.

[87] Barker, S. J., and K. K. Tremper, "Pulse Oximetry: Applications and Limitations," in *Advances in*

Oxygen Monitoring, Vol. 25, K. K. Tremper and S. J. Bakker, eds., Boston: Little, Brown and Cie, 1987, pp. 155–175.

[88] Shellock, F. G., S. M. Myers, and K. J. Kimble, "Monitoring Heart Rate and Oxygen Saturation With a Fiber Optic Pulse Oximeter During MR Imaging," *Amer. J. of Roentgenology,* Vol. 158, 1992, pp. 663–664.

[89] West, I. P., R. Holmes, and J. R. Jones, "Optical Fiber Based Pulse Oximeter for Monitoring in Magnetic Resonance Scanners," *Proc. SPIE–Int. Soc. Opt. Eng.,* Vol. 2360, 1994, pp. 94–97.

[90] Fantini, S., M. A. Franceschini, J. S. Maier, S. A. Walker, and E. Gratton, "Frequency Domain Multi-Source Optical Spectrometer and Oximeter," *Proc. SPIE–Int. Soc. Opt. Eng.,* Vol. 2326, 1994, pp. 108–116.

[91] Renault, G., D. Duboc, and M. Degeorges, "In Situ Laser Fluorimetry in Cardiology: Preliminary Results and Perspectives," *J. Appl. Cardiol.,* Vol. 2, 1987, pp. 91–95.

[92] Mayevsky, A., and B. Chance, "Methods and Apparatus for Monitoring Functions of Brain and Others Tissue," *PCT. Int. Appl. Wo* 92, 12705, 1992.

[93] Beuthan, J., O. Minet, and G. Muller, "Observations of the Fluorescence Response of the Coenzyme NADH in Biological Samples," *Opt. Lett.,* Vol. 18, 1993, pp. 1098–1109.

[94] Bocher, T., J. Beuthan, O. Minet, I. Schmitt, B. Fuchs, and G. Müller, "Fiber Optical Sampling of NADH Concentration in Guinea-Pig Heart During Ischemia," *Proc. SPIE–Int. Soc. Opt. Eng.,* Vol. 2324, 1994, pp. 166–176.

[95] Leiner, M. J. P., and O. S. Wolfbeis, "Fiber Optic pH Sensors," Chap. 8 in *Fiber Optic Chemical Sensors and Biosensors, Vol. 1,* O. S. Wolfbeis, ed., Boca Raton: CRC Press, 1991, pp. 359–384.

[96] Baldini, F., "Recent Progress in Fiber Optic pH Sensing," *Proc. SPIE–Int. Soc. Opt. Eng.,* Vol. 1368, 1991, pp. 184–190.

[97] Leiner, M. J. P., and P. Hartmann, "Theory and Practice in Optical pH Sensing," *Sensors and Actuators,* Vol. B 11, 1993, pp. 281–289.

[98] Wolthuis, R., D. MacCrae, E. Saaski, J. Hartl, and G. Mitchell, "Development of a Medical Fiber Optic pH Sensor Based on Optical Absorption," *IEEE Trans. Biomed. Eng.,* Vol. 39, 1992, pp. 531–537.

[99] Lehmann, H., G. Schwotzer, P. Czerney, and G. J. Mohr, "Fiber Optic pH Meter Using NIR Dye," *Sensors and Actuators,* Vol. B 29, 1995, pp. 392–400.

[100] Lübbers, D. W., and N. Opitz, "The pCO_2/pO_2 Optode: A New Probe for Measurement of pCO_2 and pO_2 in Gases or Liquids," *Z. Naturforsch.,* Vol. 30C, 1975, pp. 532–533.

[101] Zhujun, Z., and W. R. Seitz, "A Carbon Dioxide Sensor Based on Fluorescence," *Anal. Chim. Acta,* Vol. 160, 1984, pp. 305–309.

[102] Peterson, J. I., R. V. Fitzerald, and D. K. Buckhlod, "Fiber Optic Probe for in Vivo Measurement of Oxygen Partial Pressure," *Anal. Chem.,* Vol. 56, 1984, pp. 62–67.

[103] MacCraith, B. D., C. M. McDonagh, G. O'Keeffe, E. T. Keyes, J. G. Vos, B. O'Kelly, and J. F. McGilp, "Fibre Optic Oxygen Sensor Based on Fluorescence Quenching of Evanescent Wave Excited Ruthenium Complexes in Sol-Gel Derived Porous Coatings," *Analyst,* Vol. 118, 1993, pp. 385–388.

[104] MacCraith, B. D., G. O'Keeffe, C. M. McDonagh, and A. K. McEvoy, "LED-Based Fibre Optic Oxygen Using Sol-Gel Coating," *Electron. Lett.,* Vol. 30, 1994, pp. 888–889.

[105] Meier, B., T. Werner, I. Klimant, and O. S. Wolfbeis, "Novel Oxygen Sensor Material Based on Ruthenium Bipyridyl Complex Encapsulated in Zeolite Y: Dramatic Differences in the Efficiency of Luminescence Quenching by Oxygen on Going From Surface-Adsorbed to Zeolite Encapsulated Fluorophore," *Sensors and Actuators,* B 29, 1995, pp. 240–245.

[106] Klimant, I., P. Belser, and O. S. Wolfbeis, "Novel Metal Organic Ruthenium (II) Diimin Complexes for Use as Longwave Excitable Luminescent Oxygen Probes," *Talanta,* Vol. 41, 1994, pp. 985–991.

[107] Klimant, I., and O. S. Wolfbeis, "Novel Oxygen Sensitive Materials Based on Silicone Soluble Ruthenium Complexes," *Anal. Chem.,* Vol. 67, 1995, pp. 3160–3166.

[108] Draxler, S., M. E. Lippitsch, I. Klimant, H. Kraus, and O. S. Wolfbeis, "Effects of Polymer Matrices

on the Time Resolved Luminescence of a Ruthenium Complex Quenched by Oxygen," *J. Phys. Chem.*, 99, 1995, pp. 3162–3167.

[109] Rosenzweig, Z., and R. Kopelman, "Development of a Submicrometer Optical Fiber Oxygen Sensor," *Anal. Chem.*, Vol. 67, 1995, pp. 2650–2654.

[110] Papkovsky, D. B., "New Oxygen Sensors and Their Applications to Biosensing," *Sensors and Actuators*, Vol. B 29, 1995, pp. 213–218.

[111] Mitchell, G. L., J. C. Hartl, D. MacCrae, R. A. Wolthuis, E. W. Saaski, K. C. Garcin, and H. R. Williard, "Viologen Based Fiber Optic Oxygen Sensors: Optics Development," *Proc. SPIE–Int. Soc. Opt. Eng.*, Vol. 1587, 1992, pp. 16–20.

[112] Del Bianco, A., F. Baldini, M. Bacci, I. Klimant, and O. S. Wolfbeis, "A New Kind of Oxygen Sensitive Transducer Based on an Immobilized Metallo Organic Compound," *Sensors and Actuators*, Vol. B 11, 1993, pp. 347–352.

[113] Wolfbeis, O. S., L. J. Weiss, M. J. P. Leiner, and W. E. Ziegler, "Fiber Optic Fluorosensor for Oxygen and Carbon Dioxide," *Anal. Chem.*, Vol. 60, 1988, pp. 2028–2030.

[114] Siggaard-Andersen, O., I. H. Gorthgen, P. D. Wimberley, J. P. Rasmussen, and N. Fogh-Andersen, "Evaluation of the Gas-Stat Fluorescence Sensors for Continuous Measurement of pH , " pCO_2 and pO_2 During Cardiopulmonary Bypass and Hypothermia, *Scand. J. Clin. Lab. Invest*, Vol. 48, 1988, pp. 77–84.

[115] Leiner, M. J. P., "Optical Sensors for in Vitro Blood Gas Analysis," *Sensors and Actuators*, Vol. B 29, 1995, pp. 169–173.

[116] Gehrich, J. L., D. W. Lübbers, N. Opitz, D. R. Hansmann, W. W. Miller, J. K. Tusa, and M. Yafuso, "Optical Fluorescence and Its Application to an Intravascular Blood Gas Monitoring System," *IEEE Trans. Biomed. Eng.*, Vol. 33, 1986, pp. 117–132.

[117] Miller, W. W., M. Yafuso, C. F. Yan, H. K. Hui, and S. Arick, "Performance of an in-Vivo Continuous Blood Gas Monitor With Disposable Probe," *Clin. Chem.*, Vol. 33, 1987, pp. 1538–1542.

[118] Barker, S. J., and J. Hyatt, "Continuous Measurement of Intra-arterial pH, pCO_2 and pO_2 in the Operating Room," *Anesth. Analg.*, Vol. 73, 1991, pp. 43–48.

[119] Martin, R. C., S. F. Malin, D. J. Bartnik, A. M. Schilling, and S. C. Fourlong, "Performance and Use of Paracorporeal Fiber Optic Blood Gas Sensors," *Proc. SPIE–Int. Soc. Opt. Eng.*, Vol. 2131, 1994, pp. 426–436.

[120] Boiarski, A. A., "Integrated Optic System for Monitoring Blood Gases," U.S. Pat. 4,854,321, 1989.

[121] Gottlieb, A., S. Divers, and H. K. Hui, "In Vivo Applications of Fiber Optic Chemical Sensors," in *Biosensors With Fiberoptics*, D. L. Wise and L. B. Wingard, eds., Clifton, NJ: Humana Press, 1991, pp. 325–366.

[122] Walt, D. R., "Fiber Optic Sensors for Continuous Clinical Monitoring," *Proc. IEEE*, Vol. 80, 1992, pp. 903–911.

[123] Hui, H. K., S. Divers, T. Lumsden, T. Wallner, and S. Weir, "An Accurate, Low Cost Easily Manufacturable Oxygen Sensor," *Proc. SPIE–Int. Soc. Opt. Eng.*, Vol. 1172, 1989, pp. 233–238.

[124] Shapiro, B. A., R. D. Cane, C. M. Chomka, L. E. Bandala, and W. T. Peruzzi, "Preliminary Evaluation of an Intra-arterial Blood Gas Systems in Dogs and Humans," *Crit. Care Med.*, Vol. 17, 1989, pp. 455–460.

[125] Barker, S. J., J. Hyatt, K. K. Tremper, J. L. Gehrich, S. Arick, S. Gerschultz, and K. Safdari, "Fiber Optic Intra-arterial pHa, paO_2 and $paCO_2$ in the Operating Room," *Anesth. Analg.*, Vol. 68, 1989, p. S16.

[126] Posch, H. E., M. J. P. Leiner, and O. S. Wolfbeis, "Towards a Gastric pH Sensor: An Optrode for the 0–7 Range," *Fresenius Z. Anal. Chem.*, Vol. 334, 1989, pp. 162–165.

[127] Schulman, S. G., S. Chen, F. Bai, M. J. P. Leiner, L. Weis, and O. S. Wolfbeis, "Dependance of the Fluorescence of Immobilized 1-Hydroxypyrene-3,6,8-Trisulfonate on Solutions pH : Extension of the Range of Applicability of a pH Fluorosensor," *Anal. Chim. Acta*, Vol. 304, 1995, pp. 165–170.

[128] Netto, E. J., J. I. Peterson, M. McShane, and V. Hampshire, "A Fiber Optic Broad Range pH Sensor System for Gastric Measurements," *Sensors and Actuators*, Vol. B 29, 1995, pp. 157–163.

[129] Baldini, F., P. Bechi, S. Bracci, F. Cosi, and F. Pucciani, "In Vivo Optical Fibre pH Sensor for Gastro-Esophageal Measurements," *Sensors and Actuators*, Vol. B 29, 1995, pp. 164–168.

[130] Bechi, P., F. Pucciani, F. Baldini, F. Cosi, R. Falciai, R. Mazzanti, A. Castagnoli, A. Passeri, and S. Bocherini, "Long-Term Ambulatory Enterogastric Reflux Monitoring: Validation of a New Fiberoptic Technique," *Digest. Diseases and Sci.*, Vol. 38, 1993, pp. 1297–1306.

[131] Baldini, F., P. Bechi, and S. Bracci, "Optical Fiber Sensors in Forgut Functional Diseases," *Proc. SPIE–Int. Soc. Opt. Eng.*, Vol. 2231, 1995, pp. 72–79.

[132] Barnikol, W. K. R., T. Gaertner, N. Weiler, and O. Burkhard, "Microdetector for Rapid Changes of Oxygen Partial Pressure During the Respiratory Cycle in Small Laboratoy Animals," *Rev. Sci. Instrum.*, Vol. 59, 1988, pp. 1204–1208.

[133] Karpf, H., H. W. Kroneis, H. J. Marsoner, H. Metzler, and N. Gravenstein, "Fast Responding Oxygen Sensor for Respiratorial Analysis," *Proc. SPIE–Int. Soc. Opt. Eng.*, Vol. 1172, 1989, pp. 295–304.

[134] Keller, H. P., and D. W. Lübbers, "Control of Pedicle and Microvascular Tissue Transfer by Photometric Reflection Oximetry," *Adv. Exp. Med. Biol.*, Vol. 220, 1987, pp. 187–190.

[135] Kessler, M., D. K. Harrison, and J. Höper, "Tissue Oxygen Measurement Techniques," in *Microcirculatory Technology*, C. H. Baker and W. L. Nastuk, eds., New York: Academic Press, 1986, pp. 391–425.

[136] Manil, J., R. H. Bourgain, M. Van Waeyenberge, F. Colin, E. Blockeel, B. De Mey, J. Coremans, and R. Paternoster, "Properties of the Spontaneous Fluctuations in Cortical Oxygen Pressure," *Adv. Exp. Med. Biol.*, Vol. 169, 1984, pp. 231–239.

[137] Abel, H. H., D. Klüssendorf, and H. P. Koepchen, "New Approach to the Neurovegetative State in Man," in *Innovations in Physiological Anaesthesia and Monitoring*, R. Droh and R. Spintge, eds., Berlin: Springer-Verlag, 1989, pp. 21–24.

[138] Holst, G. A., T. Köster, E. Voges, and D. W. Lübbers, "Flox: An Oxygen-Flux-Measuring System Using a Phase Modulation Method to Evaluate the Oxygen-Dependent Fluorescence Lifetime," *Sensors and Actuators*, Vol. B 29, 1995, pp. 231–239.

[139] Ono, K., M. Kanda, J. Hiramoto, K. Yotsuya, and N. Sato, "Fiber Optic Reflectance Spectrometry System for in Vivo Tissue Diagnosis," *Appl. Opt.*, Vol. 30, 1991, pp. 98–105.

[140] Wyatt, J. S., M. Cope, D. T. Delpy, P. Van Der Zee, S. Arridge, A. D. Edwards, and E. O. R. Reynolds, "Measurement of Optical Path Length for Cerebral Near Infrared Spectroscopy in Newborn Infants," *Dev. Neurosci.*, Vol. 12, 1990, pp. 140–144.

[141] Sevick, E. M., B. Chance, J. Leigh, S. Nioka, and M. Maris, "Quantitation of Time- and Frequency-Resolved Optical Spectra for the Determination of Tissue Oxygenation," *Anal. Biochem.*, Vol. 195, 1991, pp. 330–351.

[142] Clarke, R. H., "Laser Systems for Non Invasive Blood Analysis Based on Reflectance Ratio Detection," U.S. Pat. 5,054,487, 1991.

[143] Peuchant, E., C. Salles, and R. Jensen, "Determination of Serum Cholesterol by Near-Infrared Reflectance Spectrometry," *Anal. Chem.*, Vol. 59, 1987, pp. 1816–1819.

[144] Clivaz, X., F. Marquis-Weible, R. P. Salathé, R. P. Novak, and H. H. Gilgen, "High Resolution Reflectometry in Biological Tissues," *Opt. Lett.*, Vol. 17, 1992, pp. 4–6.

[145] Walker, J. M., and E. B. Gingold, eds., *Molecular Biology and Biotechnology*, London: Royal Soc. of Chemistry, 1988.

[146] Rehm, H. J., and G. Reed, eds., *Biotechnology*, 2nd ed., Weinheim: VCH, 1991.

[147] Scheper, T., and K. F. Reardon, "Sensors in Biotechnology," Chap. 22 in *Sensors: A Comprehensive Survey, Vol. 2*, W. Göpel, J. Hesse, and J. N. Zemel, eds., Weinheim: VCH, 1991, pp. 1023–1048.

[148] Boudrant, J., G. Corrieu, and P. Coulet, eds., *Capteurs et Mesures en Biotechnologie*, Paris: Lavoisier, Tech. Doc., 1994.

[149] Wolfbeis, O. S., "Fiber Optic Sensors for Chemical Parameters of Interest in Biotechnology," in

Biosensor Intern. Workshop GBM, Vol. 10, Monograph, R. D. Schmid, G. G. Guilbault, J. Karube, H. L. Schmidt, and L. B. Wingard, eds., Weinheim: VCH, 1987, pp. 197–206.

[150] Scheper, T., W. Brandes, C. Grau, H. G. Hundeck, B. Reinhardt, F. Rüther, F. Plötz, C. Schelp, K. Schügerl, K. H. Schneider, F. Giffhorn, B. Rehr, and H. Sahm, "Applications of Biosensor Systems for Bioprocess Monitoring," *Anal. Chim. Acta*, Vol. 249, 1991, pp. 25–34.

[151] Boisdé, G., "Les Fibres Optiques et Guides d'Ondes en Biotechnologie," Chap. 11 in *Capteurs et Mesures en Biotechnologie*, J. Boudrant, G. Corrieu, and P. Coulet, eds., Paris: Lavoisier, Tech. Doc., 1994, pp. 405–442.

[152] Kennedy, M. J., M. S. Thakur, D. I. C. Wang, and G. N. Stephanopoulos, "Estimation Cell Concentration in Presence of Suspended Solids: A Light Scatter Technique," *Biotechnol. Bioeng.*, Vol. 40, 1992, pp. 875–888.

[153] Agar, D. W., "Microbial Growth Rate Measurement Technique," in *Comprehensive Biotechnology*, Vol. 4, Moo-Young, ed., London: Pergamon Press, 1985, pp. 305–327.

[154] Scheper, T., and K. Schügerl, "Characterization of Bioreactors by In-Situ Fluorometry," *J. Biotechnol.*, Vol. 3, 1986, pp. 221–229.

[155] Gikas, P., and A. G. Livingston, "Use of ATP to Characterize Biomass Viability in Freely Suspended and Immobilized Cell Bioreactors," *Biotechnol. Bioeng.*, Vol. 42, 1993, pp. 1337–1351.

[156] Anders, K. D., G. Wehnert, O. Thordsen, T. Scheper, B. Rehr, and H. Sahm, "Biotechnological Applications of Fiber-Optic Sensing: Multiple Uses of a Fiber Optic Fluorimeter," *Sensors and Actuators*, Vol. B 11, 1993, pp. 395–403.

[157] Peck, M. W., and D. P. Chynoweth, "On-line Fluorescence Monitoring of the Methanogenic Fermentation," *Biotechn. Bioeng.*, Vol. 39, 1992, pp. 1151–1160.

[158] Blum, L. J., and P. R. Coulet, eds., *Biosensor: Principles and Applications*, New York: Marcel Dekker, 1991.

[159] Coulet, P. R., L. J. Blum, and S. M. Gautier, "Luminescence Based Fibre Optic Probes," *Sensors and Actuators*, Vol. B 11, 1993, pp. 57–61.

[160] Gautier, S. M., L. J. Blum, and P. R. Coulet, "Multi-Function Fibre-Optic Sensor for the Bioluminescent Flow Determination of ATP or NADH," *Anal. Chim. Acta*, Vol. 235, 1990, pp. 243–253.

[161] Blum, L. J., S. M. Gautier, A. Berger, P. E. Michel, and P. E. Coulet, "Multicomponent Organized Bioactive Layers for Fiber Optic Luminescent Sensors," *Sensors and Actuators*, Vol. B 29, 1995, pp. 1–9.

[162] Holobar, A., B. H. Weigl, W. Trettnak, R. Benes, H. Lehmann, N. V. Rodriguez, A. Wollschlager, P. O'Leary, P. Raspor, and O. S. Wolfbeis, "Experimental Results on an Optical pH Measurement System for Bioreactors," *Sensors and Actuators*, Vol. B 11, 1993, pp. 425–430.

[163] Weigl, B. H., A. Holobar, W. Trettnak, I. Klimant, H. Kraus, P. O'Leary, and O. S. Wolfbeis, "Optical Triple Sensor for Measuring pH , Oxygen and Carbon Dioxide," *J. Biotechnol.*, Vol. 32, 1994, pp. 127–138.

[164] Agayn, V. I., and D. R. Walt, "Fiber Optic Sensor for Continuous Monitoring of Fermentation pH," *Biotechnol.*, Vol. 11, 1993, pp. 726–729.

[165] Agayn, V. I., and D. R. Walt, "Monitoring of Fermentation Parameters With Fiber Optics," *Proc. SPIE–Int. Soc. Opt. Eng.*, Vol. 2068, 1994, pp. 179–184.

[166] Wolfbeis, O. S., "Fiber Optic Probe for Kinetic Determination of Enzyme Activity," *Anal. Chem.*, Vol. 58, 1986, pp. 2874–2876.

[167] Freeman, M. K., and L. G. Bachas, "Fiber Optic Biosensor With Fluorescence Detection Based on Immobilized Alkaline Phosphatase," *Biosens. Bioelectron.*, Vol. 7, 1992, pp. 49–55.

[168] Taga, K., S. Weger, R. Göbel, and R. Kellner, "Colorimetric Activity Assays of Enzyme Modified MIR Fibers," *Sensors and Actuators*, Vol. B 11, 1993, pp. 553–559.

[169] Gautier, S. M., L. J. Blum, and P. R. Coulet, "Dehydrogenase Activity Monitoring by Flow Injection Analysis Combined With Luminescence Based Fibre Optic Sensors," *Anal. Chim. Acta*, Vol. 266, 1992, pp. 331–338.

[170] Zhujun, Z., W. R. Seitz, and S. O'Connell, "Amylase Substrate Based on Fluorescence Energy Transfer," *Anal. Chim. Acta*, Vol. 236, 1990, pp. 251–256.

[171] Li, H., and O. S. Wolfbeis, "Determination of Urease Activity by Flow Injection Analysis Using an Ammonium Selective Optrode as the Detector," *Anal. Chim. Acta*, Vol. 276, 1993, pp. 115–119.

[172] Wolfbeis, O. S., "Optrodes for Measuring Enzyme Activity and Inhibition," *NATO Adv. Sci. Ser., Part E: Appl. Sci.*, Kluwer, Vol. 252, 1993, pp. 335–344.

[173] Dremel, B. A. A., B. P. H. Schaffar, and R. D. Schmid, "Determination of Glucose in Wine and Fruit Juice Based on a Fibre Optic Glucose Biosensor and Flow Injection Analysis," *Anal. Chim. Acta*, Vol. 225, 1989, pp. 293–301.

[174] Wolfbeis, O. S., and H. E. Posch, "A Fibre Optic Ethanol Biosensor," *Frezenius Z. Anal. Chem.*, Vol. 332, 1988, pp. 255–257.

[175] Xie, X., A. A. Suleiman, G. G. Guilbault, Z. Yang, and Z. Sun, "Flow Injection Determination of Ethanol by Fiber Optic Chemiluminescence Measurement," *Anal. Chim. Acta*, Vol. 266, 1992, pp. 325–329.

[176] Arnold, M. A., "Fiber Optic Based Biocatalytic Biosensors," Chap. 20 in *Chemical Sensors and Microinstrumentation*, R. W. Murray, E. Dessy, W. R. Heineman, J. Janata, and W. R. Seitz, eds., Washington, D.C., ACS Symp. Series, Vol. 403, 1989, pp. 303–317.

[177] Scheper, T., and A. F. Bueckmann, "A Fiber Optic Biosensor Based on Fluorometric Detection Using Confined Macromolecular Nicotinamide Adenine Dinucleotide Derivatives," *Biosens. Bioelectron.*, Vol. 5, 1990, pp. 125–135.

[178] Schelp, C., T. Scheper, A. F. Bückmann, and K. F. Reardon, "Two Fibre Optic Sensors With Confined Enzymes and Co-enymes: Development and Application," *Anal. Chim. Acta*, Vol. 255, 1991, pp. 223–229.

[179] Trettnak, W., and O. S. Wolfbeis, "A Fully Reversible Fiber Optic Lactate Biosensor Based on the Intrinsic Fluorescence of Lactate Monooxygenase," *Fresenius Z. Anal. Chem.*, Vol. 334, 1989, pp. 427–430.

[180] Dremel, B. A. A., W. Yang, and R. D. Schmid, "On-line Determination of Lactic Acid During Kefir Fermentation Based on a Fibre Optic Lactic Acid Biosensor and Flow Injection Analysis," *Anal. Chim. Acta*, Vol. 234, 1990, pp. 107–112.

[181] Berger, A., and L. J. Blum," "Enhancement of the Response of a Lactate Oxidase /Peroxidase Based Fiber Optic Sensor by Compartmentalization of the Enzyme Barrier," *Enzyme Microb. Technol.*, Vol. 16, 1994, pp. 979–984.

[182] Dremel, B. A. A., R. D. Schmid, and O. S. Wolfbeis, "Comparison of Two Fibre Optic L-Glutamate Biosensors Based on the Detection of Oxygen or Carbon Dioxide and Their Application in Combination With Flow Injection Analysis to the Determination of Glutamate," *Anal. Chim. Acta*, Vol. 248, 1991, pp. 351–359.

[183] Cattaneo, M. V., and J. H. Luong, "Monitoring Glutamine in Animal Cell Cultures Using a Chemiluminescence Fiber Optic Biosensor," *Biotechnol. Bioeng.*, Vol. 41, 1993, pp. 659–665.

[184] Kulp, T. J., I. Camins, S. M. Angel, C. Munkholm, and D. R. Walt, "Polymer Immobilized Enzyme Optrodes for the Detection of Penicillin," *Anal. Chem.*, Vol. 59, 1987, pp. 2847–2853.

[185] Fuh, M. R. S., L. W. Burgess, and G. D. Christian, "Single Fiber Optic Fluorescence Enzyme Based Sensor," *Anal. Chem.*, Vol. 60, 1988, pp. 433–435.

[186] Polster, J., W. Höbel, A. Papperger, and H. L. Schmidt, "Fundamentals of Enzyme Substrate Determinations by Fiber Optics Spectroscopy," *Proc. SPIE–Int. Soc. Opt. Eng.*, Vol. 1172, 1989, pp. 273–286.

[187] Xie, X., A. A. Suleiman, and G. G. Guilbault, "A Fluorescence Based Fiber Optic Biosensor for the Flow Injection Analysis of Penicillin," *Biotechnol. Bioeng.*, Vol. 39, 1992, pp. 1147–1150.

[188] Luo, S., and D. R. Walt, "Avidin Biotin Coupling as a General Method for Preparing Enzyme Based Fiber Optic Sensors," *Anal. Chem.*, Vol. 61, 1989, pp. 1069–1072.

[189] Schügerl, K., T. Scheper, B. Hitzmann, and C. Müller, "Intelligent Sensor Systems for Bioprocess Monitoring," *J. SICE* (Japan), Vol. 34, 1995, pp. 18–24.

[190] Moreno, M. C., A. Martinez, P. Millan, and C. Camara, "Study of a pH Sensitive Optical Fibre Sensor Based on the Use of Cresol Red," *J. Molecular Structure*, Vol. 143, 1986, pp. 553–556.

[191] He, H., G. Uray, and O. S. Wolfbeis, "Optical Sensor for Salicilic Acid and Aspirin Based on a New Lipophilic Carrier for Aromatic Carboxylic Acids," *Fresenius, Z. Anal. Chem.*, Vol. 343, 1992, pp. 313–318.

[192] He, H., H. Li, G. Uray, and O. S. Wolfbeis," Non Enzymatic Optical Sensor for Penicillins," *Talanta*, Vol. 40, 1993, pp. 453–457.

[193] Sepaniak, M. J., B. J. Tromberg, and T. Vo-Dinh, "Fiber Optic Affinity Sensors in Chemical Analysis," *Prog. Anal. Spectrosc.*, Vol. 11, 1988, pp. 481–509.

[194] Vo-Dinh, T., G. D. Griffin, and M. J. Sepaniak, "Fiberoptics Immunosensors," Chap. 17 in *Fiber Optic Chemical Sensors and Biosensors, Vol. 1*, O. S. Wolfbeis, ed., Boca Raton: CRC Press, 1991, pp. 217–257.

[195] Vo-Dinh, T., "Biosensors and Antibody Probes for Environmental and Biomedical Applications," *Proc. SPIE–Int. Soc. Opt. Eng.*, Vol. 2293, 1995, pp. 132–138.

[196] Robinson, G. A., "Optical Immunosensors: An Overview," in *Advances in Biosensors, Vol. 1*, A. P. F. Turner, ed., London: JAI Press, 1991, pp. 229–256.

[197] Alarie, J. P., M. J. Sepaniak, and T. Vo-Dinh, "Evaluation of Antibody Immobilization Techniques for Fiber Optic Based Fluoroimmunosensing," *Anal. Chim. Acta*, Vol. 229, 1990, pp. 169–176.

[198] Rabbany, S. Y., A. W. Kusterbeck, R. Bredehorst, and F. S. Ligler, "Binding Kinetics of Immobilized Antibodies in a Flow Immunoassay," *Sensors and Actuators*, B 29, 1995, pp. 72–78.

[199] Balgi, G., D. E. Leckband, and J. M. Witsche, "Transport Effects on the Kinetics of Protein-Surface Binding," *Biophys. J.*, Vol. 68, 1995, pp. 2251–2260.

[200] Feldman, S. F., and E. E. Uzgiris, "Determination of the Kinetic Response and Absolute Sensitivity of a Fiber Optic Immunoassy," *Proc. SPIE–Int. Soc. Opt. Eng.*, Vol. 2068, 1994, pp. 139–144.

[201] Huber, W., R. Barner, C. Fattinger, J. Hübscher, H. Koller, F. Müler, D. Schlatter, and W. Lukosz, "Direct Optical Immunosensing (Sensitivity and Selectivity)," *Sensors and Actuators*, Vol. B 6, 1992, pp. 122–126.

[202] Miller, W. G., and F. P. Anderson, "Antibody Properties for Chemically Reversible Biosensor Applications," *Anal. Chim. Acta*, Vol. 227, 1989, pp. 135–143.

[203] Astles, J. R., and W. G. Miller, "Reversible Fiber Optic Immunosensor Measurements," *Sensors and Actuators*, Vol. B 11, 1993, pp. 73–78.

[204] Alarie, J. P., J. R. Bowyer, M. J. Sepaniak, A. M. Hoyt, T. Vo-Dinh, "Fluorescence Monitoring of a Benzo(a)Pyrene Metabolite Using a Regenerable Immunochemical Based Fiber Optic Sensor," *Anal. Chim. Acta*, Vol. 236, 1990, pp. 237–244.

[205] Millot, M-C., T. Vals, F. Martin, B. Sebille, and Y. Levy, "Surface Plasmon Resonance Response of a Polymer Coated Biochemical Sensor," *Proc. SPIE Int. Soc. Opt. Eng.*, Vol. 2331, 1995, pp. 34–39.

[206] Bright, F. V., T. A. Betts, and K. S. Litwiler, "Regenerable Fiber Optic Based Immunosensor," *Anal. Chem.*, Vol. 62, 1990, pp. 1065–1069.

[207] Betts, T. A., G. C. Catena, J. Huang, K. S. Litwiler, J. Zhang, J. Zagrobelny, and F. V. Bright, "Fiber Optic Based Immunosensors for Haptens," *Anal. Chim. Acta*, Vol. 246, 1991, pp. 55–63.

[208] Bright, F. V., K. S. Litwiler, T. G. Vargo, and J. A. Gardella, "Enhanced Performance of Fibre Optic Immunoprobes Using Refunctionalized Fluoropolymers as the Substratum," *Anal. Chim. Acta*, Vol. 262, 1992, pp. 323–330.

[209] Sloper, A. N., and M. T. Flanagan, "Scattering in Planar Surface Waveguide Immunosensors," *Sensors and Actuators*, Vol. B 11, 1993, pp. 537–542.

[210] Nellen, P .M., and W. Lukosz, "Integrated Optical Input Grating Couplers as Chemo- and Immunosensors," *Sensors and Actuators*, Vol. B 1, 1990, pp. 592–596.

[211] Clerc, D., and W. Lukosz, "Integrated Optical Output Grating Coupler as Biochemical Sensor," *Sensors and Actuators*, Vol. B 18–19, 1994, pp. 581–586.

[212] Lukosz, W., "Integrated Optical Chemical and Direct Biochemical Sensors," *Sensors and Actuators*, Vol. B 29, 1995, pp. 37–50.

[213] Laybourn, P. J. R., Y. Zhou, R. M. De la Rue, W. Cushley, C. McScharry, and J. V. Magill, "An Integrated Optical Immunosensor," *NATO ASI Ser.*, Ser. E, Vol. 252, 1993, pp. 463–470.

[214] Pariaux, O., and P. Sixt, "Sensitivity Optimization of a Grating Coupled Evanescent Wave Immunosensor," *Sensors and Actuators*, Vol. B 29, 1995, pp. 289–292.

[215] Vo-Dinh, T., T. Nolan, Y. F. Cheng, M. J. Sepaniak, and J. P. Alarie, "Phase Resolved Fiber Optics Fluoroimmunosensor," *Appl. Spectrosc.*, Vol. 44, 1990, pp. 128–132.

[216] Vo-Dinh, T., J. P. Alarie, F. Hyder, and M. J. Sepaniak, "Laser Based Fiber Optic Immunosensors for DNA-Adduct Measurements," *Polycyclic. Aromat. Compd.*, Vol. 3, 1993, pp. 765–772.

[217] Stamm, C., and W. Lukosz, "Integrated Optical Difference Interferometer as Biochemical Sensor," *Sensors and Actuators*, Vol. B 18–19, 1994, pp. 183–187.

[218] Pliska, P., and W. Lukosz, "Integrated Optical Acoustic Sensors," *Sensors and Actuators*, Vol. A 41–42, 1994, pp. 93–97.

[219] Eenink, R. G., H. E. De Bruijn, R. P. H. Kooyman, and J. Greve, "Fibre Fluorescence Immunosensor Based on Evanescent Wave Detection," *Anal. Chim. Acta*, Vol. 238, 1990, pp. 317–321.

[220] Hale, Z. M., R. Marks, C. R. Lowe, and F. P. Payne, "The Single Mode Tapered Optical Fiber Immunosensor: I Characterization With Model Analytes," *Proc. SPIE–Int. Soc. Opt. Eng.*, Vol. 2131, 1994, pp. 484–494.

[221] Slovacek, R. E., S. C. Furlong, and W. F. Love, "Feasibility Study of a Plastic Evanescent Wave Sensor," *Sensors and Actuators*, Vol. B 11, 1993, pp. 307–311.

[222] Slovacek, R. E., S. C. Furlong, and W. F. Love, "Application of a Plastic Evanescent Wave Sensor to Immunological Measurements of CKMB," *Sensors and Actuators*, Vol. B 29, 1995, pp. 67–71.

[223] Daneshvar, M. I., G. A. Casay, M. Lipowska, G. Patonay, and L. Strekowski, "Investigation of a Near Infrared Fiber Optic Immunosensor," *Proc. SPIE–Int. Soc. Opt. Eng.* Vol. 2068, 1994, pp. 128–138.

[224] Golden, J. P., L. C. Shriver-Lake, N. Narayanan, G. Patonay, and F. S. Ligler, "A Near IR Biosensor for Evanescent Wave Immunoassays," *Proc. SPIE–Int. Soc. Opt. Eng.*, Vol. 2138, 1994, pp. 241–245.

[225] Lin, J-N., W. C. Mahoney, A. A. Luderer, R. A. Brier, T. W. Sharp, and V. A. McGuire, "Ultra Sensitive Evanescent Wave Fluoroimmunosensors Using Polystyrene Integrated Lens Optical Fiber," *Proc. SPIE–Int. Soc. Opt. Eng.*, Vol. 2138, 1994, pp. 204–215.

[226] Collino, R., J. Therasse, P. Binder, E. Chaput, J. P. Boilot, and E. Levy, "Thin Films of Functionalized Amorphous Silica for Immunosensors Application," *J. Sol-Gel Sci. and Technol.*, Vol. 2, 1994, pp. 823–826.

[227] Wang, Y., and D. R. Bobitt, "Binding Characteristics of Avidin and Surface Immobilized Octylbiotin Implications for the Development of Dynamically Modified Optical Fiber Sensors," *Anal. Chim. Acta*, Vol. 298, 1994, pp. 105–112.

[228] Bluestein, B. I., M. Craig, R. Slovacek, L. Stundtner, C. Urciuoli, I. Walczak, and A. Luderer, "Evanescent Wave Immunosensors for Clinical Diagnostics," in *Biosensors With Fiberoptics*, D. L. Wise and L. B. Wingard, eds., Clifton, NJ: Humana Press, 1991, pp. 181–223.

[229] Hara, T., K. Tsukagoshi, A. Arai, and Y. Imashiro, "A Highly Sensitive Fiber Optic Immunosensor Using a Metal Complex Compound as a Chemiluminescent Catalyst," *Bull. Chem. Soc. Jpn.*, Vol. 62, 1989, pp. 2844–2848.

[230] Poirier, M. A., T. Lopes, and B-R. Singh, "Use of an Optical Fiber Based Biosensor to Study the Interaction of Blood Proteins With Solid Surfaces," *Appl. Spectrosc.*, Vol. 48, 1994, pp. 867–870.

[231] O'Neill, P. M., J. E. Fletcher, G. Christopher, C. G. Stafford, P. B. Daniels, and T. Bacarese-Hamilton, "Use of an Optical Biosensor to Measure Prostate Specific Antigen in Whole Blood," *Sensors and Actuators*, Vol. B 29, 1995, pp. 79–83.

[232] Starodub, N. F., P. Y. Arenkov, A. E. Rachkov, and V. A. Berezin, "Fiber Optic Immuno-Sensors for Detection of Some Drugs," *Sensors and Actuators*, Vol. B 13–14, 1993, pp. 728–731.

[233] Locascio-Brown, L., A. L. Plant, V. Horvath, and R. A. Durst, "Liposome Flow Injection Immunoassay: Implications for Sensitivity, Dynamic Range, and Antibody Regeneration," *Anal. Chem.*, Vol. 62, 1990, pp. 2587–2593.

[234] Ogert, R. A., J. E. Brown, B-R. Singh, L.C. Shriver-Lake, and F. S. Ligler, "Detection of Clostridium Botulinum Toxin a Using a Fiber Optic Based Biosensor," *Anal. Biochem.*, Vol. 205, 1992, pp. 306–312.

[235] Ogert, R. A., J. Burans, T. O'Brien, and F. S. Ligler, "Comparative Analysis of Toxin Detection in Biological and Environmental Samples," *Proc. SPIE–Int. Soc. Opt. Eng.*, Vol. 2068, 1994, pp. 151–158.

[236] Reck, B., K. Himmelspach, N. Opitz, and D. W. Lübbers, "Possibilities and Limitations of Continuous Thyroxine Measurement in an Optode Using the Principle of Homogeneous Fluoroimmunoassay," *Analyst*, Vol. 113, 1988, pp. 1423–1426.

[237] Horvath, J. J., M. Gueguetchkeri, D. Penumatchu, A. Gupta, and H. H. Weetall, "A New Method for the Detection and Measurement of Polyaromatic Carcinogens and Related Compounds by DNA Intercalation," *A. C. S. Symp. Ser.*, Vol. 613, 1995, pp. 44–60; and references herein.

[238] Pandey, P. C., and H. H. Weetall, "Detection of Aromatic Compounds Based on DNA Intercalation Using Evanescent Wave Biosensor," *Anal. Chem.*, Vol. 67, 1995, pp. 787–792.

[239] Seidel, C., K. Rittinger, J. Cortes, R. S. Goody, M. Koellner, J. Wolfrum, and K. O. Gruelich, "Characterization of Fluorescence Labeled DNA by Time Resolved Fluorescence Spectroscopy," *Proc. SPIE–Int. Soc. Opt. Eng.*, Vol. 1452, 1991, pp. 105–116.

[240] Vo-Dinh, T., K. Houck, and D. L. Stokes, "Surface Enhanced Raman Gene Probes," *Anal. Chem.*, Vol. 66, 1994, pp. 3379–3383.

[241] Stimpson, D. I., J. V. Hoijer, W. T. Hsieh, C. Jou, J. Gordon, T. Theriault, R. Gamble, and J. D. Baldenschwieler, "Real Time Detection of DNA Hybridization and Melting on Oligonucleotide Assays by Using Optical Waveguides," *Proc. Natl. Acad. Sci.*, Vol. 92, 1995, pp. 6379–6383.

[242] Karymov, M. A., A. A. Kruchinin, Y .A,Tarantov, I. A. Balova, L. A. Remisova, and Y. G. Vlasov, "Fixation of DNA Directly on Optical Surfaces for Molecular Probe Biosensor Development," *Sensors and Actuators*, Vol. B 29, 1995, pp. 324–327.

[243] Piunno, P. A. E., U. J. Krull, R. H. E. Hudson, M. J. Damha, and H. Cohen, "Fiber Optic Biosensor for Fluorimetric Detection of DNA Hybridization," *Anal. Chim. Acta*, Vol. 288, 1994, pp. 205–214.

Chapter 12

Environmental Monitoring, Process Control, Gas Measurements, and Sensor Networks

12.1 ENVIRONMENTAL MONITORING

The development of optical communication technology started a major interest in environmental monitoring with optical sensors [1]. Up to now, fiber-optic sensors have been applied to groundwater flow determination and measurement of pollutants in rivers, oceans, ground and drinking water, in air (toxic pollutants), and in the soil (nutrients, pesticides, herbicides). New laboratory spectrometers, portable fluorimeters, and photometers have been described [2], particularly for the assessment of organic chloride compounds, gasoline, aluminium, cyanide, and sulfur dioxide. Phytoplankton and chlorophyll distributions, polycyclic aromatic hydrocarbons, phenols in water, and pollutants such as NO_x, SO_2, and HCN have also been measured [3]. This section summarizes recent work in this field.

12.1.1 Tracer Dyes in Water

Many fluorescent tracers can be used for groundwater flow determination. Their most important characteristics are their spectral properties and the influence of pH [4]. Natural fluorescent pigments (such as chlorophyll-a) can also serve as tracers in fluorescence spectroscopy for assessment of phytoplankton (algae) growth [5]. In most cases, the sensing elements in hydrology are passive optodes using plain fibers or fiber bundles (see Figure 2.11) to direct light into the sample and to collect the fluorescent or scattered light.

12.1.2 Organic Pollutants

12.1.2.1 Polycyclic Aromatic Hydrocarbons

PAHs are one of the most commonly found environmental pollutants and occur in different forms such as particles (aerosol) or dissolved in solution [6,7]. Remote measurement for water and soil can be made directly in situ and in real time by fluorescence spectroscopy [8]. However, in natural waters, the fluorescence emission spectra of different PAHs overlap and the excitation wavelengths cover a broad range due to variations in pH and in chemical composition of the medium. Pulsed ($\approx$2 ns) time-resolved fluorescence (Figure 12.1) enhances the specificity of measurement by

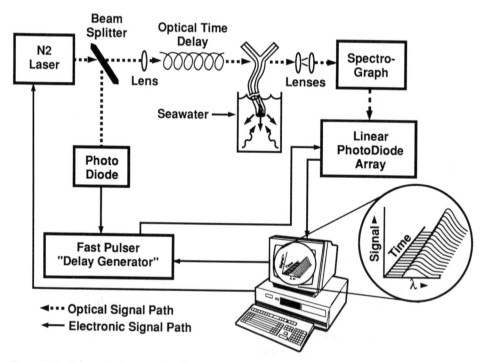

Figure 12.1   Schematic diagram of a fiber-optic time-resolved fluorimeter system for the decay time determination of polycyclic aromatic hydrocarbons in sea water. (*Source:* [9]. Reprinted with permission.)

analyzing the decay times [9]. In aerated sea water, these decay times are significantly shorter than the known values in solvents. Hence, a matrix analysis of excitation-emission wavelengths and associated decay times in the 4- to 50-ns range has been suggested for identifying unknown compounds in real samples. Cyclodextrins can be combined with PAHs to increase the sensitivity [10].

Novel instrumentation offers enhanced sensitivities; for example, a laser-excited synchronous fluorimeter has superior scanning precision, an extended wavelength range, and lifetime discrimination [11]. This prototype has a LOD for all compounds of less than 1 µg · L^{-1}, which is more than an order of magnitude better than current commercial instruments.

12.1.2.2 Halogenated Compounds

The general formula is $C_nH_mX_x$, where X is the halogenated compound (normally chloride). The best known are chloroform $CHCl_3$ ($n = m = 1$, $x = 3$), trichloroethylene C_2HCl_3 (TCE: $n = 2$, $m = 1$, $x = 3$), and carbon tetrachloride CCl_4 ($n = 1$, $m = 0$, $x = 4$). They can be determined by absorbance or fluorescence changes of pyridine or derivatives in a basic medium (Fujiwara reaction). $CHCl_3$ has been measured by absorbance in the presence of a mixture of pyridine and tetrabutylammonium chloride at aqueous concentrations of 10 to 500 ppb in geothermal wells [12]. The TCE curve is nonlinear but reproducible and allows a LOD of 25 ppb. A portable remote fluorimeter with a capillary optode has been built for gaseous detection of $CHCl_3$ in a concentration range of 4 to 50 µg · L^{-1} (equivalent to 30 to 400 ppb in aqueous solution) with a lower LOD of 1.2 µg · L^{-1} [13]. Continuous measurements of TCE and $CHCl_3$ are possible by renewing the reagent from a reservoir by means of a miniature pumping system as shown in Figure 12.2 [14].

Other techniques have been proposed for halogenated compounds. These include changes in the refractive index of coated fibers [15], a core-based absorption sensor with a silicone polymer acting as light pipe and detection membrane [16], and a fiber-optic Raman probe with a CCD as detector [17]. In addition, chlorinated hydrocarbons have been monitored by evanescence spectroscopy between 1,600 and 1,800 nm [18]. Also, a silver halide fiber coated with an appropriate polymer transparent in the mid-IR region, above 10 µm, has been coupled to a commercial FTIR spectrometer [19]. The latter sensor serves as a threshold alarm sensor in the 1- to 50-ppb concentration range. Polyisobutylene (PIB) can be a good polymer for TCE detection, and beveling the fiber end has been suggested for increasing the sensitivity of this technique [20].

A general technique has also been developed employing an integrated Mach-Zehnder structure with a membrane, made of a polysiloxane derivative, sensitive to gaseous hydrocarbons (chlorinated or not) on one arm of the interferometer [21]. However, this technique is not selective.

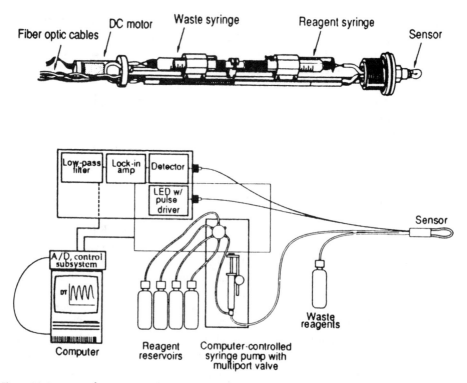

Figure 12.2   Sensor for monitoring halogenated compounds with a reagent delivery system providing fresh reagent for repeat measurements without increasing the pressure within the sensor body. The sensor is shown in the upper part of the figure; the fiber-optic sensor system is depicted below. (*Source:* [14]. Reprinted with permission.)

12.1.2.3 Aromatic Compounds and Other Solvents in Waters

Benzene, toluene, and xylene are common solvents and pollutants in drinking water. Optical fibers can be coated with special claddings that make them locally sensitive to hydrocarbons through refractive index variations related to analyte concentration. A typical response for benzene, toluene, xylene, and their mixtures is in the parts per million to parts per billion range (alternatively expressed in torr for the gas phase), using solid-state sensors or reservoir probes [22]. Commercial systems are available: Petrosense®, a portable device; Petrosense® CMS, a continuously monitoring system with up to 16 sensors; and Aersens™, a fluorescence unit.

Organic vapors of benzene, toluene, xylene, and gasoline compounds can also be measured by enhanced fluorescence of an immobilized fluorophore such as Nile red; measurements have been made in groundwater, soil samples, and continuous monitoring on a jet fuel contamination site [23]. NIR-SERS, using an excited diode laser as source, has been employed for the analysis of monosubstituted aromatic compounds such as 3-chloropyridine [24]. Various other solvents (tetrahydrofurane, ethyl acetate, ethanol) are determined by the absorbance change in the presence of triphenylmethane dyes [25]. Polynitroaromatic compounds, such as mono-, di-, or tri-nitrotoluene (TNT), can be detected at sub-part-per-million levels after a long exposure of several days to air or water [26,27]. Detection sensitivities of 8 ppb TNT were achieved using a competitive fluorescence immunoassay performed on the surface of a fiber-optic probe [28].

12.1.3 Metals Ions and Anions

Metal ions, especially heavy metals, are considered critical pollutants in water and can be assessed by optical-fiber sensing [29]. Table 12.1 summarizes a number of typical examples and various detection techniques. The types and performance of these sensors vary widely.

Other methods have also been reported. For instance, a simple sensor with plastic fibers has been demonstrated for ammonium determination in water [41], with a detection limit of 3 μM (80 nM in ammonia). Lead (II) has been measured at sub-part-per-million levels in soils by laser-induced breakdown spectroscopy [42]. In this technique, a high-power pulsed laser focused sample generates an induced plasma by vaporization and ionization and spontaneous emission at discrete wavelengths. This emission is characteristic of the elemental species present in the plasma.

Detection of anions is also possible with optical sensors. Three typical examples are:

1. A fluoride sensor (LOD = 2.6×10^{-5} M) based on the enhanced fluorescence of a fluoride ternary complex from an immobilized zirconium-calcein blue chelate [43].
2. Detection of nitrates in drinking water (with a required sensitivity <30 ppm) based on the change in absorbance of a coextraction system in a membrane [44], or on the variation of the fluorescent excitation and emission of a potential-sensitive dye for which the influence of pH is small [45]. The latter sensor operates over a 3- to 1,000-ppm range. Fluorimetric determination of nitrites and nitrates in the last 20 years has been reviewed recently [46].
3. Cyanide detection by the change of metalloporphyrin and corrins in the presence of myeglobin [47]. The sensor is reproducible and reusable in the 10^{-6} to 10^{-5} M concentration range.

12.1.4 Biological Oxygen Demand

BOD_5 (a BOD measurement made in waste water under specified standard conditions) is an indication of slow biochemical oxidation and assessment of organic pollutants.

Table 12.1
Determination of Some Metal Ions by Optodes for Environmental Monitoring

Metal Ion	Chromophore	Immobilization Support	Optical Technique	LOD (in M)	Comments	References
Ag (I)	LAD-3	Octadecyl silica beads	Absorbance	1.0×10^{-6}	Cation exchange and neutral ionophore	[30]
Al (III)	Eriochrome cyanine	XAD2 resin	Reflectance	1.0×10^{-5}	Regenerable and reproducible sensor	[31]
Cd (II), Hg (II)	Tetraporphyrin	Functionalized epoxy	Absorbance	3.0×10^{-6}	Reversible	[32]
Cu (II)	Zincon	Hydrogel on polyester	Absorbance	1.0×10^{-6}	Single shot test	[33]
Fe (II)	Calcein blue	XAD4 resin	Fluorescence	5.0×10^{-6}	Flow cell	[34]
Fe (III)	Chromazurol S	Silica gel	Reflectance	2.0×10^{-6}	Reversible	[35]
Fe (III)	Pyrocatechol derivative	Sephadex gel	Reflectance	2.0×10^{-6}	Cell	[36]
Hg (II)	Cu-calcein		F. Competitive binding	3.6×10^{-9}	Pretreatment of sample	[37]
Mn (II)	Diethylaniline		F. Zero angle photon spectrometer	10^{-9}	Flow cell for seawater	[38]
Pb (II)	Dithizone	XAD4 resin	Reflectance	10^{-8}	Flow cell	[39]
U (VI)	Chromoionophore (ETH 5315)	Plasticized PVC	Absorbance	10^{-7}	Cation exchange and neutral ionophore	[40]

Note: Abbreviations are as follows: LOD = limit of detection; LAD = lipophilic anionic dye; F = fluorescence; PVC = polyvinyl chloride; ETH = Swiss Federal Institute of Technology.

The determination of dissolved oxygen in water after five days' incubation (index 5) requires the use of sealed bottles. A scanned system of multiple fiber-optic sensors has been developed based on the luminescence quenching of oxygen-sensitive films with an immobilized ruthenium (II) complex [48]. The oxygen concentration $[O_2]$ in the range 0.1 to 15 mg $\cdot$ L^{-1} is related to an empirical equation similar to the Stern-Volmer function:

$$\log (I/I_0) = 2.718 + \alpha \cdot T \cdot [O_2] \qquad (12.1)$$

where T is the temperature (between 15°C and 45°C) and α is an experimental constant. For biotechnological applications, this sensor is usable in batches after steam sterilization.

A new fiber-optic BOD sensor using a microorganism (trichosporon cutaneum) relies on the measurement of oxygen consumed by yeast in the presence of a fluorescent ruthenium complex in a PVC membrane (Figure 12.3) [49]. This method gives a rapid estimation, in 5 to 10 mn, of the BOD value, which can be correlated to the conventional BOD_5 method. The sensor is inexpensive, does not consume oxygen, and allows in situ monitoring with a dynamic range up to 110 mg $\cdot$ L^{-1} related to the glucose/glutamate BOD standard (Figure 12.4).

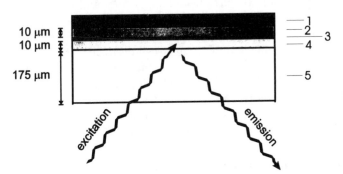

Figure 12.3 Cross section of a multilayer sensing membrane for BOD determination: (1) polycarbonate cover; (2) immobilized yeast layer; (3) 1-µm charcoal layer acting as an optical isolator; (4) oxygen-sensitive fluorescent layer; and (5) inert and gas-impermeable polyester support. The emission from the oxygen-sensitive layer is collected by a fiber bundle (not shown) underneath the polyester support. (*Source*: [49]. Reprinted with permission.)

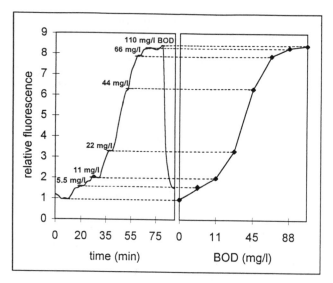

Figure 12.4 Optical signal changes and the response function of the BOD sensor when in contact with standard solutions of increasing BOD. (*After:* [49], with permission.)

12.1.5 Measurements in Seawater

Much work has been devoted to pH, pCO_2, and pO_2 sensors for physiological (see Chapter 11) and environmental applications [2]. Seawater is a special case because the pH and the exchange of CO_2 between the atmosphere and the ocean are interdependent, with other interferences from temperature and salinity variations. Thus, pH [50,51] is measured by the absorbance change of phenol red on amberlite resin, and pCO_2 (0- to 150-torr range) by the variation of the fluorescence of HPTS (hydroxypyrene trisulfonate) immobilized on cellulose [52]. Trials in the sea have been reported [53] as well as the role of CO_2 in global warning [54]. Furthermore, oxygen saturation in seawater has been determined by fluorescence quenching with a portable LED-based sensor and battery-operated instrumentation [55]. Also, dissolved free amino acids, which are important nutrients in marine microbial food webs, can be determined by time-resolved fluorescence using a FTC [56].

12.1.6 Pesticides and Herbicides

Monitoring the presence and pollution levels of pesticide and herbicide residues is essential to prevent contamination of water, soils (for agricultural purposes), and foods. Two sensing schemes are possible: with enzymatic reactions or with fluoroimmunoassays.

In the first scheme, pesticides act as inhibitors of the cleavage reaction of acetylcholine by acetylcholine esterase [57]. An advantage of this approach is the use of solid-state LEDs and photodiodes. Pyridostigmin and paraoxon (an organophosphorous pesticide) have been detected with a LOD of 1 and 0.5 μM, respectively. Paraoxon is hydrolyzed in the presence of the alkaline phosphatase enzyme and detected by chemiluminescence from conjugated polythiophene copolymer and fluorescent phycobiliprotein immobilized on an optical fiber [58].

In the second scheme, immunosensors are suitable for detecting other pesticides, such as dinitrophenole (DNP) and atrazine in the parts per billion range. An atrazine immunosensor, using a polymeric delivery system (see Section 2.6.3), is based on a competitive fluorescence energy transfer assay with FITC as donor and rhodamine-labeled atrazine as acceptor [59]. In another assay, fluorescein-labeled and nonlabeled atrazine (0.5 to 200 nM) compete for the binding sites of antiatrazine antibodies immobilized on an optical fiber [60].

12.1.7 Air Pollution

The most common air pollutants are NO_x, CO, SO_2, SH_2, HCN, alkanes, and volatile organic compounds (e.g., benzene, toluene, and xylene), which have been reviewed under gas detection [61]. Additional recent examples are presented in Section 12.3.

12.2 PROCESS CONTROL

The general concepts of optical monitoring in solution for process control applications have been summarized recently [62]. Optical-fiber sensing has been explored experimentally in three main areas of process control: biotechnology (see Chapter 11 and Section 12.2.2), nuclear process control, and explosive hydrocarbon gas monitoring (e.g., methane, and propane), a part of Section 12.3.

12.2.1 Nuclear Process Control

Remote on-line monitoring of uranium with glass fiber bundles was first done in 1960 [63], and later for plutonium over longer distances (>20m) with silica monofibers [64]. The main types of chemical sensors in this field have been reviewed [65]. The problem of sensitivity to nuclear radiation requires a careful choice of the type and material of the optical fiber so that it can withstand radiation damage without increasing its attenuation. Other technical difficulties include remote manipulation of components within the hot area, specialized optical connectors, and feed-throughs passing through thick walls, and remote replacement of optodes, as well as difficulties of operation (e.g., calibration). In particular, sensing elements must have long working lifetimes, and no active electronics can be placed within the hot zone; only passive optodes (FTCs or probes) have been used outside the laboratory in the hot area or pilot plant.

Figure 12.5 shows a typical scheme for remote on-line spectrometry in nuclear plants, in which the sampling and process zones are not distinct. Under normal current practice, no sensor is placed directly inside the Purex (purification and extraction of uranium and plutonium) process plant, and sample control is accomplished in the plant laboratory by transferring samples through air ducts. In a pilot plant, flow cells are generally placed on a secondary bypass flow or a ring of pulsed columns [66].

At the present time, the Chemex process (isotopic separation in an aqueous medium) has been abandoned in favor of isotopic separation of atomic vapor by laser. In the latter process, the control of the stability of laser dyes is possible with small optical path length cells (<1 mm).

Uranium, plutonium, and others actinides (e.g., neptunium, americium) have absorptions in the visible or NIR range. Hence, the intrinsic spectral properties of the process sample can serve to measure the concentrations of compounds such as Pu (III), Pu (IV), U (VI), and U (IV). In other cases, such as Pu (VI) and Np (V), which have a high molar absorption coefficient, a prior oxidation of lower valence ions is necessary before measurement in a FTC.

Table 12.2 summarizes the wavelengths of the main absorption peaks and the molar absorption coefficients for these compounds.

The simultaneous presence of several compounds (uranium and plutonium), together with different valence states, means that the measured spectra can be complicated. In addition, the Beer-Lambert law is not observed at high concentrations of uranium and nitrates, as shown in Figure 12.6. This is due to the presence of complexes such as UO_2^{2+}, $UO_2NO_3^+$, and $UO_2(NO_3)_2$ in nitric acid solution. A generalized empirical experimental function of the absorbance (A) for a constant path length can be written

$$A = [U] \cdot (a + b \cdot [NO_3^-] + c \cdot [NO_3^-]^2) \qquad (12.2)$$

where [U] and [NO_3^-] are the concentration of uranyl and nitrate ions, respectively, and a, b, and c are calibration constants. Similar functions are applicable to other main ions of the process.

Simultaneous determination of nitrates and uranium or plutonium from the absorption spectra has been well demonstrated [68,69]. This measurement could also be obtained from the fluorescence spectra [70]. The spectra in the solvent phase are generally more easily recognizable, but are difficult to monitor directly in the process because of the influence of radiation on these solvents.

In addition, other effects due to acidity (H^+) and the presence of hydrazine, hydroxylamine, and fine particles add to the complexity of direct absorbtion measurements, and the determination of the contribution of each compound requires a matrix method of analysis of the whole spectrum over a wide wavelength range. This is generally obtained with a fiber-optic spectrophotometer and CCD diode array (a

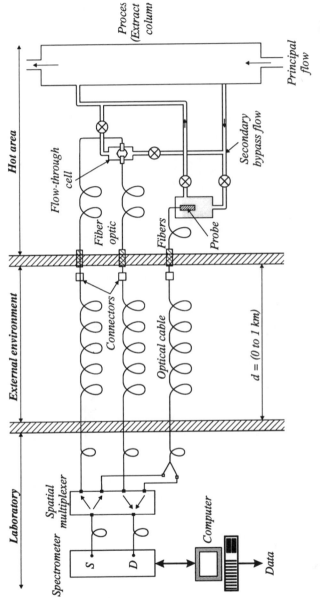

Figure 12.5 Schematic diagram showing remote measurements in a nuclear process plant using optical fibers. FTCs are placed in a secondary bypass flow circuit and must be capable of being interchanged by remote (robotic) manipulation.

Table 12.2
Main Spectral Characteristics of Nuclear Compounds in Process Control Applications

Compound	ε/λ	ε/λ	ε/λ	ε/λ	ε/λ	ε/λ	ε/λ	ε/λ	ε/λ
Am (III)	**420/503**								
Np (IV)	11/428	10/475	23/504	10/590	127/723	43/743	**160/950**		
Np (V)	7/467	22/617	9/770	**390/980**					
Np (VI)	7/557	**45/1,200**							
Pu (III)	9/430	30/566	29/602	12/664	8/800	8/885	10/1,030	12/1,120	
Pu (IV)	25/410	18/425	**65/477**	14/550	26/670	8/720	10/800	8/860	**8/1,100**
Pu (V)	15/570	5/850	**20/1,140**						
Pu (VI)	10/460	10/510	8/622	6/760	**455/830**	15/950	14/970		
U (III) in HCl medium	15/424	80/450	150/518	90/613	10/670	26/726	**170/872**	165/893	48/970
U (IV) in HCl medium	32/482	25/538	55/632	**65/653**	10/860				
U (IV) in HNO_3 medium	8/432	18/480	10/545	**33/648**	28/672	12/715	7/765		
U (VI) in HCl medium	**12/412**								
U (VI) in HNO_3 medium	**8/415**								
U (VI) in TPB (30%)-dodecane	10/403	**13/414**	12/426	9/436	6/451	4/468	2/485	47/1,100	

Notes: The main absorption peaks are reported as ε/λ, where ε is the molar absorptivity (in $L \cdot mol^{-1} \cdot cm^{-1}$) at a wavelength λ (in nanometers). The higher ε values at λ for each compound are shown in bold type. These values may be changed by the complexation effect. For instance, the values 8/415 for U (VI) in a nitrate medium corresponds to the mean value for UO_2^{2+}, $UO_2NO_3^+$, and $UO_2(NO_3)_2$ complexes with the values of 7/415, 14/425, and 22/430 for ε/λ [67]. Normally, the molar absorptivities in a $HClO_4$ medium are larger. Compounds are Am = americium; Np = neptunium; Pu = plutonium; U = uranium; TBP = tributyl phosphate. Am, Pu, and Np are quoted for a nitrate medium.

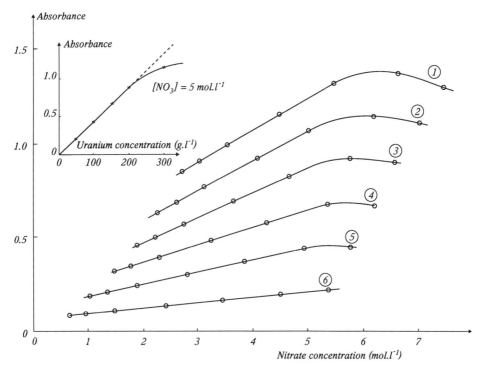

Figure 12.6   Calibration curves for on-line simultaneous determination of uranium and nitrates in nuclear plants. A complexation effect is evidenced for high concentrations of uranyl and NO_3^- ions. Uranium concentrations: (1) 300 g · L^{-1}; (2) 250 g · L^{-1}; (3) 200 g · L^{-1}; (4) 150 g · L^{-1}; (5) 100 g · L^{-1}; (6) 30 g · L^{-1}.

commercial device with 512, 1,024, or 1,728 pixels). Reference wavelengths may be chosen and selected according to the particular optical-fiber instrument and the specific solution being analyzed. Multivariate regression analysis [69,71] and chemometrics [72] are frequently employed to interpret the data.

New designs of photometers and sample cells have been built for specific process applications, such as Telephot® in France and Spectran in Germany [73], and specialized cells with short path lengths of 1 mm (for high-concentration measurements) or long lengths up to 1m (for trace detection of mg · L^{-1} of uranium in the waste from extraction columns). Remote time-resolved, laser-induced fluorimetry is a very interesting technique (Figure 12.7) and employs passive optodes for uranium trace determination with a LOD of 2.5×10^{-10} M [74]. Spectrofluorimeters equipped with optical fibers are commercially available.

Furthermore, the suitability of fiber optics for laser Raman spectrometry has been demonstrated for uranyl nitrate with capillary tubes as optodes for increasing the sensitivity [75].

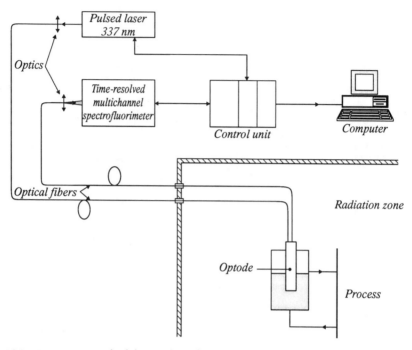

Figure 12.7 Remote time-resolved, laser-induced fluorimetry with fiber optics in a nuclear process plant.

Thermal lens spectroscopy (TLS) is a method in which heat is injected into a sample from a high-power laser, such as a dye laser. The emission spectrum is a function of the temperature. Temperature control can be accomplished with a photodiode. In dual-beam TLS, the amount of heat is measured by defocusing a second laser beam, usually a He-Ne laser. An absorption spectral scan is made by tuning the dye laser output wavelength (usually in a range of 30 to 60 nm). Transmitting the laser radiation by optical fiber allows greater flexibility in the instrumentation. This technique is well suited for rare-earth and actinide compounds that have "forbidden" transitions between the 4f or 5f electronic energy levels. The measurement of neodymium (LOD = 8×10^{-5} M) using a power density of 11 W $\cdot$ cm^{-2} [76] has been adapted to plutonium (VI) with a LOD of 3×10^{-6} M [77].

12.2.2 Other Sensors in Biotechnology

Two other areas are briefly described in this section: solvent monitoring (e.g., ethanol) and measurements in the food industry. Many other process control applications (such as fuel control, monitoring of fluidized beds, flocculation, or composite curing processes) use fiber-optic physicochemical sensors but are not considered here.

Monitoring of ethanol during the fermentation process has been made by direct spectroscopic measurements in the NIR (700 to 1,100 nm) by means of a photodiode array spectrometer [78]. Raman laser fiber-optic spectroscopy [79] has been applied to ethanol (and glucose or fructose) determination during fermentation [80]. FTIR and evanescence have been combined [81] for acetone and ethanol determination with a LOD of 5 and 3 vol%, respectively.

In the food industry, reflectance techniques for determining color indexes are widely employed for many products [82]. Chemiluminescence flow injection analysis has been applied to glucose measurements in drinks [83]. There are also a large number of applications of IR spectrometry to on-line process control [84]. A multichannel fiber-optic FTIR, with a variable path length transmission cell in the NIR and an ATR probe in the mid-IR, has been tried for carbohydrate (e.g., fructose, sucrose) and alcohol determination in fruit syrup [85]. Furthermore, the noninvasive determination of sugar through the skin of fruit and humidity in flour and other solid products by NIR spectrometry is easily automated [86] and is widely applied [87,88].

12.3 GAS MONITORING

Dissolved gas measurement in biosensing (oxygen, carbon dioxide, ammonia) was discussed in Chapter 11, and the measurement of some solvent vapors in environmental monitoring was discussed in Section 12.1. Three main techniques are considered in this section: direct spectroscopy, sensitive films, and chemically reactive films; this complements earlier reviews [61].

12.3.1 Direct Spectrometry

The intrinsic absorption of many gases can be measured directly with optical fibers serving as simple light guides (see Figure 12.8).

In particular, alkanes, such as methane with a lower explosion limit (LEL) and upper explosion limit (UEL) of about 5% and 15% by volume, respectively, are important to monitor because of their safety risk (see Table 12.3). Their weak absorption coefficients, which require very long cavity lengths with multiple pass cells, lead to many difficulties of poor signal strength, construction of the cavity, and temperature sensitivity.

The first experiments (1979 to 1983) demonstrated the remote determination of CH_4 and other alkanes with silica optical-fiber links up to 20 km long [89]. The width of their absorption bands in the NIR are broader and weaker than at longer wavelengths (mid-IR). Hence, the detection limits (LOD) obtained are around 2,000 ppm (about 4% of the LEL) at 1.33 µm [90], 400 ppm at 1.66 µm [91], and

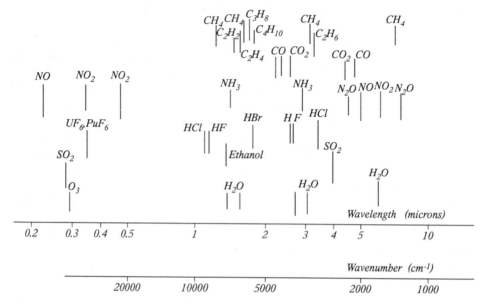

Figure 12.8 Main absorption bands of common gases that can be measured by optical-fiber sensors.

Table 12.3
LEL and UEL of Some Hydrocarbon Gases

Gas	LEL (% by volume)	UEL (% by volume)	λ (nm)	Loss (dB)
Acetylene	2.50	80.00	1,520	0.87
Ethane	3.00	12.5	1,685	0.44
Ethylene	2.75	28.60	1,619	0.71
Methane	5.00	15.00	1,666	1.10
Propane	2.12	9.35	1,696	0.41

Note: The optical losses correspond to a cavity of 1m at wavelength λ and a band pass $\Delta\lambda = 1$ nm.

300 ppm at 3.39 μm by means of As_2S_3 glass fibers [92] and fluoride glass fibers [93]. Propane has been measured at 1.68 μm [94] and butane at 1.68 to 1.8 μm [95]. LEDs and diode lasers are well adapted for frequency-modulated spectrometry [96] and a LOD of 0.3 ppm at 1.66 μm has been obtained for CH_4 [97]. This technique is suitable for simultaneous detection of methane (1.66 μm) and acetylene (1.53 μm) with two single-mode diode lasers modulated at different frequencies [98]. The LOD is 5 and 3 ppm, respectively, for a signal integration time of 1 sec and a 4-km-long optical fiber.

Applications have been reported for time-domain multiplexed sensors in coal mines [99].

Ammonia gas has a strong dipole moment usable in real-time correlation spectroscopy (see Section 8.11). Hence, it has a pronounced Stark splitting of its energy levels. In Stark modulation detection, an electrical field is applied to the gas cell, causing a synchronous modulation of the detected signal. The experimental configuration for ammonia sensing is shown in Figure 12.9 [100]. The measured LOD is about 0.16% by volume and a sensitivity of 130 ppm has been predicted. The LOD for methane is approximately 50 ppm (about 0.1% of LEL) with pressure modulation and about 300 ppm with phase modulation detection. The Stark modulation technique has been extended to water vapor sensing [101].

Single-frequency InGaAsP distributed-feedback laser diodes have an excellent tuning accuracy and spectral resolution for CH_4 and CO_2 measurement at 1.64 μm [102]. With this source, H_2S and CO_2 have also been assessed at 1.575 μm by wavelength modulation spectroscopy and harmonic detection [103]. The LOD is 10 and 100 ppm, respectively. With a different technique, remote detection of C_2H_4 at parts-per-million levels has been achieved with hollow fibers coupled to a CO_2 laser, with a wavelength at around 949 cm^{-1} that is tuned by a diffraction grating and appropriate cavity length [104].

A combination of FTIR and chalcogenide fibers permits measurement of CO_2 at 2,360 cm^{-1}, CO at 2,170 cm^{-1}, and CH_4 around 3,015 cm^{-1} with a LOD of 0.2, 0.05, and 0.3 vol%, respectively [105]. This method has also been applied at 2,932 cm^{-1} to a mixture of petroleum hydrocarbons in soils (LOD = 1.6 ppm) by coupling an extraction system (usually with CO_2) of organic compounds from solid samples to an FTIR [106]. Halide fibers can be used in the 1,350- to 700-cm^{-1} range for crude oils and distillates in petroleum plants [107]. In addition, CO_2, CO, and H_2O have been

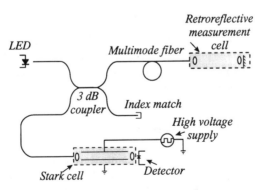

Figure 12.9 Schematic representation of correlation spectrocopy for gas measurements by electric field (Stark) modulation. Experimental configuration for an ammonia detector. (*After:* [100], with permission.)

assessed with KRS-5 fibers and a water-cooled sensor in a furnace above 1,000°C [108], and the technique extended to an ATR flow cell for gaseous mixtures of acetone, hexane, and trichloro- or trifluoroethane [109].

12.3.2 Change in Optical Properties of a Sensitive Film

In this category, optical fibers and waveguides act as a support for a sensitive film whose optical properties (refractive index, evanescence characteristics) change under the influence of the gas being measured.

A fiber cladding can be made of a sensitive polymer coating in which the refractive index changes reversibly with gas concentration [110]. Fiber transmission is monitored continuously in order to detect small changes. These intrinsic sensors are well adapted for distributed sensing and OTDR networks (see Section 12.4). They can be sensitized at specific places along the fiber [111] or along its whole length. A hydrophobic core-cladding interface using heteropolysiloxanes is produced after silanization of the fiber [112]. The selectivity and sensitivity depend on the type of cladding. For methane and heptane in air, a LOD of 7.2 and 0.02 vol%, respectively, has been achieved.

Hydrocarbon gases may be determined by the evanescent field absorption technique with special geometry fibers. A tapered fiber has been proposed for distributed sensing of methane [113]. From previous theoretical work on planar waveguides [114] it has been shown that a D-shaped optical fiber with a thin high-index overlay or porous overcoat can enhance the sensitivity by two orders of magnitude [115]. The main parameters are the core diameter, the index difference between core and cladding, and the core-to-flat distance of the D-fiber [116,117]. At 1.66 μm, the sensitivity of a 50- to 100-cm length of D-fiber is equivalent to a 1-cm spectroscopic absorption cell. A resolution of 1,000 ppm of methane should be achievable.

In another area, the change of refractive index of a polymer layer exposed to CO_2 and SO_2 [118] and to solvent vapors, such as pentane, hexane, and alcohol [119], has been investigated with integrated interferometric devices. The absolute interferometric phase can be correlated to spectral observations over a range of 70 nm corresponding to the width of the laser diode source [120]. However, a given polymer is not specific to a single gas, and detection of gas mixtures requires multicomponent analysis.

12.3.3 Chemically Reactive Film Sensors

Many gas sensors are based on a specific chemical reagent immobilized in a layer that also serves as a permeable membrane. Table 12.4 gives some typical examples of Cl_2, CO, HCl, HCN, NO_x, and NH_3 measurements.

In this manner, the relative humidity in air has been measured by the change of reflectance of entrapped dyes on a Nafion® film [137] and by the luminescent

Table 12.4
Examples of Gaseous Compounds Measured From Chemical Reactions

Gas	LOD	Chemical interaction	Immobilization support	Optical technique	Comments	References
Cl_2	45 ppb	O-toluidine	Nylon strip	Reflectance	Flow cell	[121]
Cl_2	ppm	Porphyrin/redox	Silicone rubber film	Fluorescence		[122]
Cl_2	ppm	Quinoline ester/redox	PTFE membrane	F Quenching		[123]
CO	0.05 vol%	Pd(II)	Impregnated paper	Reflectance	Flow cell	[124]
HCl	<10 ppm	Thymol blue	Polyvinyl alcohol	Fluorescence	Fluorescent fiber	[125]
HCN	1 ppb	Chloramine T and pyridine	XAD7 resin	Absorption	Nonreversible	[126]
NO	10 Pa	Cobalt-porphyrin	LB film	Absorbance	100–250°C range	[127]
NO_x	150 ppb	Aquacyanobiamide	Reagent in solution	Absorbance	Gas-permeable membrane	[128]
NO_x	Few vpm	Copper-phthallocyanine	LB film	SPR		[129]
NO_x	100 ppm	Sulfonated phthallocyanine	Spun film	Ellipsometry		[130]
NO_x	<1.0 ppm	Tetraphenyl porphine	LB film	F quenching		[131]
NO_2	<1.0 ppm	Phthallocyanine	Thin film	SPR		[132]
NH_3	2 ppb	Neutral carrier and H^+ indicator	Thin plasticized membrane	Absorbance		[133]
NH_3	<1.0 ppm	Bromocresol purple	Sol-gel on strip waveguide	Evanescence	Integrated system	[134]
NH_3	12 ppb	Lipophilic indicator	Silicone membrane	Fluorescence		[135]
NH_3	25 nM	Bromothymol blue	Indicator solution	Absorbance	Gas-permeable tubing	[136]

Note: Abbreviations are as follows: LOD = limit of detection; LB = Langmuir Blodgett; F = fluorescence; SPR = surface plasmon resonance; Pa = pressure in pascals.

lifetime after absorption on thin LB films containing porphyrins [138]. Hydrogen in the 20-ppb range can be measured by a palladium-coated single-mode fiber [139]. Adsorption of hydrogen produces dimensional changes in the palladium film that induce strains in the fiber, and the strain is measured by a Mach-Zehnder interferometer. With very thin (10 nm) palladium films, a range of 0.1% to 100% can be attained by measuring the rate of reflectivity change [140]. In another approach to multiplexed or quasidistributed sensing, an intrinsic oxygen sensor (5% to 100%) has been proposed using a ruthenium complex as fluorophore and EWS [141].

12.3.4 Other Techniques

Other techniques have been reported, such as:

1. The use of a glass porous support for detection of organic vapors [142], and of a Ta_2O_5 waveguide with an integrated optical coupler for humidity detection [143];
2. The condensation of organic vapors onto a silver surface and detection by SPR [144];
3. The change of the thermoluminescent spectrum during the process of heating Al_2O_3 powder caused by vapors, such as acetone and ethanol, adsorbed by the powder [145]. The LOD is less than 1 ppm for an adsorption time of 10 minutes. A relationship between the luminescent intensities during the heating and cooling processes can be used for mixed ethanol-butanol gas [146].

12.4 NETWORKS

12.4.1 General Approach

The combination of appropriate optical sensors and the signal transmitting capability of optical fibers in a local-area network (LAN) is the basis for fiber-optic sensor networks [147,148]. The choice of the best network topology (a linear array, ring, tree, star, and ladder architectures) depends on many factors, such as the number of nodes or sensing points and, more importantly, the way in which the sensors are modulated: spatial-division multiplexing (SDM), time-division multiplexing (TDM), frequency-division multiplexing (FDM), polarization-division multiplexing (PDM), or wavelength-division multiplexing (WDM), and various combinations of these methods. In a typical multiplexing scheme, the light flux is divided into several fiber links by a multiplexer (MUX), transmitted to the sensors, and returned to one or several detectors through a demultiplexer (DEMUX).

SDM and WDM have been described in Chapter 9. In a TDM sensor network, short pulses of light are launched into a transmissive (ladder) or reflective (linear array or star) network. The return pulses have traveled different distances in the fiber and they are therefore separated by given time intervals. In OTDR, the reflected and backscattered light is a function of the time and the distance traveled in the fiber. In distributed sensors [149], the optical fiber is sensitive along its whole length. In quasidistributed and multiplexed sensors, the fiber is sensitive at specific points. When the propagation delay is converted into a beat frequency, the technique is called OFDR (optical frequency-domain reflectometry). POTDR is OTDR with polarization modulation. In FDM, every sensor is assigned to a frequency channel and the sensor signal may be modulated in amplitude or frequency (or phase for coherent sensors).

In interferometric sensors, various passive or active techniques such as phase-generated carrier, path-matched differential interferometry, and coherence multiplexing can be used for the demodulation scheme [147].

12.4.2 Networks in Chemical Sensing

In chemical sensing, SDM has been applied to spectrophotometry with up to a maximum of 16 points in parallel for process control assessment of rare-earth and nuclear compounds. A prototype of an eight-channel optical waveguide sensor with an eight-LED array and one or several detectors has been tested for pH monitoring [150]. A multiplexed fiber-optic spectrofluorimeter has been designed to operate up to 18 biosensors [151]. Four other approaches show the potential for networking with fiber optics:

1. Geometric multiplexing of 100 diode laser sources can be coupled into a single multimode fiber [152].
2. The detection of multiplexed fluorescent signals of an electrophoresis capillary array illuminated by optical fibers has been demonstrated with a CCD camera for looking at multiple capillaries through a microscope objective. This scheme is expected to be capable of multiplexing 1,024 capillaries used in DNA sequencing [153].
3. Multiple optical fiber arrays can be used for measuring luminous intensities. Either multiple separate sources form an image on the surface of a CCD detector or multiple spectra from different fibers are focused onto the entrance slit of a grating [154].
4. Sensor arrays have been developed with polymer matrices sensitive to various analytes (pH, hydrocarbons, uranyl, and aluminium ions) in a hexagonal pattern. The fluorescence measurement by image detection has been applied to continuous groundwater monitoring [155]. Also, optical-fiber imaging can provide images and spatially resolve localized changes in the concentration of multianalytes such as O_2, CO_2, and pH [156]. This important result, while not a sensor network in itself, illustrates the potential of multiple sensing.

For the WDM technique, a miniaturized wavelength dispersion device consisting of a GRIN rod prism grating has been tested in absorbance and luminescence with 11 input/output fibers covering a range of 90 nm coming from a single source [157].

For the TDM technique, multipoint fluorescence measurements have been achieved with an $(N \times N)$ fiber-optic star coupler (with $N = 3$, 4, or 8) and fibers of 40m to 140m in length [158]. Differentiation of the signal from different sampling points is made by selecting the time of flight from each optical-fiber line.

OTDR techniques have been investigated within an important project of a Japanese government research program (1979 to 1984), sponsored by the Ministry of International Trade and Industry (MITI), and applied to petroleum refineries. Another example of a centralized chemical sensor network consists of a central unit transmitting and receiving information in the form of coded light signals and pulses [159]. The project at the Baltimore Veterans' Administration Hospital (see Figure 12.10) aims to make remote clinical measurements of in vivo tests and use sensor dipsticks on blood and urine samples taken from the patients [160].

In conclusion, the industrial applications of sensor networks are limited at present. However, an interest in multiplexing and distributed sensing is beginning to grow in chemistry [161]. Spectrometric monitoring in process control, chemical remote sensing in hospitals, LEL detection by distributed sensors for CH_4, arrays of

HOSPITAL FIBER OPTIC SENSOR/RADIOGRAPHY NETWORK

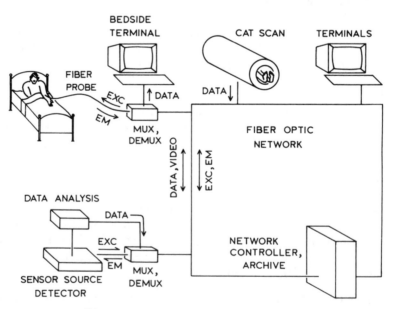

Figure 12.10 Integration of fiber-optic sensors into a hospital local-area network. (*Source:* [160]. Reprinted with permission.)

integrated sensors, and multianalyte sensors recorded by CCD cameras are all areas that will be developed in the future.

References

[1] Hirschfeld, T., T. Deaton, F. P. Milanovich, and S. M. Klainer, "The Feasibility of Using Fiber Optics for Monitoring Ground Water Contaminants," *Opt. Eng.*, Vol. 22, 1983, pp. 527–531.

[2] Klainer, S. M., K. Goswami, D. K. Dandge, S. J. Simon, N. R. Herron, D-L. Eastwood, and L. A. Eccles, "Environmental Monitoring Applications of Fiber Optic Chemical Sensors," Chap. 12 in *Fiber Optic Chemical Sensors and Biosensors, Vol. 2*, O. S. Wolfbeis, ed., Boca Raton: CRC Press, 1991, pp. 83–122.

[3] Trettnak, W., M. Hofer, and O. S. Wolfbeis, "Applications of Optochemical Sensors for Measuring Environmental and Biochemical Quantities," Chap. 18 in *Sensors: A Comprehensive Survey, Vol. 3*, W. Göpel, J. Hesse, and J. N. Zemel, eds., Weinheim: VCH, 1991, pp. 931–967.

[4] Viriot, M. L., and J. C. André, "Fluorescent Dyes: A Search for New Tracers for Hydrology," *Analusis*, Vol. 17, 1989, pp. 97–111.

[5] Zung, J. B., R. L. Woodlee, M-R.S. Fuh, and I. M. Warner, "Fiber Optic Based Multidimensional Fluorometer for Studies of Marine Pollutants," *Proc. SPIE–Int. Soc. Opt. Eng.*, Vol. 990, 1988, pp. 49–54.

[6] Niessner, R., W. Robers, and A. Krupp, "Fiber Optical Sensor System Using a Tunable Laser for Detection of PAHs on Particles and in Water," *Proc. SPIE–Int. Soc. Opt. Eng.*, Vol. 1172, 1989, pp. 145–156.

[7] Panne, U., F. Lewitzka, and R. Niessner, "Fibre Optical Sensors for Detection of Atmospheric and Hydrospheric Polycyclic Aromatic Hydrocarbons," *Analusis*, Vol. 20, 1992, pp. 533–542.

[8] Lieberman, S. H., S. M. Inman, G. A. Theriault, S. S. Cooper, P. G. Malone, Y. Shimizu, and P. W. Lurk, "Fiber Optic Based Chemical Sensors for in Situ Measurement of Metals and Aromatic Organic Compounds in Seawater and Soil Systems," *Proc. SPIE–Int. Soc. Opt. Eng.*, Vol. 1269, 1990, pp. 175–184.

[9] Inman, S. M., P. Thibado, G. A. Theriault, and S. H. Lieberman, "Development of a Pulsed Laser, Fiber Optic Based Fluorimeter: Determination of Fluorescence Decay Times of Polycyclic Aromatic Hydrocarbons in Sea Water," *Anal. chim. Acta*, Vol. 239, 1990, pp. 45–51.

[10] Panne, U., and R. Niessner, "A Fiber Optical Sensor for the Determination of Polycyclic Aromatic Hydrocarbons by Time Resolved, Laser Induced Fluorescence," *Vom Wasser*, Vol. 79, 1992, pp. 89–99.

[11] Stevenson, C. L., and T. Vo-Dinh, "Analysis of Polynuclear Aromatic Compounds Using Laser Excited Synchronous Fluorescence," *Anal. Chim. Acta*, Vol. 303, 1995, pp. 247–253.

[12] Angel, S. M., and M. N. Ridley, "Dual Wavelength Absorption Optrode for Trace Level Measurement of Thrichloroethylene and Chloroform," *Proc. SPIE–Int. Soc. Opt. Eng.*, Vol. 1172, 1989, pp. 115–122.

[13] Herron, N. R., S. J. Simon, and L. Eccles, "Remote Detection of Organochlorides With a Fiber Optic Based Sensor. III: Calibration and Field Evaluation of an Improved Chloroform Fiber Optic Chemical Sensor," *Anal. Instrum.*, Vol. 18, 1989, pp. 107–126.

[14] Milanovich, F. P., S. B. Brown, B. W. Colston, P. F. Daley, and K. C. Langry, "A Fiber Optic Sensor System for Monitoring Chlorinated Hydrocarbon Pollutants," *Talanta*, Vol. 41, 1994, pp. 2189–2194.

[15] Oxenford, J. L., S. M. Klainer, T. M. Salinas, L. Todechiney, J. A. Kennedy, D. K. Dandge, and K. Goswami, "Development of a Fiber Optic Chemical Sensor for the Monitoring of Trichloroethylene in Drinking Water," *Proc. SPIE–Int. Soc. Opt. Eng.*, Vol. 1172, 1989, pp. 108–114.

[16] Klunder, G. L., and R. E. Russo, "Core Based Intrinsic Fiber Optic Absorption Sensor for the Detection of Volatile Organic Compounds," *Appl. Spectrosc.*, Vol. 49, 1995, pp. 379–385.

[17] Bilodeau, T. G., K. J. Ewing, I. P. Kraucunas, J. Jaganathan, G. M. Nau, I. D. Aggarwal, F. Reich, and S. Mech, "Fiber Optic Raman Probe Detection of Chlorinated Hydrocarbons in Standard Soils," *Proc. SPIE–Int. Soc. Opt. Eng.*, Vol. 2068, 1994, pp. 258–270.

[18] Bürck, J., J-P. Conzen, B. Beckhaus, and H. J. Ache, "Fiber Optic Evanescent Wave Sensor for in Situ Determination of Non-polar Organic Compounds in Water," *Sensors and Actuators*, Vol. B 18/19, 1994, pp. 291–295.

[19] Göbel, R., R. Krska, S. Neal, and R. Kellner, "Performance Studies of an IR Fiber Optic Sensor for Chlorinated Hydrocarbons in Water," *Fresenius J. Anal. Chem.*, Vol. 350, 1994, pp. 514–519.

[20] Walsh, J. E., B. D. MacCraith, M. Meaney, J. G. Vos, F. Regan, A. Lancia, and S. Artjushenko, "Mid-infrared Fibre Sensor for the In-situ Detection of Chlorinated Hydrocarbons," *Proc. SPIE–Int. Soc. Opt. Eng.*, Vol. 2508, 1995, in press.

[21] Gauglitz, G., and J. Ingenhoff, "Integrated Optical Sensors for Halogenated and Non-halogenated Hydrocarbons," *Sensors and Actuators*, Vol. B 11, 1993, pp. 207–212.

[22] Klainer, S. M., J. R. Thomas, and J. C. Francis, "Fiber Optic Chemical Sensors Offer a Realistic Solution to Environmental Monitoring Needs," *Sensors and Actuators*, Vol. B 11, 1993, pp. 81–86.

[23] Barnard, S. M., and D. R. Walt, "Fiber Optic Organic Vapor Sensor," *Environ. Sci. Technol.*, Vol. 25, 1991, pp. 1301–1304.

[24] Angel, S. M., M. N. Ridley, K. Langry, T. J. Kulp, and M. L. Myrick, "New Developments and Applications of Fiber Optic Sensors," Chap. 23 in *Chemical Sensors and Microinstrumentation*, R. W. Murray, E. Dessy, W. R. Heineman, J. Janata, and W. R. Seitz, eds., ACS Symp. Series, Vol. 403, 1989, pp. 345–363.

[25] Dickert, F. L., S. K. Schreiner, G. R. Mages, and H. Kimmel, "Fiber Optic Dipping Sensor for Organic Solvents in Wastewater," *Anal. Chem.*, Vol. 61, 1989, pp. 2306–2309.

[26] Zhang, Y., and W. R. Seitz, "Single Fiber Absorption Measurements for Remote Detection of 2,4,6-Trinitrotoluene," *Anal. Chim. Acta*, Vol. 221, 1989, pp. 1–9.

[27] Reagen, W. K., A. L. Schulz, J. C. Ingram, G. D. Lancaster, and A. E. Grey, "Device and Method for Detection of Nitroaromatic Compounds in Water," U.S. Pat. 5,306,642, 1994.

[28] Shriver-Lake, L. C., K. A. Breslin, P. T. Charles, D. W. Conrad, J. P. Golden, and F. S. Ligler, "Detection of TNT in Water Using an Evanescent Wave Fiber Optic Biosensor," *Anal. Chem.*, Vol. 67, 1995, pp. 2431–2435.

[29] Seitz, W. R., "Optical Ion Sensing," Chap. 9 in *Fiber Optic Chemical Sensors and Biosensors, Vol. 2*, O. S. Wolfbeis, ed., Boca Raton: CRC Press, 1991, pp. 1–17.

[30] Hisamoto, H., E. Nagakawa, K. Nagatsuka, Y. Abe, S. Sato, D. Siowanta, and K. Suzuki, "Silver Ion Selective Optodes Based on Novel Thio Ether Compounds," *Anal. Chem.*, Vol. 67, 1995, pp. 1315–1321.

[31] Ahmad, M., and R. Narayanaswamy, "Fibre Optic Reflectance Sensor for the Determination of Aluminium (III) in Aqueous Environment," *Anal. Chim. Acta*, Vol. 291, 1994, pp. 255–260.

[32] Czolk, R., J. Reichert, and H. J. Ache, "An Optical Sensor for the Detection of Heavy Metal Ions," *Sensors and Actuators*, Vol. B 7, 1992, pp. 540–543.

[33] Oehme, I., B. Prokes, I. Murkovic, T. Werner, I. Klimant, and O. S. Wolfbeis, "LED Compatible Copper(II) Selective Optrode Membrane Based on Lipophilized Zincon," *Fresenius J. Anal. Chem.*, Vol. 350, 1994, pp. 563–567.

[34] Noiré, M. H., and B. Duréault, "A Ferrous Ion Optical Sensor Based on Fluorescence Quenching," *Sensors and Actuators*, Vol. B 29, 1995, pp. 386–391.

[35] Pulido, P., J. M. Barrero, M. C. Perez-Conde, and C. Camara, "Evaluation of Three Supports for an Optical Fiber Ferric Ion Sensor," *Quim. Anal. (Barcelona)*, Vol. 12, 1993, pp. 48–52.

[36] Jianzhong, L., and Z. Zhujun, "A Fiber Optic Iron Sensor With DEAE Sepharex as a Subtrate," *Anal. Lett.*, Vol. 27, 1994, pp. 2431–2442.

[37] Zhujun, Z., Z. Yunke, and H. Quan, "Determination of Trace Mercury Using Competitive Binding Reaction of Copper and Mercury on Calcein by Fiber Optics Fluorometry," *Xiyou Jinshu*, Vol. 6, 1987, pp. 212–216; *Chem. Abst.*, Vol. 108, 137571e.

[38] Klinkhammer, G. P., "Fiber Optic Spectrometers for In-situ Measurements in the Oceans: The ZAPS Probe," *Mar. Chem.*, Vol. 47, 1994, pp. 13–20.

[39] De Oliveira, W. A., and R. Narayanaswamy, "A Flow Cell Optosensor for Lead Based on Immobilized Dithizone," *Talanta*, Vol. 39, 1992, pp. 1499–1503.

[40] Lerchi, M., E. Reitter, and W. Simon, "Uranyl Ion Selective Optode Based on Neutral Ionophore," *Fresenius J. Anal. Chem.*, Vol. 348, 1994, pp. 272–276.

[41] Reichert, J., W. Sellien, and H. J. Ache, "Development of a Fiber Optic Sensor for the Detection of Ammonium in Environment Waters," *Sensors and Actuators*, Vol. A 25/27, 1991, pp. 481–482.

[42] Theriault, G. A., S. H. Lieberman, and D. S. Knowles, "Laser Induced Breakdown Spectroscopy for Rapid Delineation of Metals in Soils," *Proc. 4th. Symp. Field Screening Methods for Hazardous Waste and Toxic Chemicals*, Las Vegas, 22–24 Feb. 1995.

[43] Russell, D. A., and R. Narayanaswamy, "An Optical Fibre Sensor for Fluoride," *Anal. Chim. Acta*, Vol. 220, 1989, pp. 75–81.

[44] Tan, S. S. S., P. C. Hauser, N. A. Chaniotakis, G. Suter, and W. Simon, "Anion Selective Optical Sensors Based on a Co-extraction of Anion-Proton Pairs Into a Solvent Polymeric Membrane," *Chimia*, Vol. 43, 1989, pp. 257–261.

[45] Mohr, G. J., and O. S. Wolfbeis, "Solid State Nitrate Sensor Based on Potential Sensitive Fluorescent Dyes," *Anal. Chim. Acta*, 1996, in press.

[46] Viriot M. L., B. Mahieuxe, M. C. Carré, and J. C. André, "Fluorimetric Determination of Nitrate and Nitrite," *Analusis*, Vol. 23, 1995, pp. 312–329.

[47] Freeman, M. K., and L. G. Bachas," Fiber Optic Probes for Cyanide Using Metalloporphyrins and a Corrin," *Anal. Chim. Acta*, Vol. 241, 1990, pp. 119–125.

[48] Li, X. M., F. C. Ruan, W. Y. Ng, and K. Y. Wong, "Scanning Optical Sensor for the Measurement of Dissolved Oxygen and BOD," *Sensors and Actuators*, Vol. B 21, 1994, pp. 143–149.

[49] Preininger, C., I. Klimant, and O. S. Wolfbeis, "Optical Fiber Sensor for Biological Oxygen Demand," *Anal. Chem.*, Vol. 66, 1994, pp. 1841–1846.

[50] Monici, M., R. Boniforti, G. Buzzigoli, D. De Rossi, and A. Nannini, "Fibre Optic pH Sensor for Seawater Monitoring," *Proc. SPIE–Int. Soc. Opt. Eng.*, Vol. 798, 1987, pp. 294–300.

[51] Serra, G., A. Schirone, and R. Boniforti, "Fibre Optic pH Sensor for Seawater Monitoring Using a Single Dye," *Anal. Chim. Acta*, Vol. 232, 1990, pp. 337–344.

[52] Goswami, K., J. A. Kennedy, D. K. Dandge, and S. M. Klainer, "A Fiber Optic Chemical Sensor for Carbon Dioxide Dissolved in Sea Water," *Proc. SPIE–Int. Soc. Opt. Eng.*, Vol. 990, 1989, pp. 225–232.

[53] Goyet, C., D. R. Walt, and P. G. Brewer, "Development of a Fiber Optic Sensor for Measurement of pCO_2 in Sea Water: Design, Criteria and Sea Trials," *Deep Sens. Res. A. Oceanogr. Res. Pap.*, Vol. 39, No. 6A, 1992, pp. 1015–1026.

[54] MacAllister, M., and D. R. Walt, "Fiber Optic Sensor for Oceanic Carbon Dioxide," *Proc. SPIE–Int. Soc. Opt. Eng.*, Vol. 2068, 1994, pp. 213–215.

[55] Gruber, W. R., I. Klimant, and O. S. Wolfbeis, "Instrumentation for Optical Measurement Of Dissolved Oxygen Based on Solid State Technology," *Proc. SPIE–Int. Soc. Opt. Eng.*, Vol. 1885, 1993, pp. 448–457.

[56] Wing, M. R., E. J. Stromvall, and S. H. Lieberman, "Real Time Determination of Dissolved Free Amino Acids and Primary Amines in Sea Water by Time Resolved Fluorescence," *Mar. Chem.*, Vol. 29, 1990, pp. 325–338.

[57] Trettnak, W., F. Reininger, E. Zinterl, and O. S. Wolfbeis, "Fiber Optic Remote Detection of Pesticides and Related Inhibitors of the Enzyme Acetylcholine Esterase," *Sensors and Actuators*, Vol. B 11, 1993, pp. 87–93.

[58] Ayyagari, M. S. R., H. Gao, B. Bihari, K. G. Chittibabu, J. Kumar, K. A. Marx, D. L. Kaplan, and S. K. Tripathy, "Molecular Self Assembly on Optical Fiber Based Fluorescence Sensor," *Proc. SPIE–Int. Soc. Opt. Eng.*, Vol. 2068, 1994, pp. 168–178.

[59] Barnard, S. M., and D. R. Walt, "Antibody Based Fiber Optic Sensors for Environmental and Process Control Applications," *Proc. SPIE–Int. Soc. Opt. Eng.*, Vol. 1368, 1990, pp. 86–97.

[60] Oroszlan, P., G. L. Duveneck, M. Ehrat, and H. M. Widmer, "Fiber Optic Atrazine Immunosensor," *Sensors and Actuators*, Vol. B 11, 1993, pp. 301–305.

[61] Wolfbeis, O. S., "Gas Sensors," Chap. 11, in *Fiber Optic Chemical Sensors and Biosensors, Vol. 2*, O. S. Wolfbeis, ed., Boca Raton: CRC Press, 1991, pp. 55–82.

[62] Wolfbeis, O. S., and G. Boisdé, "Applications of Optochemical Sensors for Measuring Chemical Quantities," Chap. 17 in *Sensors: A Comprehensive Survey, Vol. 3*, W. Göpel, J. Hesse, and J. N. Zemel, eds., Weinheim: VCH, 1991, pp. 867–930.

[63] Colvin, D. W., "A Colorimeter for the In-line Analysis of Uranium and Plutonium," Report DP-461, Savannah River Lab., 1960.

[64] Perez, J. J., G. Boisdé, M. Goujon de Beauvivier, G. Chevalier, and M. Isaac, Automatisation de la Spectrophotometrie du Plutonium, *Analusis*, Vol. 8, 1980, pp. 344–351.

[65] Boisdé, G., F. Blanc, P. Mauchien, and J. J. Perez, "Fiber Optic Chemical Sensors in Nuclear Plants," Chap. 14 in *Fiber Optic Chemical Sensors and Biosensors, Vol. 2*, O. S. Wolfbeis. ed., Boca Raton: CRC press, 1991, pp. 135–149.

[66] McKay, C. N. N., C. L. Mills, and T. W. Kyffin, "On-line Fibre Optic Spectrophotometry in a Plutonium Pulsed Column Pilot Plant," Report DNE-R-23, Dounreay, 1990.

[67] Corriou, J. P., and G. Boisdé, "Comparison of Numerical and Physicochemical Models for Spectrophotometric Monitoring of Uranium Concentration," *Anal. Chim. Acta*, Vol. 190, 1986, pp. 255–264.

[68] Boisdé, G., F. Blanc, and J. J. Perez, "Chemical Measurements With Optical Fibers for Process Control," *Talanta*, Vol. 35, 1988, pp. 75–82 and references herein.

[69] Carey, W. P., L. E. Wangen, and J. T. Dyke, "Spectrophotometric Method for the Analysis of Plutonium and Nitric Acid Using Partial Least Squares Regression," *Anal. Chem.*, Vol. 61, 1989, pp. 1667–1669.

[70] Couston, L., D. Pouyat, C. Moulin, and P. Decambox, "Speciation of Uranyl Species in Nitric Acid Medium by Time Resolved Laser Induced Fluorescence," *Appl. Spectrosc.*, Vol 49, 1995, pp. 349–353.

[71] Baldwin, D. L., and R. W. Stromatt, "Plutonium, Uranium, Nitrate Measurements in Purex Process Stream by Remote Fiber Optic Diode Array Spectrophotometer," Report PNL-SA-15318, Richland, 1988.

[72] O'Rourke, P. E., "Chemometrics/On-line Measurements," *J.N.M.M.*, Vol. 18, 1989, pp. 85–94.

[73] Bürck, J., K. Krämer, and W. König, "Lichtleiteradaptation des Interferenzfilterphotometers Spectran für den Einsatz als In-line Monitor in der Purex Prozesskontrolle," Report KFK 4672, Karlsruhe, 1990.

[74] Moulin, C., S. Rougeault, D. Hamon, and P. Mauchien, "Uranium Determination by Remote Time Resolved Laser Induced Fluorescence," *Appl. Spectrosc.*, Vol. 47, 1993, pp. 2007–2012.

[75] Gantner, E., and D. Steinert, "Applications of Laser Raman Spectrometry in Process Control Using Optical Fibers," *Fresenius J. Anal. Chem.*, Vol. 338, 1990, pp. 2–8.

[76] Rojas, D., R. J. Silva, and R. E. Russo, "Thermal Lens Spectroscopy Using a Diode Laser and Optical Fibers," *Rev. Sci. Instrum.*, Vol. 63, 1992, pp. 2989–2993.

[77] Wruck, D. A., R. E. Russo, and R. J. Silva, "Thermal Lens Spectroscopy of Plutonium Using a Laser Diode and Fiber Optics," *J. Alloys and Compounds*, Vol. 213/214, 1994, pp. 481–483.

[78] Cavinato, A. G., D. M. Mayes, Z. Ge, and J. B. Callis, "Noninvasive Method for Monitoring Ethanol in Fermentation Processes Using Fiber Optic Near-Infrared Spectroscopy," *Anal. Chem.*, Vol. 62, 1990, pp. 1977–1982.

[79] Dao, N. Q., and M. Jouan, "The Raman Laser Fiber Optics (RLFO) Method and Its Applications," *Sensors and Actuators*, Vol. B 11, 1993, pp. 147–160.

[80] Gomy, C., M. Jouan, and N. Q. Dao, "Méthode et Analyse Quantitaive par Spectrometrie Raman Laser Associée aux Fibres Optiques pour le Suivi d'une Fermentation Alcoolique," *Anal. Chim. Acta*, Vol. 215, 1988, pp. 211–226.

[81] Heo, J., M. Rodrigues, S. J. Saggesse, and G. H. Sigel, "Remote Fiber Optic Chemical Sensing Using Evanescent Wave Interactions in Chalcogenide Glass Fibers," *Appl. Opt.*, Vol. 30, 1991, pp. 3944–3951.

[82] Marszalec, E., H. Kopola, and R. Myllylä, "Non Destructive Testing of the Quality of Naturally White Food Products," *Sensors and Actuators*, Vol. B 11, 1993, pp. 503–509.

[83] Blum, L. J., "Chemiluminescence Flow Injection Analysis of Glucose in Drinks With a Bienzyme Fiberoptic Biosensor," *Enzyme Microb. Technol.*, Vol. 15, 1993, pp. 407–411.

[84] Scotter, C., "Use of Near Infrared Spectroscopy in the Food Industry With Particulary Reference to Its Applications to In/On-line Food Process," *Food Control*, July 1990, pp. 142–149.

[85] Kemsley, E. K., R. H. Wilson, and P. S. Belton, "Potential of Fourier Transform Infrared Spectroscopy and Fiber Optics for Process Control," *J. Agric. Food Chem.*, Vol. 40, 1992, pp. 435–438.

[86] Bellon, V., and G. Boisdé, "Remote Near Infrared Spectrometry in the Food Industry With the Use of Silica and Fluoride Glasses Fibers," *Proc. SPIE–Int. Soc. Opt. Eng.*, Vol. 1055, 1989, pp. 350–358.

[87] Kawano, S., H. Watanabe, and M. Iwamoto, "Determination of Sugar Content in Intact Peaches by Near Infrared Spectroscopy With Fiber Optics in Interactance Mode," *Engei Gakkai Zasshi*, Vol. 61, 1992, pp. 445–451; *Chem. Abst.*, Vol. 117, 232379t.

[88] Bellon, V., J. L. Vigneau, and M. Leclercq, "Feasibility and Performances of a New Multiplexed, Fast and Low-Cost Fiber Optic Spectrometer for On-line Measurement of Sugar in Fruits," *Appl. Spectrosc.*, Vol. 47, 1993, pp. 1079–1083.

[89] Chan, K., H. Ito, and H. Inaba, "All Optical Fiber Based Remote Sensing System for Near Infrared Absorption of Low Level Methane Gas," *J. Lightwave Technol.*, Vol. LT-5, 1987, pp. 1706–1711.

[90] Chan, K., H. Ito, and H. Inaba, "Absorption Measurement of $v_2 + 2v_3$ Band of CH_4 at 1.33 μm Using an InGasAsP Light Emitting Diode," *Appl. Opt.*, Vol. 22, 1983, pp. 3802–3804.

[91] Chan, K., H. Ito, and H. Inaba, "Remote Sensing System for Near Infrared Differential Absorption of CH_4 Gas Using Low Loss Optical Fiber Link," *Appl. Opt.*, Vol. 23, 1984, pp. 3415–3420.

[92] Saito, M., M. Takizawa, K. Ikegawa, and H. Takami, "Optical Remote Sensing System for Hydrocarbon Gases Using Infrared Fibers," *J. Appl. Phys.*, Vol. 63, 1988, pp. 269–272.

[93] Pruss, D., "Application Of IR-Glass Fibers for Remote Spectroscopy," *Mater. Sci. Forum*, Vol. 32/33, 1988, pp. 321–330.

[94] Chan, K., H. Ito, and H. Inaba, "All Optical Remote Monitoring of Propane Gas Using a 5-km-Long Low Loss Optical Fiber Link and an InGaP Light Emitting Diode in the 1.66 μm Region," *Appl. Phys. Lett.*, Vol. 45, 1984, pp. 220–222.

[95] Fukunaga, H., S. Yabe, Y. Arakawa, and F. Inaba, "Method and Apparatus for IR Absorption Determination of Butane Gas," *Jpn. Kokai Tokkyo Koho*, JP 63/11840, 1988; *Chem. Abst.*, Vol. 109, 103832w.

[96] Mohebati, A., and T. King, "Fibre Optic Remote Gas Sensor With Diode Laser FM Spectroscopy," *Proc. SPIE–Int. Soc. Opt. Eng.*, Vol. 1172, 1989, pp. 186–193.

[97] Uehara, K., and H. Tai, "Remote Detection of Methane With a 1.66 μm Diode Laser," *Appl. Opt.*, Vol. 31, 1992, pp. 809–814.

[98] Tai, H., K. Yamamoto, M. Uchida, S. Osawa, and K. Uehara, "Long Distance Simultaneous Detection of Methane and Acetylene by Using Diode Lasers Coupled With Optical Fibers," *IEEE Phot. Technol. Lett.*, Vol. 4, 1992, pp. 804–807.

[99] Zientkiewicz, J. K., "Self Referenced Fiber Optic Methane Detection System," *Proc. SPIE–Int. Soc. Opt. Eng.*, Vol. 992, 1989, pp. 182–187.

[100] Edwards, H. O., and J. P. Dakin, "Gas Sensors Using Correlation Spectroscopy Compatible With Fibre Optic Operation," *Sensors and Actuators*, Vol. B 11, 1993, pp. 9–19.

[101] Dakin J. P., H. O. Edwards, and B. H. Weigl, "Progress With Optical Gas Sensors Using Correlation Spectroscopy," *Sensors and Actuators*, Vol. B 29, 1995, pp. 87–93.

[102] Weldon, V., P. Phelan, and J. Hegarty, "Methane and Carbon Dioxide Sensing Using a DFB Laser Diode Operating at 1.64 µm," *Electron. Lett.*, Vol. 29, 1993, pp. 560–561.

[103] Weldon, V., J. O'Gordman, P. Phelan, J. Hegarty, and T. Tanbun-Ek, "H_2S and CO_2 Gas Sensing Using DFB Laser Diode Emitting at 1.57 µm," *Sensors and Actuators*, Vol. B 29, 1995, pp. 101–107.

[104] Worell, C. A., I. P. Giles, and N. A. Adatia, "Remote Gas Sensing With Mid Infra Red Hollow Waveguide," *Electron. Lett.*, Vol. 28, 1992, pp. 615–617.

[105] Saggesse, S. J., M. R. Shahriari, and G. H. Sigel, "Evaluation of an FTIR/Fluoride Optical Fiber System for Remote Sensing of Combustion Products," *Proc. SPIE–Int. Soc. Opt. Eng.*, Vol. 1172, 1989, pp. 2–12.

[106] Heglund, D. L., D. C. Tilotta, S. B. Hawthorne, and D. J. Miller, "Simple Fiber Optic Interface for On-line Supercritical Fluid Extraction Fourier Transform Infrared Spectrometry," *Anal. Chem.*, Vol. 66, 1994, pp. 3543–3551.

[107] Ge, Z., C. W. Brown, and J. J. Alberts, "Infrared Fiber Optic Sensor for Petroleum," *Environ. Sci. Technol.*, Vol. 29, 1995, pp. 878–882.

[108] Maeda, M., N. Takahashi, and Y. Kuwano, "Optical Sensing Technique for in Situ Determination of Gas Components at Elevated Temperature by Infrared Spectroscopy," *Sensors and Actuators*, Vol. B 1, 1990, pp. 215–217.

[109] Taga, K., B. Mizaikoff, and R. Kellner, "Fiber Optic Evanescent Field Sensors for Gaseous Species Using MIR Transparent Fibers," *Fresenius J. Anal. Chem.*, Vol. 348, 1994, pp. 556–559.

[110] Archenault, M., H. Gagnaire, J. P. Goure, and N. Jaffrezic-Renault, "A Simple Intrinsic Optical Fibre Chemical Sensor," *Sensors and Actuators*, Vol. B 8, 1992, pp. 161–166.

[111] Williams, D. E., "Optical Fiber Sensor and Its Use," U.K. Pat. GB 2,210,685, 1989.

[112] Ronot, C., H. Gagnaire, J. P. Goure, N. Jaffrezic-Renault, and T. Pichery, "Optimization and Performance of a Specifically Coated Intrinsic Optical Fibre Sensor for the Detection of Alkane Compounds," *Sensors and Actuators*, Vol. A 41/42, 1994, pp. 529–534.

[113] Tai, H., H. Tanaka, and T. Yoshino, "Fiber Optic Evanescent Wave Methane Gas Sensor Using Optical Absorption for the 3.392 µm Line of a He-Ne Laser," *Opt. Lett.*, Vol. 12, 1987, pp. 437–439.

[114] Stewart, G., J. O. W. Norris, D. F. Clark, and B. Culshaw, "Evanescent Wave Chemical Sensors: A Theoretical Evaluation," *Int. J. Optoelectr.*, Vol. 6, 1991, pp. 227–238.

[115] Stewart, G., F. A. Muhammad, and B. Culshaw, "Sensitivity Improvement for Evanescent Wave Gas Sensors," *Sensors and Actuators*, Vol. B 11, 1993, pp. 521–524.

[116] Muhammad, F. A., and G. Stewart, "D-Shaped Optical Fibre Design for Methane Gas Sensing," *Electron. Lett.*, Vol. 28, 1992, pp. 1205–1206.

[117] Culshaw, B., F. Muhammad, G. Stewart, S. Murray, D. Pinchbeck, J. Norris, S. Cassidy, M. Wilkinson, D. Williams, I. Crisp, R. Van Ewyk, and A. MacGhee, "Evanescent Wave Methane Detection Using Optical Fibres," *Electron. Lett.*, Vol. 28, 1992, pp. 2232–2234.

[118] Brandenburg, A., R. Edelhäuser, and F. Hutter, "Integrated Optical Gas Sensors Using Organically Modified Silicates as Sensitive Films," *Sensors and Actuators*, Vol. B 11, 1993, pp. 361–374.

[119] Fabricius, N., G. Gauglitz, and J. Ingenhoff, "A Gas Sensor Based on an Integrated Optical Mach-Zehnder Interferometer," *Sensors and Actuators*, Vol. B 7, 1992, pp. 672–676.

[120] Gauglitz, G., and J. Ingenhoff, "Integrated Optical Sensors for Halogenated and Non Halogenated Hydrocarbons," *Sensors and Actuators*, Vol. B 11, 1993, pp. 207–212.

[121] Momin, S. A., and R. Narayanaswamy, "Optosensing of Chlorine Gas Using a Dry Reagent Strip and Diffuse Reflectance Spectrophotometry," *Anal. Chim. Acta*, Vol. 244, 1991, pp. 71–79.

[122] Baron, M. G., R. Narayanaswamy, and S. C. Thorpe, "A Kineto-Optical Method for the Determination of Chlorine Gas," *Sensors and Actuators*, Vol. B 29, 1995, pp. 358–362.

[123] Kar, S., and M. A. Arnold, "Fiber Optic Chlorine Probe Based on Fluorescence Decay of N-(6-Methoxyquinolyl)-Acetoethyl Ester," *Talanta*, Vol. 42, 1995, pp. 663–670.

[124] De Oliveira, W. A., and P. R. Saliba, "Optosensing of Carbon Monoxide Through Paper Impregnated With Palladium (II) Chloride," *J. Braz. Chem. Soc.*, Vol. 5, 1994, pp. 39–42.

[125] Muto, S., A. Ando, Tatsuo, T. Ochiai, H. Ito, H. Sawada, and A. Tanaka, "Simple Gas Sensor Using Dye-Doped Plastic Fibers," *Jpn. J. Appl. Phys. Part 1*, Vol. 28, 1989, pp. 125–127.

[126] Bentley, A. E., and J. F. Alder," Optical Fibre Sensor for Detection of Hydrogen Cyanide in Air. Part 1: Reagent Characterization and Impregnated Bead Detector Performance," *Anal. Chim. Acta*, Vol. 222, 1989, pp. 63–73.

[127] Eguchi, K., T. Hashiguchi, K. Sumiyoshi, and H. Arai, "Optical Detection of Nitrogen Monoxide by Metal Porphine Dispersed in an Amorphous Silica Matrix," *Sensors and Actuators*, Vol. B 1, 1990, pp. 154–157.

[128] Freeman, M. K., and L. G. Bachas, "Fiber Optic Sensor for NOx," *Anal. Chim. Acta*, Vol. 256, 1992, pp. 269–275.

[129] Zhu, D. G., M. C. Petty, and M. Harris, "An Optical Sensor for Nitrogen Dioxide Based on a Copper Phthalocyanine Langmuir-Blodgett Film," *Sensors and Actuators*, Vol. B 2, 1990, pp. 265–269.

[130] Martenson, J., J. Arwin, and I. Lundström, "Thin Films of Phthalocyanines Studied With Spectroscopic Ellipsometry: An Optical Gas Sensor?" *Sensors and Actuators*, Vol. B 1, 1990, pp. 134–137.

[131] Baron, M. G., R. Narayanaswamy, and S. C. Thorpe, "Luminescent Porphyrin Thin Films for NO_x Sensing," *Sensors and Actuators*, Vol. B 11, 1993, pp. 195–199.

[132] Wright, J. D., A. Cado, S. J. Peacock, V. Rivalle, and A. M. Smith, "Effects of Nitrogen Dioxide on Surface Plasmon Resonance of Substituted Phthalocyanine Films," *Sensors and Actuators*, Vol. B 29, 1995, pp. 108–114.

[133] West, S. J., S. Ozawa, K. Seiler, S. S. S. Tan, and W. Simon, "Selective Ionophore Based Optical Sensors for Ammonia Measurement in Air," *Anal. Chem.*, Vol. 64, 1992, pp. 533–540.

[134] Klein, R., and Voges, "Integrated-Optic Ammonia Sensor," *Sensors and Actuators*, Vol. B 11, 1993, pp. 221–225.

[135] Werner, T., I. Klimant, and O. S. Wolfbeis, "Optical Sensor for Ammonia Based on the Inner Filter Effect of Fluorescence," *J. Fluorescence*, Vol. 4, 1994, pp. 41–44.

[136] Kar, S., and M. A. Arnold, "Cylindrical Sensor Geometry for Absorbance Based Fiber Optic Ammonia Sensors," *Talanta*, Vol. 41, 1994, pp. 1051–1058.

[137] Sadaoka, Y., M. Matsuguchi, and Y. Sakai, "Optical Fibre and Quartz Oscillator Type Gas Sensors: Humidity Detection by Nafion® Film With Crystal Violet and Related Compounds," *Sensors and Actuators*, Vol. A 25/27, 1991, pp. 489–492.

[138] Papkovsky, D. B., G. V. Ponomarev, S. F. Chernov, A. N. Ovchinnikov, and I. N. Kurochkin, "Luminescence Lifetime Based Sensor for Relative Air Humidity," *Sensors and Actuators*, Vol. B 22, 1994, pp. 57–61.

[139] Butler, M. A., and D. S. Ginley," Hydrogen Sensing With Palladium Coated Optical Fibers," *J. Appl. Phys.*, Vol. 64, 1988, pp. 3706–3712.

[140] Butler, M. A., and R. B. Buss, "Kinetics of the Micromirror Chemical Sensor," *Sensors and Actuators*, Vol. B 11, 1993, pp. 161–166.

[141] MacCraith, B. D., G. O'Keeffe, A. K. MacEvoy, C. MacDonagh, J. F. McGilp, J. D. O'Mahony, and M. Cavanagh, "Light Emitting Diode Based Oxygen Sensing Using Evanescent Wave Excitation of a Dye Doped Sol-Gel Coating," *Opt. Eng.*, Vol. 33, 1994, pp. 3861–3866.

[142] Novak, T. J., and R. A. MacKay, "Vycor Porous Glass (Thirsty Glass) as a Reaction Medium for Optical Waveguide Based Chemical Vapor Detectors," *Spectrosc. Lett.*, Vol. 21, 1988, pp. 127–145.

[143] Kunz, R. E., "Gradient Effective Index Waveguide Sensors," *Sensors and Actuators*, Vol. B 11, 1993, pp. 167–176.

[144] Vukusic, P. S., G. P. Bryan-Brown, and J. R. Sambles, "Surface Plasmon Resonance on Gratings as a Novel Means for Gas Sensing," *Sensors and Actuators*, Vol. B 8, 1992, pp. 155–160.

[145] Utsunomiya, K., M. Nakagawa, T. Tomiyama, I. Yamamoto, Y. Matsuura, S. Chikamori, T. Wada, N. Yamashita, and Y. Yamashita, "Discrimination and Determination of Gases Utilizing Adsorption Luminescence," *Sensors and Actuators*, B 11, 1993, pp. 441–445.

[146] Nakagawa, M., "A New Chemiluminescent Sensor for Discriminating and Determining Constituents in Mixed Gas," *Sensors and Actuators*, B 29, 1995, pp. 94–100.

[147] Kist, R., "Point Sensor Multiplexing Principles," Chap. 14 in *Optical Fiber Sensors: Systems and Applications, Vol. 2*, B. Culshaw and J. P. Dakin, eds., Boston: Artech House, 1989, pp. 511–574.

[148] Ferdinand, P., *Capteurs à Fibres Optiques et Reseaux Associés*, Paris: Lavoisier, Tec. Doc., 1992.

[149] Dakin, J. P., "Distributed Optical Fiber Sensor Systems," Chap. 15 in *Optical Fiber Sensors: Systems and Applications, Vol. 2*, B. Culshaw and J. P. Dakin, eds., Boston: Artech House, 1989, pp. 575–598.

[150] Smardzewski, R. D., "Multi-element Optical Waveguide Sensor General Concept and Design," *Talanta*, Vol. 35, 1988, pp. 95–101.

[151] Lipson, D., N. G. Loebel, K. D. McLeaster, and B. Liu, "Multifiber, Multiwavelength, Fiber Optic Fluorescence Spectrometer," *IEEE Trans. Biomed. Eng.*, Vol. 39, 1992, pp. 886–892.

[152] Fan, T. Y., "Efficient Coupling of Multiple Diode Laser Arrays to an Optical Fiber by Geometric Multiplexing," *Appl. Opt.*, Vol. 30, 1991, pp. 630–632.

[153] Taylor, J. A., and E. A. Yeung, "Multiplexed Fluorescence Detector for Capillary Electrophoresis Using Axial Optical Fiber Illumination," *Anal. Chem.*, Vol. 65, 1993, pp. 956–960.

[154] Piccard, R. D., and T. Vo-Dinh, "A Multi-Optical-Fiber Array With Charge-Coupled Device Image Detection for Parallel Processing of Light Signals and Spectra," *Rev. Sci. Instrum.*, Vol. 62, 1991, pp. 584–594.

[155] Healey, B. G., S. Chadha, D. R. Walt, F. P. Milanovich, J. Richards, and S. Brown, "Development of Sensor Arrays for Continuous Ground Water Monitoring," *Proc. SPIE–Int. Soc. Opt. Eng.*, Vol. 2360, 1994, pp. 101–102.

[156] Pantano, P., and D. R. Walt, "Analytical Applications of Optical Imaging Fibers," *Anal. Chem.*, Vol. 67, 1995, pp. 481A–487A.

[157] Fuh, M-R. S., and L. W. Burgess, "Wavelength Division Multiplexer for Fiber Optic Sensor Readout," *Anal. Chem.*, Vol. 59, 1987, pp. 1780–1783.

[158] Steffen, R. L., and F. E. Lytle, "Multipoint Measurements in Optically Dense Media by Using Two-Photon Excited Fluorescence and a Fiber Optic Star Coupler," *Anal. Chim. Acta*, Vol. 215, 1988, pp. 203–210.

[159] Martin, H., "A Light-Wave Guide Serving as a Gas and/or Liquid Sensor and an Arrangement for Supervising or Affecting One Peripherical Unit Among Several Such Units From a Central Unit," *PCT Int. Appl.*, WO 94 00,750, 1994.

[160] Thompson, R. B., "Fiber Optic Sensors Advance Chemical Analysis From Afar," *Circuits and Devices*, May 1994, pp. 14–21.

[161] Lieberman, R. A., "Distributed and Multiplexed Chemical Fiber Optic Sensors," *Proc. SPIE–Int. Soc. Opt. Eng.*, Vol. 1586, 1991, pp. 80–91.

Glossary

4-MU	4-methyl-umbelliferone
4PAC	4-pyridyl-amino-coumarin
4TMU	4-trifluoromethyl-umbelifferone
5-MSA	5-maleimidyl salicylic acid
7-AM-4-	7-amino-4-methyl-
Ab	antibody
ABMB	4-aminobenzoic Meldola blue
Ag	antigen
AMCA	7-amino-4-methyl coumarin-3-acetic acid
APC	allophycocyanin
APD	avalanche photodiode
APMB	4-aminophenol Meldola blue
APTES	aminopropyltriethoxysilane
ASBD-F	(aminosulfonyl)-7-fluoro-2,1,3-benzooxadiazole
ATP	adenosine triphosphate
ATR	attenuated total internal reflectance
BBPA	bis(1-butylpentyl)adipate
BCB	bromocresol blue
BCG	bromocresol green
BCP	bromocresol purple
BD	benzodiazole
BMM	4-bromomethyl-7-methoxy-
BNAL	n-Butyl-o-(1-naphthylaminocarbonyl) lactate
BOD	biological oxygen demand
BP	Benzo[a]pyrene
BPB	bromophenol blue
BPR	bromophenol red
BSA	bovine serum albumin
BTB	bromothymol blue
CARS	coherent anti-Stokes Raman spectrometry

CB	cascade blue
CCD	charge-coupled device
CDF	carboxy dimethyl fluorescein
CDI	carbonyldiimidazole
CHEMFET	chemically sensitive field effect transistor
mCP	meta-cresol purple
CPR	chlorophenol red
CR	cresol red
CTSP	o-Cresol tetrachloro-sulfonephthalein
Da	dalton
DAZ	dansyl aziridine
DBDA	6,6′,8 hydroxy-(1,1′-diphenyl)-3,3′ diacetic acid
DBVT	di-ter.-butyl-o-o'-bis-(2,2-dichlorovinyl)-tartate
DCC	dicyclohexyl carbodiimide
DEMUX	demultiplexer
DHPDS	1,3-dihydroxypyrene-6,8-disulfonic acid
DMDCS	dimethyldichlorosilane
DNA	Deoxyribonucleic acid
DNP	dinitrophenole
DOP	bis(2-ethylhexyl) phthalate
DOS	bis(2-ethylhexyl) sebacate
DPDCS	diphenyldichlorosilane
DTCS	n-decyltrichlorosilane
DZ 49	lipophilic pH indicator
EAC	electroactive compound
EDTA	ethylenediaminetetracetic
EIA	enzyme immunoassay
EITC	eosine-5-isothiocyanate
ELISA	enzyme-linked immunosorbent assay
ETH	Swiss Federal Institute of Technology, Zurich
ETH 1001	-(R,R′)-N,N′-[bis(11-ethoxycabonyl)undecyl]-N,N′, 4,5-tetramethyl-3,6-dioxaoctane diamide
ETH 157	N,N'-dibenzyl-N-N'-diphenyl-1,2-(phenylenedioxy) diacetamide
ETH 2412	3-hydroxy-4-(4-nitrophenylazo)phenyl octadecanoate, 5-octadecanoyloxy-2-(4-nitrophenylazo)phenol
ETH 2439	9-(diethylamino)5-(4-(16-butyl-2,14-dioxo-3,15-dioxaeicosyl)phenylamino)-5H-benzo[a]phenoxazine
ETH 4120	4-octadecanoyloxymethyl-N,N,N',N'-tetracyclohexyl-1,2,phenylene dioxidi-acetamide
ETH 5294	9-(diethylamino)5-octadecanoylimino-5H-benzo[a]phenoxazine

ETH 7075	4′, 5′-dibromofluorescein octadecyl ester
EWS	evanescent wave spectroscopy
FAD	flavine adenine dinucleotide
FCFD	fluorescent capillary fill device
FD	frequency domain
FDM	frequency-division multiplexing
FEP	fluoroethylenepropylene
FIA	flow injection analysis
FITC	fluorescein-isothiocyanate
FMN	flavine mononucleotide
FOTDR	frequency OTDR
FRET	fluorescence resonance energy transfer
FT	Fourier transform
FTC	flow-through cell
FTIR	Fourier transform infrared
FTRS	Fourier transform Raman spectroscopy
GDA	glutaric dialdehyde
GDH	glucose deshydrogenase
GDPH	phosphored-GDH
GH	Goos-Hänchen
GOPTS	3-glycidoxypropyltrimethoxysilane
GOx	glucose oxidase
GREFIN	gradient of effective refractive index
GRIN	graded refractive index
HC	7-hydroxycoumarin
HCC	7-hydroxycoumarin-3-carboxylic acid
HPA	hydroxyphenyl acetic acid
HPTS	8-hydroxypyrene-1,3,6-trisulfonic trisodium salt
HRP	horseradish peroxidase
HSA	human serum albumin
IA	iodoacetamide
IC	isocyanate
ICB	intrinsic coating-based
IFE	inner filter effect
IgG	immunoglobulin G
IO	integrated optic
IOS	integrated optical sensors
IR	infrared
IRS	internal reflection spectroscopy
ISFET	ion sensitive field effect transistor
ISO	ion-sensitive optode
ITC	isothiocyanate

(K) or (NA) $T_m(CF_3)_2$PB	(Potassium) or (sodium) tetrakis (3,5-bis(trifluoromethyl)-phenyl) borate
KSR-5	thallium bromoiodide
KT_pClPB	potassium tetrakis (4-chloro-phenyl) borate
LAD	lipophilic anionic dye
laser	light amplification by stimulation emission of radiation
LB	Langmuir-Blodgett
LED	light-emitting diode
LEL	lower explosion limit
LOD	limit of detection
LP	light pattern
LS	light source
MAPS	γ-methacryloxypropyltrimethoxysilane
MEDPIN	7-decyl-2-methyl-4-(3',5'-dichlorophen-4'-one) indonaphth-1-ol
MR	methyl red
MUX	multiplexer
N-BD-Cl	4-chloro-7-nitrodibenzo-2-oxa-1,3-diazole
NA	numerical aperture
NAD	nicotinamide adenine dinucleotide
NADH	nicotinamide adenine dinucleotide hydrogenated (reduced)
NADP	phosphored-NAD
NADPH	phosphored-NADH
Nd-YAG	neodymium-yttrium-aluminum garnet
NEP	noise equivalent power
NHS	N-hydroxysuccinimide
NIR	near IR
NPOE	2-nitrophenyloctylether
NVP	N-vinylpyrrolidone
NY	nitrazine yellow
OFS	optical-fiber sensor
oNPOE	ortho-NPOE
OPA	o-phthalaldhehyde
OS	oxygen saturation
OTCS	octadecyltrichlorosilane
OTDR	optical time domain reflectometry
OTES	octadecyltriethoxysilane
PAH	polycyclic aromatic hydrocarbon
PAN	1(-2-pyridylazo)-2-naphthol
PANDA	polarization-maintaining and absorption-reducing
PAR	4(-2-pyridylazo)-resorcinol

PBP	periplasmic binding protein
PC	polycarbonate
PCB	palatine chrome black
PCS	plastic-clad silica
PD	photodiode
PDM	polarization-division multiplexing
PE	polyethylene
PES	polyethylsulfone
PHEMA	poly(hydroxyethyl methacrylate)
PI	polyimide
PIB	Polyisobutylene
PMMA	polymethylmethacrylate
PMT	photomultiplier tubes
POTDR	polarization OTDR
PR	phenol red
PS	polystyrene
PSD	potential-sensitive dye
PTFCE	Poly(tetrafluorochloroethylene)
PTFE	polytetrafluoroethylene
PVA	poly(vinyl)alcohol
PVC	polyvinyl chloride
PVCFC	polyvinylchloroformate chloride
PVDF	polyvinylidene fluoride
PVI	polyvinylimidazole
PVP	poly(vinylpyrrolidone)
PZT	Pb(lead)zirconate titanate
RBITC	rhodamine-B-isothiocyanate
redox	reducing-oxidizing
RNA	ribonucleic acid
SDM	spatial-division multiplexing
SERRS	surface-enhanced resonance Raman scattering
SERS	surface enhanced Raman spectroscopy
SLD	superluminescent diode
SMF	single-mode fiber
SNAFL	seminaphthofluorescein
SNARF	seminaphthorhodafluor
SPR	surface plasmon resonance
SPU	signal processing unit
SVRS	solochrome violet RS
SX	siloxane
TAN	1-(2'-thiazolylazo)-2-naphtol
TAR	4-(2'-thiazolylazo)-2-rearcinol

TAVS	triacetoxyvinylsilane
TB	thymol blue
TBCS	tributylchlorosilane
TBG	thyroxine binding globulin
TBPB	tetrabromophenol blue
TBPE	tetrabromophenolphthalein ethyl ester
TBPSP	3,4,5,6-tetrabromophenolsulfonephthalein
TBTB	tetrabromothymol blue
TCE	trichloroethylene
TCPSP	tetrachlorophenol-sulfonephthalein
TCTC	tricyclohexyltin chloride
TD	time domain
TDM	time-division multiplexing
TDMACl	tridodecylmethylammonium chloride
TE	transverse electric
TEM_{00}	transverse electromagnetic mode
TEOS	tetraorthoethylsilicate
Texas red	sulforhodamine 101 sulfonyl chloride
TIR	total internal reflection
TIRF	total internal reflection fluorescence
TLC	thin-layer chromatography
TLS	thermal lens spectroscopy
TM	transverse magnetic
TMSDMA	n-trimethylsilyldimethylamine
TN	turnover number
TNT	tri-nitrotoluene
TOT	trioctyl tin chloride
TOTC	trioctyltin chloride
TPTZ	tripyridyltriazine
TRITC	tetramethyl-rhodamine isothiocyanate
TTF	tetrathiafulvalene
UEL	upper explosion limit
UV	ultraviolet
WDM	wavelength-division multiplexing
XOD	xanthine oxidase
ZBLA	zirconium, barium, lanthanum, aluminum fluoride
ZBLAN	ZBLA with sodium fluoride

About the Authors

Gilbert E. Boisdé was born in the Loire Atlantique, France, on 19 April 1931 and received an engineering degree in chemistry from the Ecole Supérieure de Chimie de l'Université de Rennes (ENSCUR) in 1954. In 1957, he joined the Commissariat Energy Atomique (CEA, the French Atomic Energy Commission). He worked initially in electrochemical research on molten salts (1957 to 1968), corrosion (1968 to 1972), and automated analytical instrumentation (1972 to 1974). Since 1973, his current interests have grown in the field of fibers optics, especially in instrumentation, remote spectrometry, the behavior of fibers in nuclear environments, and physical and chemical optical sensors. With the Association Nationale de la Recherche Technologique (ANRT, National Association for Technological Research) he was a stimulating leader for a French national pilot project on optical-fiber sensors. He is author of more than 80 technical papers, 6 contributions to books, and 20 patents in the field of fiber optics. He retired in 1992, and since then has worked on this book with the support of the CEA.

Alan Harmer has 20 years of R&D experience in industrial optics and electronics, and he was the first person to make a comprehensive worldwide survey of optical fiber sensors and their applications in 1981. He is currently at the International Electrotechnical Commission in Geneva, Switzerland. Prior to that, he ran a technical consultancy group for the design and prototyping of fiber-optic sensors. He was section manager at Battelle Geneva Research Center, Switzerland, from 1974 to 1989, where he was responsible for work on optoelectronics and fiber optics, with emphasis on their use in industrial systems. His activities cover optoelectronic instrumentation and fiber-optic sensors, hybrid optoelectronics, optical-fiber fabrication, micro-optics, and industrial optical systems. He graduated from Cambridge University and holds higher degrees from Oxford University and MIT in physics and electrical engineering. He has published more than 70 technical papers, holds 20 patents, and has created novel optoelectronic inventions leading to 5 commercial developments.

Index

Absorbance measurements, 208–10
  fiber-optic absorption spectrometers, 208–9
  IR spectroscopy, 209–10
  limitations of, 208
Absorption, 161–62
  attenuation and, 162
  coefficient, 161
Active components, 255–57
  active fiber, 256–57
  effects, 255–56
  Faraday magneto-optic effect, 256
  Kerr effect, 255–56
  Pockels effect, 255
  See also Passive components
Active optodes, 28
Adsorption, 122–23
Affinity, 52
Air pollution, 347
Ammonia sensors, 300–301
  fluorescence, 301
  with renewable reagent, 34
Amplitude modulation, 200–201
Analyte
  concentration, 29
  concentration relationship, 26
  intermediate, 316–18
  intrinsic properties of, 39
  signal characteristics as function of, 207
Anions, 343
Anion sensing, 295
Antibodies
  antigen interaction, 52–53
  bioreceptor, 57
  forms of, 41

hapten interaction, 53
  polyclonal, 53
  shape recognition of, 42
  structure of, 50–52
Antigen-antibody interaction, 52–53
Applications, 285–361
  areas of, 287
  measured parameters, 287
  See also specific applications
Aromatic compounds, 342–43
Associations
  competitive, 27
  in multiple steps, 23
Attenuated total reflectance (ATR)
      measurements, 209
Attenuation, 161–62
  defining, 161–62
  of optical intensity, 162
Auxochromes, 65
Avalanche photodiodes (APD), 250
Avidity, 52–53
Azo compounds, 76–78
  examples, 77
  illustrated, 76
  pH measurement, 77

Bandpass filters, 252
Bathochromic effect, 134
Beer-Lambert equation, 213
Binding, 121
  competitive, 28–30
  constant, 22
  covalent, 125, 128
  electrostatic, 125

Biocatalytic reactions, 45–49
    coenzymes, 49
    enzymes, 45–48
    inhibitors, 48
    measurement methods, 49
Biocatalytic sensors, 49–50
    classes, 49
    enzyme loading and, 50
Biochemical sensors, 289–325
Biological oxygen demand (BOD), 343–46
    determination, 345
    sensors, 345–46
Biomass, 313–15
Bioreceptors, 57–60
    antibodies, 57
    DNA, 59
    DNA-binding proteins, 59
    enzymes, 57
    lectins, 57–58
    neuroreceptors, 59
Biosensing
    chemical sensing vs., 15
    defined, 13
    limitations of, 8–9
    for medical applications, 8
    scheme, 14
Biosensors, 289–325
    in biotechnology, 313
    creatinine, 301–2
    defined, 5
    enzyme-based, 297
    ethanol, 316
    glucose, 295–99
    regenerable, 320–21
    reversible, 298
    urea, 301–2
    See also Biosensing; Sensors
Biotechnology, 312–18
    ATP and, 313–15
    biomass and, 313–15
    enzyme activity, 316
    enzyme optodes with intermediate
        analytes, 316–18
    fiber-optic biosensors in, 313
    NADH and, 313–15
    pH and, 315–16
    process groups, 312
    process sensors, 312
Birefringent fibers, 186–88
    creation methods, 186–87
    See also Optical fibers
Blood
    chemical/biochemical substances in, 290
    composition, 289

electrolytes, 290–95
Blood gas, 304–7
    extracorporeal sensors, 307–8
    monitor, 307
Button probes, 269

Carbon-nitrogen compounds, 76–80
    azo compounds, 76–78
    ferroins, 78
    hydrazones, 78
    triazines, 78–80
    See also Indicators
Carbon-nitrogen-oxygen compounds, 80–84
    chemiluminescent compounds, 84
    diphenylcarbazone, 83
    EDTA, 83
    nitro, 82
    nitroso, 82
    oxazines, 81–82
    oxime, 82
    phenoxazines, 81–82
    rhodamines, 80–81
    symmetric dipheylcarbazide, 83
    See also Indicators
Carbon-oxygen compounds, 69–76
    hydroxy aromatic compounds, 73–76
    phthaleins, 69–71
    sulfophthaleins, 71–72
    See also Indicators
Catalysts, 43–45
    activity measurement, 44
    defined, 43
    See also Reactions
Cation exchange reactions, 95–96
Cation-selective optodes, 31
Chemically reactive film sensors, 356–58
Chemical measurement techniques, 204–8
    light modulation and, 205
    list of, 205
    signal characteristics for, 206, 207
Chemical reactions, 17–39
    by equilibrium equation, 19
    competitive binding, 28–30
    concentration changes in, 21
    direct, 22–28
    exchange, 36–39
    forward, 19
    gases compounds measured from, 357
    indirect, 28–31
    rate changes in, 21
    rate equations of, 19–21
    reverse, 20
    reversible, 17–19
    simultaneous, 31

successive, 30–31
thermodynamic description, 17–19
Chemical sensing
    active components, 255–57
    biosensing vs., 15
    networks in, 359–61
    optical techniques for, 7–9
    passive components, 252–55
    scheme, 14
Chemical sensors
    multimode fibers in, 191
    planar waveguide configurations in, 180–81
    reversibility of, 28
    See also Chemical sensing; Sensors
Chemical-to-optical transduction schemes, 6
Cholesterol, 302–3
    esterase, 303
    oxidase, 303
Chromophores, 65–66
    defined, 65
    interaction, 66
Cladding loss model, 268
Clinical diagnostics, 289–303
Cofactors, 49
Coherence, 156–58
    fiber-optic sensors and, 158
    function, 257
    length, 157
    See also Light
Coimmobilization, 136–38
Competitive associations, 27
Competitive binding, 28–30
    defined, 28
    effects of, 137
    equilibrium, 30
    See also Binding
Complementary determining residues, 52
Configurations, 118–21
    bindings for, 121
    illustrated, 119–20
Controlled-reagent sensors, 32–36
    polymeric delivery systems, 33–36
    with renewable reagent, 33
    with reservoir, 32–33
Core-based optodes, 270–71
    hybrid, 272
    See also Intrinsic optodes
Correlation spectroscopy, 228–29
Covalent binding, 125
    on organic polymers, 128
    See also Binding
Covalent-covalent attachment, 128–31
    illustrated, 129
    lipophilic membrane example, 128–29

planar-waveguide example, 129
    steps for, 128
Covalent immobilization, 125–31
    covalent binding on organic polymers, 128
    covalent-covalent attachment, 128–31
    on fatty acid monolayers, 130
    silanization, 126–27
    surface preactivation, 127–28
    uses, 125
    See also Immobilization techniques
Creatinine sensors, 301–2
Crown ethers. See Macrocyclic compounds

Dichroic mirrors, 252
Differential path length factor, 311
Diffraction, 155–56
    Fraunhofer, 155–56
    gratings, 156, 252
    through narrow slit, 155
    See also Light
Diphenylcarbazide, 83
Diphenylcarbazone, 83
Direct reactions, 22–28
    association in multiple steps, 23
    competitive association, 27
    homogeneous/heterogeneous phases, 27–28
    with immobilized reagent, 23–27
    single-step, 22
    See also Chemical reactions
Direct sensors, 267–68
    EWS, 267–68
    examples of, 36
Direct spectrometry, 353–56
Dispersion, 160–61
    effect of, 160
    material, 160–61
    minimum, 160
    in optical fiber, 161
    See also Light
DNA, 59–60
    defined, 59
    double-stranded, 323
    gene probe, 323
    hybridization, 325
    illustrated, 60
    single-stranded, 323–25
DNA-binding proteins, 59
Doped fibers, 189–91
Dye immobilization, 124
Dye lasers, 245
Dyes
    absorbance-pH function, 137
    absorption spectrum, 134
    coimmobilization of, 136–38

Dyes (continued)
  maximum absorption, 136
  potential-sensitive (PSDs), 91–94
  tracer, 339–40
  See also Indicators

Electrolytes, 290–95
  calcium, 294, 295
  potassium, 292, 294
  sodium, 292
Electromagnetic waves, 144–49
  basic equations, 144–46
  intensity of, 146
  polarization, 146–49
  velocity, 145
  See also Light
Electronic transitions, 65–66
Electrostatic binding, 125
Ellipsometry, 230
Emission lines, 159, 160
Energy transfer, 168–70
  model, 168
  nonresonant, 170
  radiative, 168
  resonant, 168–69
  spatial, 169
  spin coupling, 169, 170
Enthalpy, 17–18
  defined, 17
  interaction ranges, 19
Entropy, 17, 18
Environmental monitoring, 339–47
  air pollution, 347
  biological oxygen demand, 343–46
  metals ions and anions, 343
  organic pollutants, 340–43
  pesticides and herbicides, 346–47
  seawater measurements, 346
  tracer dyes in water, 339–40
Enzyme immunoassay (EIA), 53
Enzyme-linked immunosorbent assays (ELISA)
      method, 55
Enzymes
  activity of, 316
  amplification, 53–54, 56
  binding site, 47
  bioreceptor, 57
  concentration, 46
  defined, 45
  immobilization of, 50, 129
  immunoreactions with, 53–54
  luciferase, 47–48
  optodes, 316–18
Equilibrium constant, 20

Equilibrium equation, 19–20
Ethylenediaminetetracetic (EDTA), 83
Evanescent waves, 152
Evanescent wave sensors, 221–22
  instrumentation for, 221
  for NIR domain, 222
  with optical fiber, 222
Evanescent wave spectroscopic (EWS)
  direct optodes, 267–68
  sensors, 266
Evanescent wave spectroscopy, 188–89, 219–22
  defined, 219
  instrumentation, 221
  principle of, 220
Exchange reactions, 36–39
  cation, 95–96
  exchange of charged particle, 38
  exchange of electrons, 36
  exchange of protons, 37–38
  for neutral chemical compounds, 97
  See also Chemical reactions
Extrinsic optodes, 259, 262–66
  active, 263
  collection efficiency, 263
  fiber, 263–65
  fluorescent light in, 264
  illustrated, 261
  with immobilized compounds, 265–66
  passive, 263
  sensing cells, 262–63
  using plain fibers, 265
  See also Intrinsic optodes; Optodes

Fabry-Perot interferometer, 226, 227
Fiber-face-to-face configuration, 264
Fiber-optic absorption spectrometers, 208–9
Fiber-optic immunosensors, 107
Fibers. *See* Optical fibers
Flash lamps, 242–43
Flow injection analysis (FIA), 262
Fluorescein, 70
Fluorescence, 164–67
  ammonia sensors, 301
  atomic, 164
  characteristics of, 214–17
  in fiber-optic sensors, 167
  indicators, 221
  measurement of, 214–17
  molecular, 165
  polarization methods, 218
  quenching effect on, 216
  remote time-resolved, 264
  steady-state, 166
  Stokes, 165

time-decay curve, 167
transitions, 165
See also Light
Fluorescence resonance energy transfer
        (FRET), 216
    optical spectra showing, 169
    sensor examples, 218
Fluorescent capillary fill device (FCFD), 272–74
    commercial, 273
    defined, 272
    illustrated, 273
    inner walls, 273–74
Fluorescent indicators, 221
Fluorescent labels, 103–5
Fluorescent sensors, 217
Food industry, 353
Fourier transform infrared (FTIR)
    glass materials for, 192
    spectroscopy, 209–10
Fourier transform Raman spectroscopy
        (FTRS), 245
Fraunhofer diffraction, 155–56
Frequency-division multiplexing
        (FDM), 358, 359
Frequency domain (FD) fluorometry, 203
Frequency OTDR (FOTDR), 230
Fresnel equations, 151–53

Gas discharge tubes, 241–42
Gas lasers, 243–45
    argon-krypton, 244–45
    drawbacks to, 245
    He-Ne, 243–44
    See also Lasers
Gas monitoring, 353–58
    ammonia, 355
    chemically reactive film sensors, 356–58
    direct spectrometry, 353–56
    hydrocarbon, 356
    sensitive film properties, 356
Gastroenterology, 309–10
Geometrical optics, 149–51
    internal reflection, 150–51
    light flux, 149–50
    reflection, 150
    refraction, 150
Gibbs free energy, 18
Glass porous support, 358
Glucose biosensors, 30, 295–99
    affinity, 296
    applications for, 299
    measurement reactions, 298
    schemes for, 296
    See also Biosensors

Goos-Hänchen shift, 152, 153
GOx sensors, 298–99
Graded refractive index (GRIN) lenses, 253
Graded-refractive-index fibers, 183, 184
Grating couplers, 180, 181, 182
Grating light reflection spectroscopy, 229
Guiding monolayer structures, 274–76
    defined, 274
    illustrated, 275
Guiding multilayer structures, 276–77
    defined, 276
    illustrated, 276

Halogenated compounds, 341–42
    general formula, 341
    monitoring sensor, 342
Hapten-antibody interaction, 53
Herbicides, 346–47
Historical perspective, 1–4
Human serum albumin (HSA), 321
Hydroxy aromatic compounds, 73–76
    hydroxybenzenes, 73
    hydroxycoumarins, 75
    hydroxynaphthalenes, 73–74
    See also Indicators

Immobilization
    absorption spectrum and, 134
    coimmobilization, 136–38
    covalent, 125–31
    of enzymes, 50, 129
    of indicators, 125
    of polymers, 125
    procedure evaluation, 132
Immobilization techniques, 121–31
    adsorption, 122–23
    covalent immobilization, 125–31
    electrostatic (ionic) binding, 125
    lipophilic membranes, 124
    physical entrapment, 122
    physicochemical methods, 122–24
    sol-gel method, 123–24
    surface modification and, 122
Immobilized compounds, 132–38
    characterization, 132–33
    coimmobilization of several dyes and, 136–38
    in pH sensors, 133–36
Immobilized reagent
    interaction, 23–24
    measurement in solution with, 27
    multiple, 31
    phase ratios, 27
    support layer with, 32
    systems with, 23–27

Immunoassays, 54–56
    competitive, 54–55
    at continuous surface, 58
    direct, 54
    disadvantages of, 56
    heterogeneous, 56
    noncompetitive, 55
Immunoglobulins, 52
Immunological interactions, 50–53
    antibody structure and, 50–52
    antigen-antibody, 52–53
    hapten-antibody, 53
Immunosensors, 318–25
    characteristics of, 319
    determinations from, 324
    examples, 323–25
    fluoroimmunosensor, 322
    progress in, 322–23
    regenerable, 319–21
    reversible, 319–21
    uses, 318–19
Incomplete antigen, 53
Indicators, 66–91
    in acid-base reactions, 66–67
    azo compounds, 76–78
    carbon-nitrogen compounds, 76–80
    carbon-nitrogen-oxygen compounds, 80–84
    carbon-oxygen compounds, 69–76
    characteristics, 68
    as chemical transducer, 66
    common, 69–90
    defined, 66
    diphenylcarbazide, 82–83
    diphenylcarbazone, 83
    EDTA, 83
    ferroins, 78
    fluorescent, 221
    fundamentals of, 66–68
    hydrazones, 78
    hydroxy aromatic compounds, 73–76
    immobilization of, 125
    LEDs and, 101
    macrocyclic compounds, 86–90
    NADH, 304
    nitro, 82
    nitroso, 82
    optical operating requirements, 101–3
    oxazines, 81–82
    oxime, 82
    pH, 133
    phenoxazines, 81–82
    phosphorescent, 218–19
    phthaleins, 69–71
    polycyclic aromatic hydrocarbons, 69

redox, 90–91
    rhodamines, 80–81
    sulfophthaleins, 71–72
    sulfur compounds, 84–86
    TBPSP, 67
    triazines, 78–80
    two consecutive, 137
    useful, 68
Indirect reactions, 28–31
    competitive binding, 28–30
    successive reactions, 30–31
    See also Chemical reactions
Indirect sensors
    examples of, 36
    rate of consumption, 32
Infrared fibers, 192–95
Inhibitors, 48
Inner filter effect (IFE), 216, 218
Integrated sensors, 277–78
    defined, 4–5, 277
    examples, 277–78
    illustrated, 278
    See also Sensors
Intensity modulation, 200–201
Interaction, light with matter, 158–71
Interference, 153–54
    electric fields and, 153
    pattern from light beam, 154
Interferometers, 226–27
    Fabry-Perot, 226, 227
    illustrated, 227
    Mach-Zehnder, 226, 227, 228
    Michelson, 226, 228
    types of, 226
Interferometric sensors, 189, 227–28
    examples of, 227–28
    integrated-optic, 228
    See also Interferometers
Intermediate analyte, 5
Internal reflection, 150–51
    ATR measurements, 209
    phase changes during, 153
    See also Reflection
Intravascular sensors, 308–9
Intrinsic coating-based (ICB) optodes, 269–70
    hybrid, 272
    See also Intrinsic optodes
Intrinsic optodes, 259, 266–72
    coating-based, 269–70
    core-based, 270–71
    direct EWS, 267–68
    hollow fiber, 271
    hybrid ICB and core-based, 272
    illustrated, 261

porous fiber, 271
refractometric, 266–67
two-state fluorescence coupling system, 271
See also Extrinsic optodes; Optodes
Invasive techniques, 308–10
gastroenterology, 309–10
intravascular sensors, 308–9
respiratory analysis, 310
In vivo monitoring, 303–12
blood gas and pH measurements, 304–7
extracorporeal blood gas sensors, 307–8
invasive techniques, 308–10
NADH as indicator, 304
noninvasive techniques, 311–12
oximetry, 303–4
transcutaneous techniques, 310–11
Ion carriers, 94–101
Ionic and molecular recognition process, 13–14
Ionophores
chemical formulas, 100
chromo chemical formulas, 99
Ion sensitive optodes (ISOs), 98

Kerr effect, 171, 189, 255–56
Kinetics, 41, 43–50
biocatalytic reactions, 45–49
catalysis, 43–45
concepts, 43–45
desorption, 133
first-order reactions, 43
higher order reactions, 43
measurements, 57
optical biocatalytic sensors, 49–50
photokinetics, 45
of surface reactions, 56–57
Kretschmann configuration, 183, 223

Labeling protocols, 106–7
Labels, 103–6
classes of, 106
fluorescent, 103–5
phosphorescent, 106
transient, 106
Langmuir-Blodgett monomolecular films, 171
Lasers, 243–46
dye, 245
gas, 243–45
solid-state, 245–46
Lasing fibers, 189–91
Lectins, 57–58
LEDs, 304
cross section, 244
indicator choice and, 101
as light source, 243

structure types, 243
Light, 143–58
as combined photons and waves, 151–58
cone of, 150
coupling, 270
electromagnetic waves, 144–49
flux, 149–50
geometrical optics, 149–51
intensity, 152
interaction limitations, 8
interaction with matter, 158–71
modulation of, 8
particle nature, 143–44
pattern and intensity distribution, 177
propagation model, 173–77
from return fibers, 199
See also Optical modulation
Light sources, 241–46
dye lasers, 245
flash lamps, 242–43
gas discharge tubes, 241–42
gas lasers, 243–45
lamp sources, 241–43
light-emitting diodes, 243
solid-state lasers, 245–46
white light sources, 241–43
See also Light
Linear birefringence, 147
Lipophilic membranes, 124
Liposomes, 56
Local-area networks (LANs), 358
Lower explosion limit (LEL), 353
of hydrocarbon gases, 354
See also Upper explosion limit (UEL)
Low-loss fibers, 1
Luminescence, 164–68, 214–19
characteristics of fluorescence, 214–17
fluorescence, 164–67
measurements, 217–18
phosphorescence, 167
sensors, 218–19

Mach-Zehnder interferometer, 226, 227
defined, 226
illustrated, 227
IO, 228
molded, 228
See also Interferometers
Macrocyclic compounds, 86–90
calix(n)arenes, 87–88
configurations, 86
cryptands, 87
hemispherands, 88–89
porphyrins, 90

Macrocyclic compounds (continued)
    spherands, 87
    See also Indicators
Mass balance expressions, 29
Material dispersion, 160–61
Maxwell's equations, 175
    for plane wave, 146–47
    to TE and TM modes, 177–79
Measurements
    absorbance, 208–10
    ATR, 209
    biocatalytic reaction, 49
    blood gas, 304–7
    chemical techniques, 204–8
    glucose biosensor reactions, 298
    luminescence, 217–18
    NADH, 314
    nuclear process, 349
    pH, 77
    photoacoustic, 228–29
    Raman scattering, 228–29
    reflectance, 211–13
    in seawater, 346
    substrate, 49
MEDPIN, 78, 293–94
Membranes
    composite flat, 118
    hydrophobic/hydrophilic characteristics, 117–18
    lipophilic, 124
    permeability of, 117
    theoretical aspects of, 117–18
Metal ions, 343
    determination by optodes, 344
    See also Environmental monitoring
Michaelis-Menten constant, 47, 48
Michaelis-Menten expression, 46
Michelson interferometers
    with acoustic techniques, 228
    defined, 226
    See also Interferometers
Mie scattering, 159
Mode propagation, 174–75
Modulation. *See* Optical modulation
Modulators, 256
Molecular bioreceptors, 57–60
Molecular transitions, 162–63
Monitoring
    environmental, 339–47
    gas, 353–58
    solvent, 352–53
    in vivo, 303–12
Monochomators, 252
Monomode fibers, 184
Multianalyte sensing, 274

Multimode fibers, 183, 184–86
    capillary, 186
    in chemical sensing, 191
    illustrated, 185
    See also Optical fibers
Multiplexing, 200
Multiplicity, 163
Multiwavelength spectrometry, 251

NADH
    biotechnology and, 313–15
    as indicator, 304
    measurement of, 314
Nernst's equation, 17
Networks, 358–61
    in chemical sensing, 359–61
    general approach, 358–59
Neuroreceptors, 59
Nitro, 82
Nitroso, 82
Noninvasive techniques, 311–12
Nonlinear effects, 170–71
    harmonics, 171
    in planar waveguides, 180
    types of, 171
Nonreversible sensors, 32
Normalized frequency, 175
Nuclear process control, 347–52
    calibration curves and, 351
    Chemex process, 348
    compounds and, 350
    remote measurements in, 349
    See also Process control
Numerical aperture (NA), 173

Optical chemical sensors, 3
    techniques, 229–31
    transduction steps, 5
    See also Sensors
Optical chemoreception, 13–14
Optical chemoreceptors, 63
Optical conversion factors, 146
Optical electrical transduction, 14, 141
Optical fiber probe, 309
Optical fibers, 183–95
    asymmetrical structures, 176
    birefringent, 186–88
    capillary, 186
    doped, 189–91
    for evanescent wave spectroscopy, 188–89
    fluoride glass, 174
    graded-refractive-index, 183, 184
    historical perspective, 1–4
    infrared, 192–95

inherent properties of, 7
for interferometry, 189
lasing, 189–91
low loss transmission of, 7
materials for, 7–8, 191–95
monomode, 184
multimode, 183, 184–86
number of modes and, 179
plastic, 174, 189, 195
polarization in, 175–77
polarization-maintaining, 186–87
for sensors, 184–91
silica, 174, 191–92
single-mode, 183
step-index, 183
structure of, 174
telluride, 193
types of, 183–84
Optical frequency-domain reflectometry
    (OFDR), 359
Optical modulation, 200–204
    chemical measurement techniques and, 205
    examples, 202, 203, 204
    intensity, 200–201
    phase, 202
    polarization, 203
    schemes, 201
    time, 203–4
    wavelength, 201–2
Optical signal processing, 14, 141
Optical techniques
    in analytical chemistry, 7
    for chemical sensing, 7–9
Optical time domain reflectometry (OTDR),
    230–31, 359, 360
Optical waveguide sensors, 7–8
Optic Bio-Chemical Sensor European
    workshop, 4
Optics, 143–71
    geometrical, 149–51
    light and, 143–58
    light interaction with matter and, 158–71
    linear, 170
Optodes, 1, 259–72
    active, 28
    cation-selective, 31, 295
    classification of, 259–61
    defined, 5, 259
    enzyme, 316–18
    ethanol, 317
    extrinsic, 259, 261, 262–66
    intrinsic, 259, 261, 266–72
    ion sensitive (ISOs), 98
    membrane formulas, 99

situ, 312
surface structures, 119–20
system characteristics, 98
Optoelectrochemical sensors, 278–79
Optoelectrochemical transduction, 183
Optrodes, 259
    defined, 5
    See also Optodes
Organic pollutants, 340–43
    aromatic compounds, 342–43
    halogenated compounds, 341–42
    polycyclic aromatic hydrocarbons, 340–41
    See also Environmental monitoring
Oxazines, 81–82
Oxime, 82
Oximetry, 303–4

Paraday magneto-optic effect, 256
Partial antigen, 53
Passive components, 252–55
    bandpass filters, 252
    conventional, 252
    dichroic mirrors, 252
    diffraction gratings, 252
    fiber-optic, 253–55
    monochomators, 252
    spatial multiplexing, 254–55
    wavelength-division multiplexing, 253–54
    See also Active components
Penicillin detection, 318
Peripheral sensors, 312
Pesticides, 346–47
PH
    determination, 305
    indicators, 133
    sensitive dyes, 315
    See also PH sensors
Phase matching, 171
Phase modulation, 202
Phenoxazines, 81–82
Phosphorescence, 167
    indicators, 218–19
    See also Fluorescence
Photoacoustic measurements, 228–29
Photochemical effects, 163
Photodeposition method, 138
Photodetectors, 199, 246–52
    avalanche photodiodes (APD), 250
    defined, 246
    homojunction, 250
    noise equivalent power (NEP), 246
    photoemissive, 247
    principles, 246–47
    reverse-biased pin diodes, 248–50

Photodetectors (continued)
  sensitivity, 246
  solid-state, 247–52
  thermal, 246–47
Photoelectrode sensors, 278–79
Photoemissive detectors, 247
Photokinetics, 45
Photometers, 351
Photomultiplier tubes (PMTs), 247
Photons
  absorption, 164
  characteristics of, 144
  waves and, 151–58
  See also Light
PH sensors, 37
  optical characteristics in, 133–36
  sterilizable, 315
  See also PH
Phthaleins, 69–71
  chemical composition of, 69
  defined, 69
  fluorescein, 70
  SNAFL, 70–71
  SNARF, 71
  structural formula of, 70
  See also Indicators
Pitts-Giovanelli theory, 213
Planar sensors, 9
Planar waveguides, 177–83
  configurations of, 180–83
  devices, 272–78
  grating couplers and, 180, 181, 182
  Kretschmann configuration, 183
  nonlinearity in, 180
  wave propagation, 177–79
  See also Waveguides
Plastic fibers, 174, 189, 195
  advantages/disadvantages of, 195
  step-index, 195
  typical losses of, 194
  See also Optical fibers
Pockels effect, 255
Polarization, 146–49
  controllers, 256
  elliptical, 148
  fluorescence methods, 218
  linear, 149
  modulation, 203
  in optical fibers, 175–77
  OTDR (POTDR), 230, 359
  plane of, 147
  propagation and, 147
  See also Light
Polarization-division multiplexing (PDM), 358

Polycyclic aromatic hydrocarbons, 69, 340–41
Polymeric delivery systems, 33–36
  illustrated, 35
  parts of, 33
  performance, 33–35
Potential-sensitive dyes (PSDs), 91–94
  defined, 93
  development of, 91–92
  examples of, 94
  rhodamine dyes, 93
  See also Dyes
Process control, 347–53
  food industry, 353
  nuclear, 347–52
  solvent monitoring, 352–53
Propagation
  mode, 174–75
  polarization and, 147
  wave, 177–79
Protein
  fragment attachment steps, 131
  labeling protocols, 106–7
Quenching
  defined, 215
  effect on fluorescence, 216
  static, 215
Raman scattering, 159–60
  measurements, 213–14
  surface-enhanced resonance (SERRS), 214
  surface-enhanced spectroscopy, 214
  See also Scattering
Rate equations, 19–21
Rayleigh scattering, 159
Ray model, 173–77
Reactions
  biocatalytic, 45–49
  catalytic, 43–45
  cation exchange, 95–96
  chemical, 17–39
  coextraction, 96
  enzymatic, 45
  enzyme-controlled, 44
  first-order, 43
  higher order, 43
  immunological, 51
  surface, 56–57
Reagents
  immobilized, 23–27
  mixing, 35–36
  renewable, 33
  total concentration of, 25
Receptors, 94–101

Redox indicators, 90–91
Reflection, 150
    illustrated, 151
    internal, 150–51
    phase changes during, 153
Refraction, 150
Refractometric optodes, 266–67
Refractometry, 225–26
Regenerable biosensors, 320–21
Renewable reagent, 33
    ammonia sensor with, 34
    sensor design and, 33
Reservoir
    defined, 32
    disadvantages of, 33
    material, 32
Respiratory analysis, 310
Reverse-biased pin diodes, 248–50
Reversible sensors, 22–31
    analyte in, 22
    direct reactions, 22–28
    indirect reactions, 28–31
    nonreversible sensors becoming, 32
    for phenytoin measurement, 320
    reagent in, 22
    simultaneous reactions, 31
    types of, 28
    See also Chemical sensors
Rhodamines, 80–81

Scattering, 158–60, 211–14
    illustrated, 159
    Mie, 159
    Raman, 159–60
    Raman measurements, 213–14
    Rayleigh, 159
    reflectance measurements, 211–13
    surface-enhanced Raman spectroscopy, 214
    techniques, 212
    types of, 211
    See also Light
Seawater measurements, 346
Self-correlation function, 157
SELFOC, 253
Sensing cells, 262–63
Sensitive film, 356
Sensors
    ammonia, 34, 300–301
    bindings for, 121
    biocatalytic, 49–50
    BOD, 345–46
    configurations for, 118–21
    controlled-reagent, 32–36
    controlled-release polymer, 35

creatinine, 301–2
direct, 36
direct EWS, 267–68
dye-based, 93
evanescent wave, 221–22
EWS, 266
fabricating, 8
fluorescence techniques in, 167
fluorescent, 101, 217
fouling of, 321
glucose, 30
GOx, 298–99
ICB, 269–70
indirect, 32, 36
integrated, 4–5, 277–78
interferometric, 189
intravascular, 308–9
IO refractive index, 225–26
molecular recognition in, 38–39
nonreversible, 32
optical chemical, 3, 5
optical-fiber (OFS), 3
optical fibers for, 184–91
optical waveguide, 7–8
optoelectrochemical, 278–79
peripheral, 312
pH, 37, 133–36
photoelectrode, 278–79
planar, 9
potassium, 294
reversible, 22–31, 320
smart, 5
SPR, 223–24
steps in, 13–14
urea, 301–2
Shape recognition, 41–42, 50–60
    advantages, 41
    of antibody, 42
    defined, 41
    enzyme amplification, 56
    immunoassays, 54–56
    immunological interactions, 50–53
    immunoreactions with enzymes, 53–54
    kinetics of surface reactions and, 56–57
    liposome, 56
    molecular bioreceptors, 57–60
    of molecules, 42
    stereospecificity, 41, 42
Signal characteristics, 205–8
    as function of analyte, 207
    for measurement techniques, 206
Silanization, 126–27
    agents, 126
    defined, 126

Silanization (continued)
    example, 127
Silica fibers, 174, 191–92
    defined, 191
    typical losses, 192
    See also Optical fibers
Simultaneous reactions, 31
Single-step reaction, 22
Situ optodes, 312
Slab waveguides, 175
    TE modes in, 176
    See also Waveguides
Smart sensors, 5
Snell's law, 150, 173, 225
Sol-gel method, 123–24
    defined, 123
    dye immobilization with, 124
    glasses from, 268
    See also Immobilization techniques
Solid-state detectors, 247–52
    avalanche photodiodes (APD), 250
    defined, 247
    extrinsic, 247–48
    intrinsic, 247
    performance of, 250
    reverse-biased pin diodes, 248–50
    transitions in, 249
    See also Photodetectors
Solid-state lasers, 245–46
Solvent monitoring, 352–53
Spatial-division multiplexing (SDM), 358, 359
Spatial multiplexing, 254–55
Spectrofluorimeters, 351
Spectroscopy
    correlation, 228–29
    evanescent wave, 188–89, 219–22
    FT NIR, 214
    grating light reflection, 229
    IR, 209–10
    photoacoustic, 228
    surface-enhanced, 214
    thermal lens (TLS), 352
    TIRF, 276
Spin coupling, 169, 170
Stark effect, 228
Step-index fibers, 183, 184
Stern-Volmer equation, 216
Stokes fluorescence, 165
Substrates, 45
    defined, 41, 45
    high concentration, 46
    inhibition, 48
    measurement methods, 49
    See also Kinetics

Successive reactions, 30–31
Sulfophthaleins, 71–72
    defined, 71
    pH measurement, 72
    structure of, 71
    See also Indicators
Sulfur compounds, 84–86
    dithiocarbamates, 85
    dithizone, 84
    illustrated, 85
    thiazines, 84–85
    thiazolylazo, 85–86
    thio-oxine, 84
    thiosemicarbazide, 84
    See also Indicators
Supports, 113–18
    chemical formulas for, 115, 116
    functions of, 113
    intrinsic fluorescence, 117
    materials, 113–17
    membrane theory, 117–18
    polymer type, 117
    types of, 113, 117
Surface-enhanced resonance Raman scattering
        (SERRS), 214
Surface penetration, 127–28
Surface plasmon resonance (SPR), 57, 222–24
    guiding structure, 276
    illustrated, 224
    schematic diagram, 223
    sensors, 223–24, 277
    uses, 224

Thermal detectors, 246–47
Thermal lens spectroscopy (TLS), 352
Thermoluminescent spectrum, 358
Time-division multiplexing (TDM), 358, 360
Time domain (TD) fluorometry, 203
Time modulation, 203–4
Tracer dyes, 339–40
Transcutaneous techniques, 310–11
Transverse correlation, 156

Upper explosion limit (UEL), 353
    of hydrocarbon gases, 354
    See also Lower explosion limit (LEL)
Urea biosensors, 301–2

Verdet constant, 189

Wall effect, 308
Waveguides
    light transmission in, 174
    mode propagation in, 174–75
    planar, 177–83

slab, 175, 176
Wavelength-division multiplexing
        (WDM), 253–54, 358, 359
Wavelength modulation, 201–2
White light sources, 241–43

Young's fringes, 154
Young's slits experiment, 154, 155

The Artech House Optoelectronics Library

Brian Culshaw, Alan Rogers, and Henry Taylor, *Series Editors*

Acousto-Optic Signal Processing: Fundamentals and Applications, Pankaj Das

Amorphous and Microcrystalline Semiconductor Devices, Optoelectronic Devices, Jerzy Kanicki, editor

Bistabilities and Nonlinearities in Laser Diodes, Hitoshi Kawaguchi

Chemical and Biochemical Sensing With Optical Fibers and Waveguides, Gilbert Boisdé and Alan Harmer

Coherent and Nonlinear Lightwave Communications, Milorad Cvijetic

Coherent Lightwave Communication Systems, Shiro Ryu

Electro-Optical Systems Performance Modeling, Gary Waldman and John Wootton

Elliptical Fiber Waveguides, R. B. Dyott

The Fiber-Optic Gyroscope, Hervé Lefèvre

Field Theory of Acousto-Optic Signal Processing Devices, Craig Scott

Frequency Stabilization of Semiconductor Laser Diodes, Tetsuhiko Ikegami, Shoichi Sudo, Yoshihisa Sakai

Fundamentals of Multiaccess Optical Fiber Networks, Denis J. G. Mestdagh

Germanate Glasses: Structure, Spectroscopy, and Properties, Alfred Margaryan and Michael A. Piliavin

High-Power Optically Activated Solid-State Switches, Arye Rosen and Fred Zutavern, editors

Highly Coherent Semiconductor Lasers, Motoichi Ohtsu

Iddq Testing for CMOS VLSI, Rochit Rajsuman

Integrated Optics: Design and Modeling, Reinhard März

Introduction to Electro-Optical Imaging and Tracking Systems, Khalil Seyrafi and S. A. Hovanessian

Introduction to Glass Integrated Optics, S. Iraj Najafi

Introduction to Radiometry and Photometry, William Ross McCluney

Introduction to Semiconductor Integrated Optics, Hans P. Zappe

Laser Communications in Space, Stephen G. Lambert and William L. Casey

Optical Control of Microwave Devices, Rainee N. Simons

Optical Document Security, Rudolf L. van Renesse

Optical Fiber Amplifiers: Design and System Applications, Anders Bjarklev

Optical Fiber Sensors, Volume I: Principles and Components, John Dakin and Brian Culshaw, editors

Optical Fiber Sensors, Volume II: Systems and Applicatons, John Dakin and Brian Culshaw, editors

Optical Interconnection: Foundations and Applications, Christopher Tocci and H. John Caulfield

Optical Network Theory, Yitzhak Weissman

Optical Transmission for the Subscriber Loop, Norio Kashima

Optoelectronic Techniques for Microwave and Millimeter-Wave Engineering, William M. Robertson

Reliability and Degradation of LEDs and Semiconductor Lasers, Mitsuo Fukuda

Semiconductor Raman Laser, Ken Suto and Jun-ichi Nishizawa

Semiconductors for Solar Cells, Hans Joachim Möller

Single-Mode Optical Fiber Measurements: Characterization and Sensing, Giovanni Cancellieri

Smart Structures and Materials, Brian Culshaw

Ultrafast Diode Lasers: Fundamentals and Applications, Peter Vasil'ev

For further information on these and other Artech House titles, contact:

Artech House
685 Canton Street
Norwood, MA 02062
617-769-9750
Fax: 617-769-6334
Telex: 951-659
email: artech@artech-house.com

Artech House
Portland House, Stag Place
London SW1E 5XA England
+44 (0) 171-973-8077
Fax: +44 (0) 171-630-0166
Telex: 951-659
email: artech-uk@artech-house.com